Trace Elements *in* Laboratory Rodents

Trace Elements *in* Laboratory Rodents

Edited by

Ronald Ross Watson

CRC Press
Boca Raton New York London Tokyo

Methods in Nutrition Research

Edited by Ronald Ross Watson and Ira Wolinsky

Published Titles

Trace Elements in Laboratory Rodents, Ronald R. Watson

Forthcoming Titles

Methods for Investigation of Amino Acid and Protein Metabolism in Humans and in Animal Models, Antoine El-Khoury

Methods in Nutrition Research

A new CRC Series on Research in Nutrition

SERIES EDITORS

Ronald Ross Watson and Ira Wolinsky

Publisher/Life Sciences:	Robert B. Stern
Editorial Assistant:	Carol Messing
Project Editor:	Carrie L. Unger
Marketing Manager:	Susie Carlisle
Direct Marketing Manager:	Becky McEldowney
Cover design:	Denise Craig
PrePress:	Kevin Luong
Manufacturing:	Sheri Schwartz

Library of Congress Cataloging-in-Publication Data

Trace elements in laboratory rodents / edited by Ronald R. Watson.
p. cm.
Includes bibliographical references and index.
ISBN 0-8493-9611-5 (alk. paper)
1. Rodents as laboratory animals—Nutrition. 2. Trace elements in animal nutrition.
I. Watson, Ronald R. (Ronald Ross). II. Series.
SF407.R6T735 1996
636′.9323—dc20

96-15926
CIP

International Standard Book Number 0-8493-9611-5
Library of Congress Card Number 96-15926
Printed in the United States of America 1 2 3 4 5 6 7 8 9 0
Printed on acid-free paper

Series Preface

Methods are critical to both good data and their correct interpretation. While there are methods series for biochemistry and other disciplines, nutritional sciences have suffered from the absence of such a resource. Many small but important techniques, procedures, and carefully tested methods make experiments easier and more accurate. This series, *Methods in Nutrition Research,* is designed to gradually fill this gap in research resources. It will provide published and unpublished details of technical procedures used by experts in different areas of nutrition research. It will also describe potential pitfalls to be avoided. The series is designed for the researcher with a focus on laboratory and field nutritional research methods and how to apply them precisely. It will contain books ranging from descriptions of studies of micronutrients in animal models to macronutrients in human studies. The overall aim of the series is to carry out nutritional research as efficiently as possible, without technical errors, from known procedural methods, ones which may not be easily found in the literature. Nutritional researchers will bring together their own as well as others' experiences to provide a single source of detailed, tested methods with complete descriptions on how to apply them.

We welcome*Trace Elements in Laboratory Rodents,* the first book in this series, knowing it will fill a void in the literature on experimental methods in animal research.

Ronald Ross Watson
Ira Wolinsky
Series Editors

PREFACE

Trace Elements in Laboratory Rodents is written by researchers in nutritional studies using trace elements in rodent models. They describe the best methods to provide deficient or supplemental trace elements to the animals and how to assay them. This book also provides the technical and sometimes unpublished details as well as potential pitfalls to optimize trace element studies in small animals. It is a ***hands on, how to do*** approach with its main focus on the technical details which make good, reliable studies. Common as well as rare or recently recognized minerals are described, relating to both dietary supplementation and measurement in tissues. Some chapters focus on the use of trace elements in the whole animal, while others look at their application to cell culture. The main purpose of this book, then, is as a primary resource for studies using trace elements.

Ronald Ross Watson
Editor

THE EDITOR

Ronald Ross Watson, Ph.D., has edited 35 books including 15 on various aspects of nutrition.

Dr. Watson attended the University of Idaho, but graduated from Brigham Young University in Provo, Utah with a degree in Chemistry in 1966. He completed his Ph.D. in Biochemistry in 1971 at Michigan State University. His postdoctoral schooling was completed at the Harvard School of Public Health in Nutrition and Microbiology, with a 2-year postdoctoral research experience in Immunology. He was an Assistant Professor of Immunology at the University of Mississippi Medical Center in Jackson from 1973 to 1974 where he also did research. He was an Assistant Professor of Microbiology and Immunology at the Indiana University Medical School from 1974 to 1978 as well as an Associate Professor at Purdue University in the Department of Food and Nutrition from 1978 to 1980. In 1982 he joined the faculty at the University of Arizona in the Department of Family and Community Medicine, where he is a Professor of Research. He has published over 400 research papers and review chapters.

Dr. Watson is a member of several national and international nutrition, immunology, and cancer societies and research societies on alcoholism. His research interests have involved studying the immune function under nutritional stress including research with human subjects. His major research interest involves effects of dietary material on disease resistance and immune functions. This includes understanding the effects of alcohol and drug abuse on immunosuppression associated with AIDS, the role of vitamins A and E and beta carotene as modulators of skin cancer resistance, and immune function in people and animals. He has published over 200 journal papers and 200 reviews and chapters in these areas.

Contributors

Frederick N. Bebe, M.S.
Nutrition Research Program
Kentucky State University
Frankfort, Kentucky

Sam J. Bhathena, Ph.D.
Metabolism and Nutrient Interaction Laboratory/ARS
Beltsville Human Nutrition Research Center
U.S. Department of Agriculture
Beltsville, Maryland

Margaret Ann Bock, Ph.D.
Department of Human Nutrition
New Mexico State University
Las Cruces, New Mexico

John D. Bogden, Ph.D.
Department of Preventive Medicine and Community Health
UMDNJ —New Jersey Medical School
Newark, New Jersey

Philip Carthew, Ph.D., MRC Path.
MRC Toxicology Unit
University of Leicester
Leicester, England

Merrill J. Christensen, Ph.D.
Department of Food Science and Nutrition
Brigham Young University
Provo, Utah

Robert A. Felty, Ph.D.
Department of Chemistry
Dartmouth College
Hanover, New Hampshire

John Finley, Ph.D.
U.S. Department of Agriculture/ARS
Grand Forks Human Nutrition Research Center
Grand Forks, North Dakota

Tommaso Galeotti
Instituto di Parologia Generale
Università Cattolica
Rome, Italy

Edward Harris
Department of Biochemistry and Biophysics
Texas A & M University
College Station, Texas

Deborah L. Hasten, Ph.D.
Department of Internal Medicine
Washington University School of Medicine
St. Louis, Missouri

Maren Hegsted, Ph.D.
School of Human Ecology
Louisiana State University
Baton Rouge, Louisiana

Curtiss D. Hunt, Ph.D.
U.S. Department of Agriculture/ARS
Grand Forks Human Nutrition Research Center
Grand Forks, North Dakota

Phyllis E. Johnson, Ph.D.
U.S. Department of Agriculture/ARS
Grand Forks Human Nutrition Research Center
Grand Forks, North Dakota

Henry C. Lukaski, Ph.D.
U.S. Department of Agriculture/ARS
Grand Forks Human Nutrition Research Center
Grand Forks, North Dakota

Brooke D. Martin, B.Sc.
Department of Chemistry
Dartmouth College
Hanover, New Hampshire

G. Stephen Morris
Department of Kinesiology
Louisiana State University
Baton Rouge, Louisiana

Donald Oberleas, Ph.D.
Department of Food and Nutrition
Texas Tech University
Lubbock, Texas

C. Oguakoya, B.Sc.
Department of Chemical Biochemistry
King's College of Medicine and Dentistry
London, England

Myna Panemangalore, Ph.D.
Nutrition Research Program
Kentucky State University
Frankfort, Kentucky

T.J. Peters, D.Sc.
Department of Chemical Biochemistry
King's College of Medicine and Dentistry
London, England

Yong Chang Qian
Department of Biochemistry and Biophysics
Texas A & M University
College Station, Texas

Philip G. Reeves, Ph.D.
U.S. Department of Agriculture/ARS
Grand Forks Human Nutrition Research Center
Grand Forks, North Dakota

George D. V. van Rossum, M.A., D.Phil
Department of Pharmacology
Temple University School of Medicine
Philadelphia, Pennsylvania

Alan Shenkin, Ph.D., FRCP, FRC Path.
Department of Clinical Chemistry
University of Liverpool Hospital
Liverpool, England

R.J. Simpson, D.Phil.
Department of Chemical Biochemistry
King's College of Medicine and Dentistry
London, England

A.G. Smith, Ph.D., FRSC
MRC Toxicology Unit
University of Leicester
Leicester, England

Scott M. Smith
Nutrition and Metabolism Laboratory
Biomedical Operations and Research Branch
NASA Johnson Space Center
Houston, Texas

Diane M. Stearns, Ph.D.
Department of Chemistry
Dartmouth College
Hanover, New Hampshire

Evelyn Tiffany-Castiglioni
Department of Veterinary Anatomy and Public Health
Texas A & M University
College Station, Texas

Eric O. Uthus, Ph.D.
U.S. Department of Agriculture/ARS
Grand Forks Human Nutrition Research Center
Grand Forks, North Dakota

Moshe J. Werman, Ph.D.
Department of Food Engineering and Biotechnology
Technion —Israel Institute of Technology
Haifa, Israel

Catherine A. Wardle, M.Sc., Ph.D.
Department of Clinical Chemistry
University of Liverpool Hospital
Liverpool, England

Connie Weaver, Ph.D.
Department of Foods and Nutrition
Purdue University
West Lafayette, Indiana

Rob E. C. Wildman
Department of Food and Nutrition
Ohio University
Athens, Ohio

Karen E. Wetterhahn, Ph.D.
Department of Chemistry
Dartmouth College
Hanover, New Hampshire

Table of Contents

Acknowledgments

Research by Dr. Watson over two decades led to the genesis of this book. It was and is supported in the area of nutritional supplementation of the elderly to improve immune functions and reduce cancer by Wallace Genetics Foundation, Inc.

Dedication

To Roscoe Derrick Watson, Ph.D., Professor Emeritis of the University of Idaho and Ellen Kemp Watson, Ph.D., Washington State University, whose example of research and questioning initiated the research that led to this series and this book.

GENERAL OVERVIEW

Chapter 1

AIN-93 Purified Diets for the Study of Trace Element Metabolism in Rodents

Philip G. Reeves
U.S. Department of Agriculture/ARS
Grand Forks Human Nutrition Research Center
Grand Forks, North Dakota

CONTENTS

0-8493-9611-5/97/$0.00+$.50

I. INTRODUCTION

Diet, as a vehicle for the delivery of adequate nutrition to the experimental animal, is very important. In experimental animal studies where the mechanism of action of a particular nutrient is to be resolved, knowledge of the precise composition of the diet for the animal model is of paramount importance. Laboratory animal diets are placed into three separate categories. One is termed cereal-based, unrefined, or nonpurified diets. These diets are made primarily with natural ingredients and are classed as open or closed formula. Open formula refers to diets where the composition is known and available to the potential user and must be constructed as specified. Closed formula refers to diets where the composition is known only to the commercial manufacturer. Closed formula diets cannot be expected to have the same nutrient composition from batch to batch. Another classification of laboratory animal diets is the purified diet. This type of diet is made of refined ingredients such as isolated proteins and refined sugars and oils or fats. The third classification is the chemically defined diet. Such diets are made of chemically pure sources of amino acids, mono- or disaccharides, and purified fatty acids or triglycerides. Minerals are provided from reagent grade sources, and the vitamins are of the highest purity.

For the study of trace element nutrients, purified or chemically defined diets are the first choices. Most often, the purified diet is used. Although the composition of the major ingredients may contain small amounts of a particular trace element, clever manipulation of the ingredients will allow the formulation of a diet adequately supplied with all essential nutrients except the one in question. In 1977 and 1981, the American Institute of Nutrition (AIN) published the formula of a purified diet called AIN-76A for use in studies involving experimental rodents.[1,2] This diet could be easily manipulated for use in trace element studies. However, over many years of use, numerous problems were encountered. Perhaps the most significant one was the repeated occurrence of kidney calcification in female rats that consumed this diet. In 1989, participants in an AIN workshop came to the conclusion that revisions in the AIN-76A should be made. In 1993, replacement formulae for the AIN-76A diet were published and given a general designation of AIN-93.[3] The formulations were based on certain criteria: they can be made from purified ingredients; they conform to or exceed the nutrient requirements suggested by the National Research Council (NRC);[4,5] they can be made with readily available ingredients at a reasonable cost; the compositions are consistent and reproducible; and they can be used over a wide range of applications.

The new diet comes in two versions, one of which is called AIN-93G and is to be used during the growth phase and during reproduction. The other is called AIN-93M and is to be used during adult maintenance. Table 1 gives the general formulae of these diets. Table 2 gives the formulae of the mineral mixes for each diet, and Table 3 gives the ingredient list for the vitamin mix to be used for either diet. Table 4 lists the vitamins and minerals contained in the two diets and their targeted amounts. This table also shows the 1978 and 1995 NRC estimates of requirements for these nutrients.[4,5] In the conversion from the AIN-76A to the AIN-93 diets, some major changes were made. These changes are highlighted as follows.

TABLE 1.

Formulation of the AIN-93G Diet for Rapid Growth and the AIN-93M Diet for Maintenance of Rodents When Casein Is Used as the Protein Source

Ingredients	AIN-93G	AIN-93M
	(g/kg Diet)	
Cornstarch	397.486	465.692
Casein (≥85% protein)	200.000	140.000
Dextrinized cornstarch (90 to 94% tetrasaccharides)[a]	132.000	155.000
Sucrose	100.000	100.000
Soybean oil (no additives)	70.000	40.000
Fiber[b]	50.000	50.000
Mineral mix (AIN-93G-MX)	35.000	35.000
Vitamin mix (AIN-93-VX)	10.000	10.000
L-cystine	3.000	1.800
Choline bitartrate (41.1% choline)[c]	2.500	2.500
Tert-butylhydroquinone (TBHQ)	14.0 mg	8.0 mg

[a] Dyetrose (Dyets, Inc., Bethlehem, PA) and Lo-Dex 10 (American Maize, Co., Hammond, IN) meet these specifications. An equivalent product also may be used.

[b] Solka-Floc®, 200 FCC (FS&D Corp., St. Louis, MO) or its equivalent is recommended.

[c] Based on the molecular weight of the free base.

II. AIN-93 DIETS: GENERAL FORMULATIONS

A. Carbohydrates

The carbohydrate sources for the new diets are cornstarch, dextrinized cornstarch, which contains from 90 to 94% tetrasaccharides, and sucrose.

TABLE 2.

Mineral Mixes That Supply the Recommended Concentrations of Minerals for Diets AIN-93G and AIN-93M, Respectively

Ingredients	AIN-93G-MX	AIN-93M-MX
	(g or mg/kg of mix)	
Calcium carbonate, anhydrous, 40.04% Ca	357.00	357.00
Potassium phosphate, monobasic, 22.76% P; 28.73% K	196.00	250.00
Potassium citrate, tri-potassium, monohydrate, 36.16% K	70.78	28.00
Sodium chloride, 39.34% Na; 60.66% Cl	74.00	74.00
Potassium sulfate, 44.87% K; 18.39% S	46.60	46.60
Magnesium oxide, 60.32% Mg	24.00	24.00
Ferric citrate, 16.5% Fe	6.06	6.06
Zinc carbonate, 52.14% Zn	1.65	1.65
Sodium meta-silicate · $9H_2O$, 9.88% Si	1.45	1.45
Manganous carbonate, 47.79% Mn	0.63	0.63
Cupric carbonate, 57.47% Cu	0.30	0.30
Chromium potassium sulfate · $12H_2O$, 10.42% Cr	0.275	0.275
Boric acid, 17.5% B	81.5 mg	81.5 mg
Sodium fluoride, 45.24% F	63.5 mg	63.5 mg
Nickel carbonate, 45% Ni	31.8 mg	31.8 mg
Lithium chloride, 16.38% Li	17.4 mg	17.4 mg
Sodium selenate, anhydrous, 41.79% Se	10.25 mg	10.25 mg
Potassium iodate, 59.3% I	10.0 mg	10.0 mg
Ammonium paramolybdate · $4H_2O$, 54.34% Mo	7.95 mg	7.95 mg
Ammonium vanadate, 43.55% V	6.6 mg	6.6 mg
Powdered sucrose	221.026	209.806
Total	1000.00 g	1000.00 g

TABLE 3.

AIN-93-VX Vitamin Mix Recommended for Use with the AIN-93G and AIN-93M Diet Formulations

Vitamins	g/kg of mix
Nicotinic acid	3.000
Ca Pantothenate	1.600
Pyridoxine-HCl	0.700
Thiamin-HCl	0.600
Riboflavin	0.600
Folic acid	0.200
Biotin	0.020
Vitamin B-12 (cyanocobalamine)(0.1% in mannitol)	2.500
Vitamin E (all-*rac*-α-tocopheryl acetate) (500 IU/g)[a]	15.000
Vitamin A (all-*trans*-retinyl palmitate) (500,000 IU/g)[a]	0.800
Vitamin D-3 (cholecalciferol) (400,000 IU/g)	0.250
Vitamin K-1 (phylloquinone)	0.075
Powdered sucrose	974.655

[a] The gelatin-matrix form of these vitamins is recommended.

TABLE 4.

Target Concentrations of Minerals and Vitamins for the Basic AIN-93G and AIN-93M Rodent Diets Derived from the Total Ingredients, and Their Comparison with the 1978 and 1995 NRC Estimates of Requirements for Growing Rats

Minerals	AIN-93[a]	1978 NRC[b] (mg/kg Diet)	1995 NRC[c]
Calcium	5,000.00	5,000.00	5,000.00
Potassium	3,600.00	3,600.00	3,600.00
Phosphorus	3,000.00	4,000.00	3,000.00
Chloride	1,571.00	500.00	500.00
Sodium	1,019.00	500.00	500.00
Magnesium	507.00	400.00	500.00
Sulfur	300.00	300.00	[b]
Iron	35.00	35.00	35.00
Zinc	30.00	12.00	12.00
Manganese	10.00	50.00	10.00
Copper	6.00	5.00	5.00
Silicon	5.00		
Iodine	0.20	0.15	0.15
Molybdenum	0.15		0.15
Selenium	0.15	0.10	0.15
Chromium	1.00	0.30	
Fluoride	1.00	1.00	
Nickel	0.50		
Boron	0.50		
Lithium	0.10		
Vanadium	0.10		
Vitamins			
Nicotinic acid	30 mg	20 mg	15 mg
Pantothenic acid	15 mg	8 mg	10 mg
Pyridoxine	6 mg	6 mg	6 mg
Thiamin	5 mg	4 mg	4 mg
Riboflavin	6 mg	3 mg	3 mg
Folic acid	2 mg	1 mg	1 mg
Choline	1,000 mg	1,000 mg	750 mg
Vitamin K-1	750 μg	50 μg	1,000 μg
Biotin	200 μg		200 μg
Vitamin B-12	25 μg	50 μg	50 μg
Vitamin A	4,000 IU	4,000 IU	2,300 IU
Vitamin D-3	1,000 IU	1,000 IU	1,000 IU
Vitamin E	75 IU	30 IU	27 IU

[a] Targeted nutrient concentrations to meet or exceed the NRC requirements for laboratory rodents. Actual analysis may reveal higher amounts of some nutrients because of contaminants from certain ingredients such as zinc and some vitamins from casein that were not accounted for.

[b,c] Based on the 3rd (1978) and 4th (1995) Editions of the NRC Nutrient Requirements of Laboratory Animals.[4,5] A blank space in a column indicates that the NRC did not estimate a requirement for this nutrient.

These sources of carbohydrate were chosen for various reasons. (1) Starch was chosen to replace the sucrose originally used in the AIN-76A diet because high dietary sucrose, compared with other carbohydrates, can cause several complications in rodents. These include hyperlipidemia, hepatic lesions,[6] enhancement of nephrocalcinosis,[7] enhancement of copper requirement,[8,9] and enhancement of iron absorption.[10] (2) When starch is the major component of the diet, the diet will not pellet properly. Therefore, a more water-soluble carbohydrate source must be substituted for a portion of the cornstarch. It was previously shown by the animal diets industry that dextrinized starch could fill these requirements. Dextrinized starch is cornstarch that has been hydrolyzed to simpler sugars. The product that contains greater than 90% tetrasaccharides seems to work best. It was found that this product must be added in the correct proportions with other components of the diet; i.e., making major changes in the protein, fat, or other components may require adjustments in the amount of dextrinized starch to make the diet pellet properly. (3) A small amount of sucrose was added to the diet to provide sweetness and improve palatability, and to act as a dipersal medium for vitamins and minerals.

B. Protein

Many sources of protein have been used over the years in experimental rodent diets. Table 5 shows the amino acid composition and the mineral and vitamin contents of six of these protein sources. Casein was chosen as the source of protein for the AIN-93 diets because it has a reasonably adequate amino acid composition, is readily available, and is reasonably priced. As with most protein sources, however, there are a few shortages. The major limitation of casein is its short supply of sulfur amino acids. Compared with other protein sources, it is low in cystine (cysteine), but contains a moderate amount of methionine. Therefore, to balance the sulfur amino acids of the AIN-93 diets, L-cystine is used. This contrasts with most other diets designed for experimental use, which use DL-methionine. The assumption is that sufficient methionine will be converted to cysteine *in vivo* to prevent a deficiency of this amino acid. Casein at 200 g/kg and L-cystine at 3.0 g/kg of the AIN-93G diet will provide a sufficient concentration of each essential amino acid to meet the 1978 NRC requirements for growth.[5] Based on the 1978 report, 140 g of casein and 1.8 g of L-cystine in the AIN-93M diet will provide more than enough amino acids for maintenance of adult rodents. However, in 1995, the 4th revised edition of the NRC requirements of laboratory animals recommended an increase in many of the essential amino acids, including the sulfur amino acids.[4] The requirement level for the latter increased from 6.0 to 9.8 g/kg of diet (Table 6). At 200 g/kg of the AIN-93G diet, casein falls short of the recommended amount of sulfur amino acids. Even with the addition of L-cystine at 3.0 g/kg of diet, it is still below the recommendation for growth by about 1.6 mg/kg. To bring the AIN-93G diet up to requirements, it is recommended unofficially that 1.6 g of L-methionine be added/kg of diet in addition to 3.0 g of L-cystine. An alternative would be to increase the L-cystine concentration to 4.6 mg/kg. However, this is not recommended because studies have shown that as the concentration of L-cystine is increased in the diet of rats, there is an increased chance of a negative interaction with copper metabolism, thus lowering the copper status of animals, especially when they are fed a marginal copper diet.[11] The suggested changes are shown in Tables 7 and 8 (AIN-93G-CAS and AIN-93M-CAS).

The use of other protein sources in the diet will require alterations in the basic dietary ingredients. The sulfur amino acid content of egg white solids is relatively high. At 200 g/kg of diet, this protein source will supply the required amounts of sulfur amino acids and no further additions to the diet are required. On the other hand, the use of lactalbumin will require an addition of L-arginine and L-cystine to help meet the 1995 NRC requirements (Tables 7 and 8; AIN-93G-LAC and AIN-93M-LAC). If wheat gluten is used, 250 g of protein source/kg of diet is recommended with additional L-methionine, L-lysine, and L-threonine (Tables 7 and 8; AIN-93G-GLU and AIN-93M-GLU).

Not only do the protein sources contain variable amounts of amino acids, but they also contain other components, including minerals and vitamins, often in relatively high concentrations. In addition, there is a natural variability in mineral concentration from batch to batch (Table 9). If an experimental protocol dictates using a protein source other than casein and if the diet is to meet the AIN-93 formulation criteria, changes in the mineral mix must be considered. For example, egg white solids do not contain as much phosphorus as casein, but they contain excessive amounts of sodium, chlorine, potassium, and inorganic sulfur. Consequently, the AIN-93 mineral mixes have to be altered to increase the phosphorus content and to eliminate the sodium, chlorine, and sulfur sources. The potassium content must also be decreased. Examples of mineral mixes to be used for egg white-based AIN-93 diets are given in Tables 10 and 11 (AIN-93G-EGG-MX and AIN-93M-EGG-MX).

The substitution of other protein sources requires other changes. The use of lactalbumin in place of casein requires a single change in the mineral mix. Lactalbumin contains less phosphorus than casein; therefore, the phosphorus content of the mineral mix should be increased. Tables 10 and 11 show examples of the required changes (AIN-93G-LAC-MX and AIN-93M-LAC-MX). The use of wheat gluten will also require an increase in the amount of phosphorus in the mineral mix (Tables 10 and 11; AIN-93G-GLU-MX and AIN-93M-GLU-MX).

Of the proteins listed in Table 4, isolated soybean protein is the most difficult to adjust for changes in mineral content. The protein isolate has very high concentrations of phosphorus and sodium relative to other sources, but a low concentration of chlorine. However, a large part of the phosphorus is bound up in phytic acid, a form that may not be totally

TABLE 5.

Estimated General Composition and the Amino Acid, Mineral, and Vitamin Contents of Major Protein Sources on an As-Is Basis

	Soybean	Casein	Vitamin-Free Casein	Egg White Solids	Wheat Gluten	Lactalbumin
General Composition, g/kg						
Protein	908	869	930	824	760	860
Fat	3	11	Trace	[a]	11	36
Carbohydrate				45		
Fiber	1					
Moisture	55	100	50	85	80	50
Ash	33	19	20	46	12	15
Amino Acid, g/kg						
Alanine	36.3	22.6	24.2	51.1	27.4	42.1
Arginine	66.3	31.3	33.5	46.9	34.2	18.9
Aspartic Acid	104.4	56.6	60.5	70.8	38.0	92.9
Cystine	11.8	3.5	3.7	19.8	19.0	17.2
Glutamic Acid	188.9	181.0	193.4	111.2	251.6	151.4
Glycine	37.2	15.7	16.7	29.6	31.2	16.3
Histidine	22.7	22.6	24.2	18.1	19.0	18.1
Isoleucine	42.7	41.8	44.6	48.6	26.6	43.9
Leucine	70.8	76.6	81.8	69.2	52.4	103.2
Lysine	55.4	64.4	68.8	49.4	22.0	79.1
Methionine	11.8	22.6	24.2	31.3	9.1	21.5
Phenylalanine	47.2	43.5	46.5	50.2	35.7	30.1
Proline	48.1	101.8	108.8	29.6	82.8	53.3
Serine	45.4	47.0	50.2	58.5	35.0	31.8

Threonine	32.7	33.1	35.3	35.4	22.0	42.1
Tryptophan	13.6	10.4	11.2	12.4	8.4	16.3
Tyrosine	35.4	46.1	49.3	32.1	12.9	32.7
Valine	37.2	49.6	53.0	59.3	32.7	49.0
Minerals, mg/kg						
Calcium	2,000	280	280	890	450	590
Chlorine	1,000	500	500	10,600		
Copper	13	1.2	1.2	3.2	10	0.5
Iron	140	3.8	3.8	2.4	60	5.0
Magnesium	490	30	30	420	350	31
Manganese	13	0.4	0.4		30	0.2
Phosphorus	7,800	7,600	7,600	890	1,400	1,700
Potassium	1,500	50	50	11,200	300	160
Selenium	0.2	0.2	0.2	1.4		
Sodium	7,500	<100	<100	12,400	400	70
Zinc	42	40	40	1.6	40	55
Vitamins, mg/kg						
Biotin		0.06	0.003	0.6		0.1
Folic Acid		0.2	0.1		0.03	0.4
Niacin	12.0	7.2	0.5	7.2	60.0	5.0
Pantothenic Acid	0.6	0.2	0.1	19.6		10.0
Pyridoxine	0.2	0.2	0.1	0.2	1.0	1.0
Riboflavin	0.7	1.2	0.7	23.0	1.0	4.0
Thiamin	1.5	0.2	0.1	0.4	1.0	0.5
Vitamin B-12, µg/kg	10.0	20.0	10.0	5.0	1.0	100.0

[a] An empty space indicates that accurate information was unavailable.

available under normal feeding conditions. Although the rat has intestinal phytase activity[12–14] and the potential to break down phytate, they develop low phosphorus status when fed diets containing 3.0 g of total phosphorus/kg where 50 to 70% of phosphorus is in the form of phytate.[15–17] This suggests that if phytate is degraded, only a portion of the released phosphorus is available. The low status in these rats could be corrected by increasing dietary phosphorus to 4 g/kg. The phytate from soybean protein also reduces the absorption of zinc[18,19] and iron,[20] and it enhances the absorption of copper.[21] Davis et al.[22] found that a zinc supplement of 30 mg/kg of diet was enough to overcome the detrimental effects of feeding 25% soybean protein in the diet. Tables 10 and 11 give suggested formulations of the mineral mixes to be used if soybean protein isolates are used in the AIN-93 diets (AIN-93G-SOY-MX and AIN-93M-SOY-MX). These mineral mixes will provide 30 mg of zinc/kg of diet in addition to the endogeneous zinc from the soybean protein source itself.

C. Fat

Recent investigations suggest that at least two fatty acids, linoleic [18:2; *n*-6] and linolenic [18:3; *n*-3], are dietary essentials. Bourre et al.[23,24] fed variable concentrations of linoleic and linolenic acids in the diet of rats. By using tissue saturation of 20:4; *n*-6 and 22:6; *n*-3 as criteria for requirements, they determined that 12 g of linoleic and 2 g of α-linolenic acid/kg of diet were the minimal requirements for rats. Few dietary sources of fat will provide these amounts of essential fatty acids without a complicated mixture. Table 12 shows the composition of some of the fat sources often used in animal diets. Soybean oil in the amount of 30 g/kg of diet will provide the above requirements for linoleic and linolenic acids. However, in Bourre's[23,24] work, an amount of fat equivalent to 50 to 60 g of soybean oil was required to reach the plateau for maximal concentrations of the marker fatty acids in many tissues of growing rats. Therefore, a 15% margin of safety was added and the amount of fat source for the AIN-93G diet was set at 70 g of soybean oil/kg of diet. This amount of fat is recommended for young growing rats and reproducing females. For the maintenance diet of adult males and nonreproducing females, the recommended amount is 40 g of soybean oil/kg of diet. Studies by Lee et al.[25] suggest that a ratio of *n*-6:*n*-3 of 5 and a polyunsaturate:saturate ratio of 2 are points of greatest influence on tissue lipids and eicosanoid production. Soybean oil provides an *n*-6:*n*-3 ratio of 7 and a polyunsaturate:saturate ratio of 4.

D. Antioxidants

Fats, especially those with a high polyunsaturate content, are subject to oxidation and must be protected.[26,27] Because the AIN-93 diets

TABLE 6.

Amino Acid Composition of the AIN-93G Diet When 200 g of Casein/kg is Used as the Source of Protein, and Compared with the Amino Acid Requirements for Growth as Proposed by the National Research Council in 1978 and 1995

Amino Acids	AIN-93G	1978 NRC[b]	1995 NRC[c]	(AIN-93G)-(1995 NRC)[d]
			(g/kg Diet)	
Alanine	4.52			
Arginine	6.26	6.00	4.30	1.96
Asparagine		4.00		
Aspartic Acid	11.32			
Cystine	3.70[a]	3.00	4.9	−1.2
Glutamic Acid	36.20	40.00		
Glycine	3.14			
Histidine	4.52	3.00	2.80	1.72
Isoleucine	8.36	5.00	6.20	2.15
Leucine	15.32	7.50	10.70	4.61
Lysine	12.88	7.00	9.20	3.68
Methionine	4.52	3.00	4.90	−0.38
Phenylalanine	8.70	4.00	5.10	3.60
Proline	20.36	4.00		
Serine	9.40			
Threonine	6.62	5.00	6.20	0.41
Tryptophan	2.08	1.50	2.0	0.09
Tyrosine	9.22	4.00	5.10	4.12
Valine	9.92	6.00	7.40	2.52
Nonessentials		5.90	66.00	
Total	177.04	108.90	134.80	

[a] Includes 3.0 g of L-cystine added to the diet.
[b,c] Based on the 3rd (1978) and 4th (1995) Editions of the NRC Nutrient Requirements of Laboratory Animals.[4,5]
[d] The amount of amino acids provided in the AIN-93G diet minus the 1995 NRC requirements. A negative notation indicates that 200 g of casein/kg of diet plus the recommended amino acid supplement does not provide a sufficient amount of these amino acids to meet the 1995 NRC requirements, and they must be added as the free amino acid.

contain primarily polyunsaturated fatty acids including 18:3; *n*-3, it is recommended that the diet contain an antioxidant. It has been shown that tertiary-butylhydroquinone (TBHQ) is an effective antioxidant against oxidation of the highly polyunsaturated *n*-3 fatty acids of fish oil.[28–30] Because soybean oil contains *n*-3 fatty acids, it is recommended that TBHQ be used in the AIN-93 diets at a level of 200 mg/kg of oil. If, however, the experimental design precludes the use of antioxidants, the diet may be protected from oxidation on a limited basis by storing it in a nitrogen atmosphere in the cold. If the experimental protocol calls for the study of dietary copper or selenium and their effects on antioxidant enzyme, then adding antioxidants to the diet might be a hindrance. Purified diets contain free copper and iron and other oxidizing components could make them more prone to oxidation than a natural ingredient-based diet where these metals are organically bound.

TABLE 7.

Modified Formulation of the AIN-93G Diet to Meet the 1995 NRC Requirements for Rapid Growth Phase of Rodents When Either Casein (CAS), Soybean Isolate (SOY), Egg White Solids (EGG), Wheat Gluten (GLU) or Lactalbumin (LAC) Is Used as the Protein Source

Ingredients	AIN-93G-CAS	AIN-93G-SOY	AIN-93G-EGG	AIN-93G-GLU	AIN-93G-LAC
			(g/kg of diet)		
Cornstarch	395.886	395.406	390.486	342.786	397.386
Casein (≥85% protein)	200.000				
Soybean protein isolate (91% protein)		200.000			
Egg white solids (83% protein)			200.000		
Wheat gluten (76% protein)				250.000	
Lactalbumin (86% protein)					200.000
Dextrinized cornstarch (90 to 94% tetrasaccharides)[a]	132.000	132.000	132.000	132.000	132.000
Sucrose	100.000	100.000	100.000	100.000	100.000
Soybean oil (no additives)	70.000	70.000	70.000	70.000	70.000
Fiber[b]	50.000	50.000	50.000	50.000	50.000
Mineral mix (AIN-93G-CAS-MX)[c]	35.000				
Mineral mix (AIN-93G-SOY-MX)		35.000			
Mineral mix (AIN-93G-EGG-MX)			35.000		
Mineral mix (AIN-93G-GLU-MX)				35.000	
Mineral mix (AIN-93G-LAC-MX)					35.000
Vitamin mix (AIN-93-VX)	10.000	10.000	10.000	10.000	10.000
L-Methionine	1.600	2.540		3.000	
L-Cystine	3.000	2.540			2.100
L-Arginine					1.000
L-Lysine				3.700	
L-Threonine				1.000	
Biotin premix[d]			10.00		

Choline bitartrate (41.1% choline)[e]	2.500	2.500	2.500	2.500	2.500
Tert-butylhydroquinone (TBHQ)	14.0 mg	14.0 mg	14.0 mg	14.0 mg	14.0 mg
Total	1000.000	1000.000	1000.000	1000.000	1000.000

[a] Dyetrose (Dyets, Inc., Bethlehem, PA) and Lo-Dex 10 (American Maize, Co., Hammond, IN) meet these specifications. An equivalent product also may be used.

[b] Solka-Floc®, 200 FCC (FS&D Corp., St. Louis, MO) or its equivalent is recommended.

[c] See Table 10 for mineral mixes.

[d] Premix is made by mixing 180 mg of biotin with 999.82 g of cornstarch. This amount, when fed at 10 g/kg of diet, and in addition to that in the vitamin mix, will provide 2.0 mg of biotin/kg of diet that contains 200 g of egg white solids/kg.[68]

[e] Based on the molecular weight of the free base.

TABLE 8.

Modified Formulation of the AIN-93M Diet for Maintenance of Rodents When Either Casein (CAS), Soybean Isolate (SOY), Egg White Solids (EGG), Wheat Gluten (GLU) or Lactalbumin (LAC) Is Used as the Protein Source

Ingredients	AIN-93G-CAS	AIN-93G-SOY	AIN-93G-EGG	AIN-93G-GLU	AIN-93G-LAC
			(g/kg of diet)		
Cornstarch	464.272	463.932	457.492	427.092	465.292
Casein (≥85% protein)	140.000				
Soybean protein isolate (91% protein)		140.000			
Egg white solids (83% protein)			140.000		
Wheat gluten (76% protein)				175.000	
Lactalbumin (86% protein)					140.000
Dextrinized cornstarch (90 to 94% tetrasaccharides)[a]	155.000	155.000	155.000	155.000	155.000
Sucrose	100.000	100.000	100.000	100.000	100.000
Soybean oil (no additives)	40.000	40.000	40.000	40.000	40.000
Fiber[b]	50.000	50.000	50.000	50.000	50.000
Mineral mix (AIN-93M-CAS-MX)[c]	35.000				
Mineral mix (AIN-93M-SOY-MX)		35.000			
Mineral mix (AIN-93M-EGG-MX)			35.000		
Mineral mix (AIN-93M-GLU-MX)				35.000	
Mineral mix (AIN-93M-LAC-MX)					35.000
Vitamin mix (AIN-93-VX)	10.000	10.000	10.000	10.000	10.000
L-Methionine	1.12	1.780		2.100	
L-Cystine	2.10	1.780			1.500

Ingredient					
L-Arginine					0.700
L-Lysine				2.600	
L-Threonine				0.700	
Biotin premix[d]			10.000		
Choline bitartrate (41.1% choline)[e]	2.500	2.500	2.500	2.500	2.500
Tert-butylhydroquinone (TBHQ)	8.0 mg	8.0 mg	8.0 mg	8.0 mg	8.0 mg
Total	1000.000	1000.000	1000.000	1000.000	1000.000

[a] Dyetrose (Dyets, Inc., Bethlehem, PA) and Lo-Dex 10 (American Maize, Co., Hammond, IN) meet these specifications. An equivalent product also may be used.

[b] Solka-Floc®, 200 FCC (FS&D Corp., St. Louis, MO) or its equivalent is recommended.

[c] See Table 11 for mineral mixes.

[d] Premix is made by mixing 130 mg of biotin with 999.87 g of cornstarch. This amount, when fed at 10 g/kg of diet, and in addition to that in the vitamin mix, will provide 1.5 mg of biotin/kg of diet that contains 140 g of egg white solids/kg.[68]

[e] Based on the molecular weight of the free base.

TABLE 9.

Example of the Natural Variability in Mineral Content of Protein Sources, Casein, Soybean Protein, and Lactalbumin

Mineral	Casein		Soybean Protein		Lactalbumin	
	mg/kg	Range	mg/kg	Range	mg/kg	Range
Calcium	281 ± 61[a]	111–288	2086 ± 222	1960–2480	596 ± 238	407–1240
Copper	1.2 ± 0.6	0.5–2.6	12.9 ± 2.2	9.0–14.7	0.5 ± 0.2	0.3–0.8
Iron	3.8 ± 1.7	1.9–7.3	142 ± 11	125–156	5.0 ± 3.0	3.5–14.8
Magnesium	29 ± 8	20–37	489 ± 56	390–515	31 ± 8	23–55
Manganese	0.4 ± 0.1	0.3–0.5	13.0 ± 2.2	11.2–16.7	0.2 ± 0.05	0.1–0.3
Phosphorus	7651 ± 544	6670–8200	7778 ± 345	7300–8260	1720 ± 383	1240–2490
Potassium	<100		1470 ± 370	862–1860	160 ± 32	99–219
Sodium	<100		7368 ± 317	6990–7830	71 ± 21	50–111
Zinc	38.7 ± 6.2	25.1–48.6	42.3 ± 1.9	40.5–44.9	54.8 ± 14.4	34.9–83.5

[a] Values are means ± SD and the range of values for samples from different batches of casein (11), soybean protein (5), and lactalbumin (13) over a period from 1/91 to 10/95. Samples were analyzed at Hazleton Labs., Madison, WI, by Inductively Coupled Plasma Elemental Technology. This information was provided by Dr. Ron Rose, Harlan Teklad, Madison, WI.

TABLE 10.

Mineral Mixes That Supply the Recommended Concentrations of Elements for Modified AIN-93G Diets That Contain Either Casein (CAS), Soybean Isolate (SOY), Egg White Solids (EGG), Wheat Gluten (GLU) Or Lactalbumin (LAC) as the Protein Source

MINERALS	AIN-93G-CAS-MX	AIN-93G-SOY-MX	AIN-93G-EGG-MX	AIN-93G-GLU-MX	AIN-93G-LAC-MX
			(g or mg/kg of mix)		
Calcium carbonate, anhydrous, 40.04% Ca	357.00	102.97	83.56	104.10	102.78
Calcium phosphate, dibasic, 29.46% Ca; 22.77% P		306.17	376.40	332.57	331.36
Potassium phosphate, monobasic, 22.76% P; 28.73% K	186.00[a]				
Potassium citrate, tri-potassium, monohydrate, 36.16% K	78.87[a]	83.44	108.09	226.69	226.69
Potassium chloride, 52.44% K; 47.56% Cl		82.36[b]		9.25	
Sodium chloride, 39.34% Na; 60.66% Cl	74.00	[c]	[c]	66.74	74.00
Potassium sulfate, 44.87% K; 18.39% S	46.60	46.60	[e]	46.60	46.60
Magnesium oxide, 60.32% Mg	24.00	19.40	24.00	19.51	24.00
Ferric citrate, 16.5% Fe	6.06	1.21[d]	6.06	6.06	6.06
Zinc carbonate, 52.14% Zn	1.65	1.65	1.65	1.65	1.65
Sodium meta-silicate · $9H_2O$, 9.88% Si	1.45	1.45	1.45	1.45	1.45
Manganous carbonate, 47.79% Mn	0.63	0.63	0.63	0.63	0.63
Cupric carbonate, 57.47% Cu	0.30	0.30	0.30	0.30	0.30
Chromium potassium sulfate · $12H_2O$, 10.42% Cr	0.275	0.275	0.275	0.275	0.275
Boric acid, 17.5% B	81.50 mg	81.50 mg	81.50 mg	81.50 mg	81.50 mg
Sodium fluoride, 45.24% F	63.50 mg	63.50 mg	63.50 mg	63.50 mg	63.50 mg
Nickel carbonate, 45% Ni	31.80 mg	31.80 mg	31.80 mg	31.80 mg	31.80 mg
Lithium chloride, 16.38% Li	17.40 mg	17.40 mg	17.40 mg	17.40 mg	17.40 mg

TABLE 10. (Continued)

Mineral Mixes That Supply the Recommended Concentrations of Elements for Modified AIN-93G Diets That Contain Either Casein (CAS), Soybean Isolate (SOY), Egg White Solids (EGG), Wheat Gluten (GLU) Or Lactalbumin (LAC) as the Protein Source

MINERALS	**AIN-93G-CAS-MX**	**AIN-93G-SOY-MX**	**AIN-93G-EGG-MX**	**AIN-93G-GLU-MX**	**AIN-93G-LAC-MX**
			(g or mg/kg of mix)		
Sodium selenate, anhydrous, 41.79% Se	10.25 mg	10.25 mg	[f]	10.25 mg	10.25 mg
Potassium iodate, 59.3% I	10.00 mg	10.00 mg	10.00 mg	10.00 mg	10.00 mg
Ammonium paramolybdate · $4H_2O$, 54.34% Mo	7.95 mg	7.95 mg	7.95 mg	7.95 mg	7.95 mg
Ammonium vanadate, 43.55% V	6.60 mg	6.60 mg	6.60 mg	6.60 mg	6.60 mg
Powdered sucrose	222.936	353.314	424.606	197.506	181.566
Total, g	1000.00	1000.00	1000.00	1000.00	1000.00

[a] Values are based on the average phosphorus analysis of casein (Table 9), and differs from the original AIN-93G-MX.

[b] Soybean protein isolates contain limited amounts of chloride. Therefore, diets containing this protein source require the addition of a form of Cl to the mineral mix.

[c] Soybean protein isolates and egg white solids contain a large amount of sodium, thus diets containing these protein sources require no addition of a sodium source to the mineral mix.

[d] Soybean protein contains a relatively high amount of iron; thus the amount of iron source in this mix can be reduced.

[e,f] Egg white solids contain sufficient sulfur and selenium so that when this protein source is used at 200 g/kg of diet no extra sulfur or selenium is required in the mineral mix.

TABLE 11.

Mineral Mixes that Supply the Recommended Concentrations of Elements for Modified AIN-93M Diets that Contain Either Casein (CAS), Soybean Isolate (SOY), Egg White Solids (EGG), Wheat Gluten (GLU) or Lactalbumin (LAC) as the Protein Source

Minerals	AIN-93M-CAS-MX	AIN-93M-SOY-MX	AIN-93M-EGG-MX	AIN-93M-GLU-MX	AIN-93M-LAC-MX
			(g or mg/kg of mix)		
Calcium carbonate, anhydrous, 40.04% Ca	357.00	68.33	82.43	96.85	95.90
Calcium phosphate, dibasic, 29.46% Ca; 22.77% P		364.89	360.80	345.69	346.57
Potassium phosphate, monobasic, 22.76% P; 28.73% K	243.03[a]				
Potassium citrate, tri-potassium, monohydrate, 36.16% K	33.50[a]	85.34	152.98	217.23	226.63
Potassium chloride, 52.44% K; 47.56% Cl		85.97[b]	5.22	6.48	
Sodium chloride, 39.34% Na; 60.66% Cl	74.00	[c]	[c]	68.92	74.00
Potassium sulfate, 44.87% K; 18.39% S	46.6	46.60	[e]	46.60	46.60
Magnesium oxide, 60.32% Mg	24.00	20.42	20.90	20.80	24.00
Ferric citrate, 16.5% Fe	6.06	2.67[d]	6.06	6.06	6.06
Zinc carbonate, 52.14% Zn	1.65	1.65	1.65	1.65	1.65
Sodium meta-silicate · $9H_2O$, 9.88% Si	1.45	1.45	1.45	1.45	1.45
Manganous carbonate, 47.79% Mn	0.63	0.63	0.63	0.63	0.63
Cupric carbonate, 57.47% Cu	0.30	0.30	0.30	0.30	0.30
Chromium potassium sulfate · $12H_2O$, 10.42% Cr	0.275	0.275	0.275	0.275	0.275
Boric acid, 17.5% B	81.50 mg	81.50 mg	81.50 mg	81.50 mg	81.50 mg
Sodium fluoride, 45.24% F	63.50 mg	63.50 mg	63.50 mg	63.50 mg	63.50 mg

TABLE 11. (Continued)

Mineral Mixes that Supply the Recommended Concentrations of Elements for Modified AIN-93M Diets that Contain Either Casein (CAS), Soybean Isolate (SOY), Egg White Solids (EGG), Wheat Gluten (GLU) or Lactalbumin (LAC) as the Protein Source

Minerals	AIN-93M-CAS-MX	AIN-93M-SOY-MX	AIN-93M-EGG-MX	AIN-93M-GLU-MX	AIN-93M-LAC-MX
			(g or mg/kg of mix)		
Nickel carbonate, 45% Ni	31.80 mg	31.80 mg	31.80 mg	31.80 mg	31.80 mg
Lithium chloride, 16.38% Li	17.40 mg	17.40 mg	17.40 mg	17.40 mg	17.40 mg
Sodium selenate, anhydrous, 41.79% Se	10.25 mg	10.25 mg	[f]	10.25 mg	10.25 mg
Potassium iodate, 59.3% I	10.00 mg	10.00 mg	10.00 mg	10.00 mg	10.00 mg
AmmoniumpParamolybdate · $4H_2O$, 54.34% Mo	7.95 mg	7.95 mg	7.95 mg	7.95 mg	7.95 mg
Ammonium vanadate, 43.55% V	6.60 mg	6.60 mg	6.60 mg	6.60 mg	6.60 mg
Powdered sucrose	211.276	322.869	367.086	186.836	175.706
Total, g	1000.00	1000.00	1000.00	1000.00	1000.00

[a] Values are based on the average phosphorus analysis of casein (Table 9), and differs from the original AIN-93M-MX.

[b] Soybean protein isolates contain limited amounts of chloride. Therefore, diets containing this protein source require the addition of a form of Cl to the mineral mix.

[c] Soybean protein isolates and egg white solids contain a large amount of sodium, thus diets containing these protein sources require no addition of a sodium source to the mineral mix.

[d] Soybean protein contains a relatively high amount of iron; thus the amount of iron source in this mix can be reduced.

[e,f] Egg white solids contain sufficient sulfur and selenium so that when this protein source is used at 140 g/kg of diet no extra sulfur or selenium is required in the mineral mix.

TABLE 12.

Major Fatty Acids in Commonly Used Oils and Fats in Laboratory Animal Diets

Dietary Oil	Saturated	Monounsaturated	Polyunsaturated	Oleic	Linoleic	Linolenic
			(g/100 g of oil)			
Corn Oil	12.7[a]	24.2	58.7	24.2	58.0	0.7
Soybean Oil	14.4	23.3	57.9	22.8	51.0	6.8
Canola Oil	7.1	58.9	29.6	56.1	20.3	9.3
Safflower Oil	9.1	12.1	74.5	11.7	74.1	0.4
Coconut Oil	86.5	5.8	1.8	5.8	1.8	—
Sunflower Oil	10.5	19.5	65.7	19.5	65.7	—
Beef Tallow	49.8	41.8	4.0	36.0	3.1	—
Lard	39.2	45.1	11.2	41.2	10.2	—
Butterfat	50.5	23.4	3.0	20.4	1.8	—

[a] Values are taken from the USDA Handbook 8.

E. Fiber

Fiber is not considered a nutrient for the rat, but it may provide some beneficial effects in controlling the gut microflora populations. The AIN-93 diets have included 50 g of wood fiber/kg of diet as the fiber source. The mineral content of fiber sources can vary. Of the trace minerals in fiber, iron seems to be the highest in concentration. In general, the less costly, less purified sources contain the most minerals. The Solka Floc® (FS and D, St. Louis, MO)* source of fiber recommended for the AIN-93 diets contains about 100 mg of iron/kg. At 50 g/kg of diet this could add a moderate amount of iron to the diet. Alphacel (ICN Biochemicals, Irvine, CA) contains only about 1.3 mg of iron/kg, and Avicel cellulose fiber (FMC Corp., Philadelphia, PA) contains 1.6 mg of iron/kg. Therefore, for iron deficiency studies, the source of fiber should be a consideration.

F. Vitamins

Table 4 lists the known essential vitamins for laboratory rodents, as well as the amounts recommended for use in the AIN-93 diets. This table also provides information about the NRC requirements. The vitamins are subject to oxidation and should be protected as much as possible during storage and preparation of the diet. Because riboflavin and vitamin K-1 are particularly vulnerable to degradation by light, the diets should be prepared in reduced light and stored in a dark cold room. There are interactions among some of the vitamins and trace elements. For example, a diet that is very high in vitamin E may cause experimental animals to lack the signs of selenium deficiency. Diets with low amounts of vitamin D may affect the outcome of studies involving calcium metabolism. A vitamin B-12 deficiency anemia might be initially confused with symptoms of iron deficiency.

G. Minerals

Female rats are prone to develop kidney calcification when fed purified diets. This is especially true when they are fed the AIN-76A diet. The female rat seems to be very sensitive to the dietary Ca to P molar ratio, and the ratio in the AIN-76A is too low to prevent the initiation of nephrocalcinosis. Recent studies showed that a Ca to P ratio of 1.3 or greater works well[31] in the prevention of nephrocalcinosis, and the amounts of calcium and phosphorus for the AIN-93 diets were adjusted accordingly.[3]

---

* Mention of a trademark or proprietary product does not constitute a guarantee or warranty of the product by the U.S. Department of Agriculture and does not imply its approval to the exclusion of other products that may also be suitable.

If the calcium concentration in the diet is kept at 5.0 g/kg, then the phosphorus concentration will be 3.0 g/kg. Because casein is the protein of choice for the AIN-93 diets and it contains phosphorus, the phosphorus content of the mineral mix must be adjusted to obtain the correct dietary ratio. The dietary calcium and phosphorus concentrations should not go below the requirements, however.

Other changes in the mineral composition of the AIN-93 diets compared with the AIN-76A diet are noteworthy. The 1978 NRC requirement for manganese was set at 50 mg/kg of diet; however, more recent experimental evidence suggested that it was much lower. Therefore, the 1995 NRC has lowered the requirement to only 10 mg of manganese/kg of diet. This is in agreement with the AIN-93G and -93M diets. Although molybdenum is considered an essential mineral because of its involvement in the metabolism of sulfur compounds,[32] the 1978 NRC did not have a requirement level, nor did the AIN-76A have molybdenum included in the mineral mix. However, the 1995 NRC recommends 150 μg of molybdenum/kg of diet. This is in line with that proposed for the AIN-93 diets. Chromium was listed in the 1978 NRC as a required mineral. However, succeeding investigations could not corroborate earlier suggestions that chromium was essential for the rat;[33,34] therefore, the 1995 NRC placed it in the "potentially beneficial nutrient" category until more conclusive evidence is found. However, chromium is still included in the AIN-93 mineral mixes, but at a reduced amount compared with the AIN-76A diet.

The ultratrace elements, other than selenium, molybdenum, and iodine, represent a special class of proposed nutrients. When the formulae of the AIN-93 diets were proposed, it was recommended that a group of ultratrace elements be added to the diets. These included fluoride, boron, vanadium, nickel, lithium, and silicon (regarded as a trace element). Although to date there is no convincing and/or repeatable evidence that these elements are essential for any specific metabolic process, studies show that they may interact with other nutrients under various conditions to give the appearance of beneficial effects.[35–39] These elements are found in relative abundance in natural ingredient diets but their concentrations are often very low or nonexistent in purified diets. Therefore, minimal amounts were placed in the AIN-93 diets. It is in the best interest of the investigator to factor in and/or minimize fluctuations in the concentrations of any dietary component. The ultratrace elements are of particular concern because so little goes so far, i.e, there is a fine line between beneficiality and toxicity. At present, the best approach to accomplishing this task is to monitor the nutrient concentrations by analysis, and to take steps to reduce the concentrations if they are considered excessive, or to increase them if they are suboptimal.[3]

III. MODIFIED AIN-93 DIET FORMULATIONS FOR USE IN TRACE MINERAL DEFICIENCY STUDIES

Mineral deficiency studies can be carried out with any type of diet that is adequate in all required nutrients except the mineral in question. The formulation of these diets requires precise knowledge of the nutrient requirements of the animal species used in the study, as well as knowledge of the composition of the various dietary ingredients, including contamination with the mineral in question (see Table 9). Protein sources can be a major contributor of trace element contaminants, and the sources of the macrominerals such as calcium carbonate, calcium phosphate, and potassium phosphate also can contribute to trace element contamination. Only the purest grades of these sources should be used. The investigator should also be aware of possible interactions between the mineral in question and other components in the diet.

The following is a presentation of diet formulatons that might be used to study trace element metabolism. They were designed with the AIN-93 base formulae in mind. Some of the formulae have already been used to produce a deficiency of a particular mineral, and some have yet to be used for this purpose. These formulae are presented as possible examples of how the diets should be constructed in order to meet or exceed the requirements of all known nutrients for rodents, including the trace elements. It is the investigator's responsibility, however, to insure that each batch of diet actually contains the amount of trace element desired. This is most accurately done by direct analysis. It should also be noted that the type of caging, bedding material, and drinking water can contribute to the intake of certain trace elements. Therefore, all possible sources of contamination should be considered when conducting trace element studies in rodents.

A. Zinc

Zinc is an important trace element that is absolutely required for growth, reproduction, and maintenance of laboratory rodents. It is involved in myriad metabolic pathways primarily as a functional part of numerous enzymes; an estimate of some 300 different enzymes across all classes is sometimes given.[40] Visual signs of severe zinc deficiency occur very rapidly in the rat. These include loss of appetite and slow weight gain, both of which occur within days of first feeding a zinc-deficient diet. About 2 weeks into the deficiency, scaly and bleeding skin lesions along the paws and back may occur. Excessive loss of hair may also be observed. Biologically, serum zinc concentrations begin to fall within hours of first feeding a zinc-deficient diet. Later, the activities of enzymes will begin to be depressed. These include serum and tissue alkaline phosphatase, red cell carbonic anhydrase, intestinal carboxypeptidase, alcoholic dehydrogenase, and testicular angiotensin converting enzyme. With an extended period of deficiency, many other zinc-dependent enzyme systems will show reduced activity.[41]

Initially, zinc deficiency was difficult to produce in laboratory animals because zinc was ubiquitous in the diet ingredients, and in the early days, zinc-galvanized cages were used to house animals. With the arrival of stainless steel caging and purified diet ingredients, a zinc deficiency can easily be produced. However, one must select the diet ingredients with caution. For example, at present there is only one protein source, egg white solids, that is naturally low enough in zinc to be used in zinc-deficiency studies. Other sources such as soybean protein isolate or casein can be used, but they must be put through extensive processing to remove the zinc.[42] A diet composed of purified amino acids can also be used.

Control diets that mimic the AIN-93G and -93M diets and are suitable for use in zinc deficiency studies, are shown in Tables 7 and 8. These diets use egg white solids as the protein source and are named AIN-93G-EGG and AIN-93M-EGG to distinguish them from the original diets that use casein as the protein source. It is very important that the designated mineral mixes be used with each diet (Table 10 and 11) to insure that the correct mineral balance is obtained.

The use of egg white solids as a protein source poses special problems. Spray-dried egg white solids contain trypsin inhibitor and avidin, a compound that binds biotin and makes it unavailable to the animal. Some laboratory animals such as the guinea pig may not eat egg white-based diets unless the protein is autoclaved and dried before introduction into the diet. Autoclaving will remove up to 99% of the trypsin inhibitor. An autoclaving procedure that has been used successfully is as follows: mix 2.5 parts of dried egg white to 1 part of deionized water and autoclave the mixture in shallow pans at 121°C for 30 min. The result will be a rubbery mass. Cut the material into 1 in. cubes and dry it at 60°C. Regrind the dry material to a fine powder before use.[43,44] As a precaution, extra biotin should be added to an egg white-based diet whether treated in this manner or not.

To produce the zinc-deficient version of the diet, remove the zinc source ($ZnCO_3$) from the mineral mix (Table 10 and 11) and replace it with an equal amount of powdered sucrose. When this mix is used in the diet as prescribed, the dietary zinc concentration will be less than 1.0 mg/kg. Recent studies have shown that this diet is an excellent one for producing zinc deficiency in weanling male rats and mice (Reeves, unpublished work) and in adult male rats.[45] It should be noted that rats fed egg white-based diets with adequate zinc have normal serum zinc concentrations that are 20 to 30% higher than those of rats fed casein-based diets with a similar amount of zinc. The mechanism for this difference is unknown.

B. Copper

Copper is an important trace element that is absolutely required for numerous metabolic processes in laboratory rodents. Copper is involved in oxidase enzyme systems such as cytochrome c oxidase in mitochondria, lysyl oxidase in the cross linking of elastin and collagen, and in superoxide dismutase (SOD), which breaks down the superoxide radical.[46] Visual signs of copper deficiency are slow to occur and are not readily apparent. After about 5 weeks of consuming a severely copper-deficient diet (<1.0 mg/kg), weanling rats will begin to lose weight, and some may begin to suffer cardiac hypertrophy and hemorrhage of the large vessels of the heart, resulting in spontaneous death by dissecting aneurysms. Biochemical changes during this period are characterized by very low serum copper concentrations and serum ceruloplasmin activity, low liver copper, low red cell and liver SOD activity, and low liver cytochrome c oxidase activity.[47]

Copper-deficient diets are relatively easy to produce. There are at least three protein sources that can be used without further processing: casein, egg white solids, and lactalbumin. Control diets that meet the requirements proposed for the AIN-93G and -93M diets, and are also suitable for use in copper deficiency studies, are shown in Tables 7 and 8. These diets are named AIN-93G-EGG, AIN-93M-EGG, AIN-93G-CAS, AIN-93M-CAS, AIN-93G-LAC, and AIN-93M-LAC to distinguish them from the original AIN-93 diets. It is very important that the designated mineral mixes be used with each diet (see Tables 10 and 11) to insure that the correct mineral balance is obtained.

To produce the copper-deficient version of the diet, remove the copper source ($CuCO_3$) from the mineral mix and replace it with an equal amount of powdered sucrose. When the mix is used in the diets as prescribed, the dietary copper concentration will be less than 1.0 mg/kg. Recent studies have shown that casein-based AIN-93 diets are good for producing copper deficiency in weanling male rats[48,49] and for lowering copper status of adult male mice.[50] However, some aspects of a severe copper deficiency, including heart enlargement and anemia,[51] may take longer to develop than when other diet formulations without copper are used.[47,52] As a general rule, female rats[53] and mice[54] are less susceptible to copper deficiency than their male counterparts. Therefore, a longer experimental period may be required to produce the desired level of deficiency. Strain differences among rats and mice may also be a consideration when doing copper deficiency studies. Tissue distribution and biliary excretion of copper tend to vary among different strains of rats.[55–57]

C. Iron

Of the essential trace elements, iron is probably the most studied, in both humans and animals. Iron is essential for the synthesis and the oxygen carrying function of hemoglobin. Hemoglobin accounts for more than 60% of total body iron, and is greater than 95% of the protein content of erythrocytes. Thus, one of the primary signs of iron deficiency is anemia, which is characterized by reduced hematocrit and low serum iron and ferritin concentrations. However, in a mild deficiency, e.g., iron depletion, low serum ferritin concentration may be the only sign. These signs develop rapidly in the young growing animal. In older animals, signs of severe deficiency may take longer to develop because of possible high iron stores.

The protein sources most suitable for iron deficiency studies are casein, egg white solids, and lactalbumin. Theoretically, the concentration of iron in casein and lactalbumin diets using Solka Floc® as the fiber source would be about 6 mg/kg. If egg white solids and this fiber source were used, it would be about 5 mg/kg. However, actual analyses in my laboratory have found values of 10 and 9 mg of iron/kg in these diets, respectively. With the use of a more purified source of fiber such as Alphacel or Avicel, the values will be <5 mg/kg. The mineral source of calcium may vary in iron content and may also contribute to the iron supply in the diet. Thus, to lower dietary iron to a minimum, a purified source of fiber and a pure grade of the calcium source should be used. Typical analyses of this type of formulation in my laboratory show values of <2.0 mg of iron/kg of diet. In iron deficiency studies, Wein and Van Campen[58] fed weanling rats diets similar to AIN-76A (casein based with Solka Floc®) but with the carbohydrate source consisting of equal amounts glucose and cornstarch (similar to AIN-93) replacing the sucrose. After 3 weeks, hematocrit values were only 57% of normal, spleen iron concentrations were only 17% of normal, and growth was reduced by about 10%. The analyzed concentration of iron in the deficient diet was about 6.5 mg/kg.

D. Manganese

Manganese functions as a part of three important metalloenzymes. One is arginase, an enzyme in the urea cycle that aids in the elimination of nitrogen from the body. Another is pyruvate carboxylase, which functions in carbohydrate and lipid metabolism.[59] The third is mitochondrial superoxide dismutase, which functions to dismutate the superoxide radical to hydrogen peroxide. Manganese also functions as a nonspecific activator of several enzyme systems. Many times, however, magnesium can substitute for manganese in these systems. Signs of manganese deficiency include impaired growth, defects in lipid and carbohydrate metabolism, abnormalities of the skeleton, and reproductive failure.[60]

For many years, it was believed that the requirement for manganese for the rodent was relatively high, around 50 mg/kg of diet. The 1978 edition of the NRC Nutrient Requirements of Laboratory Animals gave this value as the requirement.[4] However, in the 1995 revision, justification was given for lowering the requirement to only 10 mg/kg.[5] Evidence was cited to suggest that as little as 5 mg/kg was sufficient, but because of animal strain differences in response to manganese deficiency, 10 mg/kg was set as the requirement. In addition, excess manganese may be detrimental to iron metabolism.[61]

In the past, diets frequently used in manganese studies with rodents were similar to the AIN-76A where casein was used as the protein source.[61–63] Casein has an inherently low concentration of manganese (0.4 mg/kg) and at 200 g/kg of diet, without a manganese supplement, the dietary concentration will be, theoretically, less than 0.1 mg/kg. However, actual measurements in a similar diet shows values of >1.0 mg of manganese/kg of diet. Egg white solids and lactalbumin would also be good sources of protein for manganese deficiency studies because of their equally low concentration of manganese (Tables 5 and 9). Soybean protein, however, contains a higher amount of manganese and at 200 mg/kg of diet, it could provide a range between 2 and 4 mg of manganese/kg of diet, which is not necessarily a deficient concentration[63] (see Table 9). Thus, the use of diets AIN-93G or -93M with casein, egg white, or lactalbumin (see Tables 7 and 8) would suffice in manganese studies with laboratory rodents.

E. Other Trace Minerals

For the study of other mineral elements such as selenium and iodine, special diet formulations are often used. For example, selenium studies are carried out with diets that contain 300 g of Torula yeast/kg as the protein source, which is extremely low in selenium.[64] However, this protein source is very low in sulfur amino acids and is moderately low in tryptophan. In addition, Torula yeast has a bizarre mineral composition with extremely high concentratons of phosphorus, potassium, and magnesium. The use of this protein source makes it difficult to balance the diet with respect to these minerals. Theoretically, the soybean protein and casein-based diets shown in Tables 7 and 8 would give a selenium concentration of approximately 0.04 mg/kg when the protein source is used at 200 g/kg. However, egg white solids could not be used because of the high selenium content. Pure amino acid-based diets can also be used.

For iodine studies, some investigators use the so-called Remington diet, which contains 180 g of wheat gluten/kg.[65] The base for this diet is ground yellow corn (Harlan Teklad, Madison, WI; #170360). The wheat gluten diet shown in Tables 7 and 8 may also work for iodine deficiency studies. Others have used ovalbumin as the protein source.[66] Whether this component is the same as egg white solids is not known. Casein-based diets have been used successfully for the study of molybdenum deficiency.[32] In the study cited, the basal concentration of molybdenum was

0.025 mg/g of diet. However, other analyses of the casein-based AIN-76A diet and the AIN-93G diet without the ultratrace element mix showed molybdenum concentrations between 0.4 and 0.6 mg/kg of diet, which is above the requirement.[32] Consequently, extreme care should be taken in choosing the dietary ingredients for molybdenum studies. For the study of any of the ultratrace elements, a pure amino acid-based diet should be considered.

IV. DIET PREPARATION

References by Baker[67] and Reeves et al.[3] outline the basic procedures for preparing purified diets for animal studies. Many research facilities do not have the equipment or the expertise to mix diets; therefore, the investigators must rely on commercial suppliers. Of those that do have the facilities, most are only equipped to make 10 to 20 kg of diet at a time. Thus, the following procedures for diet mixing apply to small, in-house operations.

The main goal in diet preparation is to produce a nutritious diet. A secondary goal, but of equal importance, is to produce a homogenous mixture of all ingredients so that the experimental animals consistently receive the same amount of nutrients throughout the study. This is especially important with respect to such components as vitamins and ultratrace minerals that are present in extremely low concentrations.

A. Ingredients and Sources

All dietary ingredients should be of the highest quality possible and chemically analyzed for mineral contaminants. For trace mineral studies, the ingredients for the mineral mixes should be analytical grade, and if the contaminate composition is unknown, they should be analyzed for the mineral in question. Use of the diet for long periods may require large purchases of single batches of ingredients to avoid lot-to-lot variation. Some of the mineral forms recommended in the AIN-93 diet formulations are hygroscopic and should be stored in a desiccator. Some of the vitamins are degraded by light and should be stored in the dark and the supply periodically replaced with fresh stock. Highly unsaturated oils are subject to oxidation and should be stored in the cold and checked frequently for peroxidation.

B. Premix Preparation

Most of the vitamins and some mineral components are present in the diet in very small quantities. To insure adequate mixing, these components should be incorporated into separate premixes before they are added to the total diet. For the AIN-93 diets, it is recommended that the vitamins be contained in an amount of powdered sucrose such that when 10 g are added/kg of diet, the required concentrations of each vitamin will be met. The vitamin premix can be prepared by adding the required amounts of each vitamin to powdered sucrose in a large plastic bottle equipped with a plastic paddle attached to the lid. The bottle is then placed on a roller (ball-mill type) and mixed. Elbow type mixers made of polycarbonate are also available for this procedure. Because the vitamin K-1 source is a thick oily material, some investigators may prefer to make a separate mix using soybean oil or dextrose as the base for this vitamin. Preparation of the vitamin mix should be carried out in reduced light.

Some of the ingredients for the mineral mix come in powdered form. For those that do not, they should be ground to a fine powder with a mortar and pestle. Then, all the powdered components of the mineral mix are placed together inside a large ball-mill made of burundum. Burundum balls, 1 in. in diameter, are placed inside, and the container rolled until the mixture is homogeneous. Stainless-steel balls are also available for this procedure, but they should never be used to mix minerals for studies with chromium. Their use in the preparation of the mineral mix for vanadium and nickel studies is discouraged. The finished product is stored in a plastic container in a desiccator.

C. Diet Mixing

There are several methods for mixing diets. One is to place the major ingredients (starch, protein, carbohydrate, fiber) in a stainless steel bowl of a Hobart-type mixer equipped with an open-blade stainless steel paddle, cover the bowl with a plastic sleeve or with cloth, and mix the ingredients until they are homogeneous (about 5 min slow speed). Dust accumulation on the sleeve or cloth should be wiped back into the bowl. Then, add one at a time the amino acid supplement, vitamin mix, mineral mix, and the oil to which the antioxidant has been added. Mix for about 2 min after each addition and an additional 15 min after the last ingredient. Stainless-steel equipment should not be used if the study concerns chromium, vadnium, or nickel.

Another procedure that may reduce the dust problem is to place casein, sugar, dextrinized starch, and soybean oil plus antioxidant in the mixing bowl and throughly mix. The remaining ingredients are then added one at a time in order of preponderance and are mixed after each addition. Over-mixing and/or high-speed mixing should be avoided because it could result in heat generation and loss of vitamins and cause fatty acid oxidation. The time for blending depends on the type and model of the mixer. Consult the technical service department of the equipment source for advice on the best performance of the model being used.

D. Diet Pelleting

Pelleting purified diets is a difficult task and should be left to the commercial suppliers who have the required equipment. Successful pelleting depends on the composition of the diet. For the most part, the composition of the diets presented in this paper have been tested and found to meet the commercial standards for diet pelleting. Water is used in the pelleting process, and tap water can contain a variety of trace minerals; therefore, only the purest source of water should be used.

E. Diet Storage

For short-term studies (no more than 3 months), diets can be stored in the dark at 4°C in opaque plastic containers with tight-fitting lids.[26] For longer term studies (up to 6 months) diets can be stored frozen. However, storage for more than 6 months, under the best of conditions, should be avoided. Deterioration, especially with regard to the unsaturated fatty acids, should be monitored periodically. Guidelines for limiting autoxidation of highly unsaturated oils in purified diets during storage and feeding have been presented by Warner et al.[27] and Fritsche and Johnston.[27]

V. PRECAUTIONS FOR DIET FORMULATIONS

From time to time, nutritional studies require alterations in one or more of the major ingredients. If changes in the source or amount of fat, protein, or carbohydrate are made, the investigator should be aware of the composition of the component being substituted and make adjustments to ensure that the required amounts of nutrients are present in the reformulation. When experimental protocols require alterations in calories derived from fat or require changes in fatty acid compositions, adjustments in fat calories should be made on an isocaloric basis, not simply by dilution or substitution. It is rather easy to change the fatty acid composition of the diet because there are many common oil and fat sources available for use in animal diets (Table 12). However, the investigator should recognize that fats and oils containing highly saturated fatty acids with chain lengths longer than 16 are poorly digested.[3] Diets with high concentrations of highly unsaturated fatty acids may also require a higher concentration of antioxidants.

The investigator should also be aware that the bioavailability of a mineral may not always be the same among different mineral sources. If mineral sources other than those recommended here are used, the bioavailability of the mineral in question should be considered. In addition, many of the mineral sources are hygroscopic and may cause the mineral mix to clump. However, substituting mineral sources to make the task of diet mixing easier may not always be the best idea. For example, silicon dioxide is a good mix component, but silicon from this source is almost totally unavailable to the animal. Therefore, the investigator or the supplier should not substitute convenience for prudence.

The investigator should also be aware that the drinking water can provide a substantial amount of certain trace elements, depending on the source of the water. Therefore, it is highly recommended that distilled or deionized water be used in trace element experiments. However, the method of distillation or deionization will depend on the mineral in question. For example, studies with boron should not use water distilled in a glass apparatus.

ACKNOWLEDGMENTS

The author wishes to acknowledge Jim Lindlauf for the review of the diet formulations, Don Yowell of Dyets, Inc., for information regarding the use levels of dextrinized starch in the formulae for pelleting, Dr. Ron Rose of Harlan Teklad, for information on the mineral analyses of protein sources, and Drs. Kathy Sukalski and Eric Uthus for critical review of the manuscript.

REFERENCES

1. American Institute of Nutrition, Second report of the ad hoc committee on standards for nutritional studies, *J. Nutr.*, 110, 1726, 1980.
2. American Institute of Nutrition, Report of the American Institute of Nutrition ad hoc committee on standards for nutritional studies, *J. Nutr.*, 107, 1340, 1977.
3. Reeves, P.G., Nielsen, F.H., and Fahey, G.C., Jr., AIN-93 purified diets for laboratory rodents: Final report of the American Institute of Nutrition ad hoc committee on the reformulation of the AIN-76A rodent diet, *J. Nutr.*, 123, 1939, 1993.
4. National Research Council, *Nutrient Requirements of Laboratory Animals, Third Revised Edition*, National Academy of Sciences, Washington, D.C., 1978.
5. National Research Council, *Nutrient Requirements of Laboratory Animals, Fourth Revised Edition*, National Academy Press, Washington, D.C., 1995.
6. Medinsky, M.A., Popp, J.A., Hamm, T.E., and Dent, J.G., Development of hepatic lesions in male Fischer-344 rats fed AIN-76A purified diet, *Toxicol. Appl. Pharmacol.*, 62, 111, 1982.
7. Bergstra, A.E., Lemmens, A.G., and Beynen, A.C., Dietary fructose vs. glucose stimulates nephrocalcinogenesis in female rats, *J. Nutr.*, 123, 1320, 1993.
8. Failla, M.L., Babu, U., and Seidel, K.E., Use of immunoresponsiveness to demonstrate that the dietary requirement for copper in young rats is greater with dietary fructose than dietary starch, *J. Nutr.*, 118, 487, 1988.
9. Fields, M., Ferretti, R.J., Smith, J.C., Jr., and Reiser, S., Effect of copper deficiency on metabolism and mortality in rats fed sucrose or starch diets, *J. Nutr.*, 113, 1335, 1983.

10. O'Dell, B.L., Fructose and mineral metabolism, *Am. J. Clin. Nutr.*, 58 Suppl. 771S, 1993.
11. Kato, N., Saari, J.T., and Schelkoph, G.M., Cystine feeding enhances defects of dietary copper deficiency by a mechanism not involving oxidative stress, *J. Nutr. Biochem.*, 5, 99, 1994.
12. Cooper, J.R. and Gowing, H.S., Mammalian small intestinal phytase (EC 3.1.3.8), *Br. J. Nutr.*, 50, 673, 1983.
13. Williams, P.J. and Taylor, T.G., A comparative study of phytate hydrolysis in the gastrointestinal tract of the golden hamster (*Mesocricetus auratus*) and the laboratory rat, *Br. J. Nutr.*, 54, 429, 1985.
14. Taylor, T.G. and Coleman, J.W., A comparative study of the absorption of calcium and the availability of phytate-phosphorus in the golden hamster (*Mesocricetus auratus*) and the laboratory rat, *Br. J. Nutr.*, 42, 113, 1979.
15. Moore, R.J., Reeves, P.G., and Veum, T.L., Influence of dietary phosphorus and sulphaguanidine levels on phosphorus utilization in rats, *Br. J. Nutr.*, 51, 453, 1984.
16. Moore, R.J. and Veum, T.L., Adaptive increase in phytate digestibility by phosphorus-deprived rats and the relationship of intestinal phytase (EC 3.1.3.8) and alkaline phosphatase (EC 3.1.3.1) to phytate utilization, *Br. J. Nutr.*, 49, 145, 1983.
17. Moore, R.J. and Veum, T.L., Effect of source and level of dietary yeast product on phytate phosphorus utilization by rats fed low phosphorus diets. *Nutr. Rep. Int.*, 25, 221, 1982.
18. O'Dell, B.L. and Savage, J.E., Effect of phytic acid on zinc availability, *Proc. Soc. Exp. Biol. Med.*, 103, 304, 1960.
19. Davies, N.T. and Nightingale, R., The effects of phytate on intestinal absorption and secretion of zinc, and whole body retention of Zn, copper, iron and manganese in rats, *Br. J. Nutr.*, 34, 243, 1975.
20. Hurrell, R., Juillerat, M.A., Reddy, M.B., Lynch, S.R., Dassenko, S.A., and Cook, J.D., Soy protein, phytate, and iron absorption in humans, *Am. J. Clin. Nutr.*, 56, 573, 1992.
21. Lee, D.Y., Schroeder, J., III, and Gordon, D.T., Enhancement of Cu bioavailability in the rat by phytic acid, *J. Nutr.*, 118, 712, 1988.
22. Davis, P.N., Norris, L.C., and Kratzer, F.H., Interference of soybean proteins with the utilization of trace minerals, *J. Nutr.*, 77, 217, 1962.
23. Bourre, J.-M., Piciotti, M., Dumont, O., Pascal, G., and Durand, G., Dietary linoleic acid and polyunsaturated fatty acids in rat brain and other organs. Minimal requirements of linoleic acid, *Lipids*, 25, 465, 1990.
24. Bourre, J.-M., Francois, M., Youyou, A., Dumont, O., Piciotti, M., Pascal, G., and Durand, G., The effect of dietary α-linolenic acid on the composition of nerve membranes, enzymatic acitivity, amplitude of electrophysiological parameters, resistance to poisons and performance of learning tasks in rats, *J. Nutr.*, 119, 1880, 1989.
25. Lee, J.H., Jukumoto, M., Nishida, H., Ikeda, I., and Sugano, M., The interrelated effects of n-6/n-3 and polyunsaturated/saturated ratios of dietary fats on the regulation of lipid metabolism in rats, *J. Nutr.*, 119, 1893, 1989.
26. Fullerton, F.L., Greenman, D.L., and Kendall, D.C., Effects of storage conditions on nutritional qualities of semipurified (AIN-76) and natural ingredient (NIH-07) diets, *J. Nutr.*, 112, 567, 1982.
27. Warner, K., Bookwalter, G.N., Rackis, J.J., Honig, D.H., Hockridge, E., and Kwolek, W.F., Prevention of rancidity in experimental rat diets for long term feeding, *Cereal. Chem.*, 59, 175, 1982.
28. Ke, P.J., Nash, D.M., and Ackman, R.G., Mackerel skin lipids as an unsaturated fat model system for the determination of antioxidative potency of TBHQ and other antioxidant compounds, *J. Am. Oil. Chem.*, 54, 417, 1977.
29. Fritsche, K.L. and Johnston, P.V., Rapid autoxidaton of fish oil in diets without added antioxidants, *J. Nutr.*, 118, 425, 1988.
30. Gonzalez, M.J., Gray, J.I., Schemmel, R.A., Dugan, L., Jr., and Welch, C.W., Lipid peroxidation products are elevated in fish oil diets even in the presence of added antioxidants, *J. Nutr.*, 122, 2190, 1992.

31. Reeves, P.G., Rossow, K.L., and Lindlauf, J., Development and testing of the AIN-93 purified diets for rodents: results on growth, kidney calcification and bone mineralization in rats and mice, *J. Nutr.*, 123, 1993.
32. Wang, X., Oberlease, D., Yang, M.T., and Yang, S.P., Molybdenum requirement of female rats, *J. Nutr.*, 122, 1036, 1992.
33. Holdsworth, E.S. and Neville, E., Effects of extracts of high- and low-chromium brewer's yeast on metabolism of glucose by hepatocytes from rats fed on high- or low-chromium diets, *Br. J. Nutr.*, 63, 623, 1990.
34. Flatt, P.R., Juntti-Berggren, L., Berggren, P.O., Gould, B.J., and Swanston-Flatt, S.K., Effects of dietary inorganic trivalent chromium (Cr^{3+}) on the development of glucose homeostasis in rats, *Br. J. Nutr.*, 63, 623, 1989.
35. Nielsen, F.H., Nickel, *Trace Elements in Human and Animal Nutrition*, Mertz, W., Ed., Academic Press, Inc., New York, 1987, p. 245.
36. Nielsen, F.H., Vanadium, *Trace Elements in Human and Animal Nutrition*, Mertz, W., Ed., Academic Press, Inc., New York, 1987, p. 275.
37. Seaborn, C.D., Mitchell, E.D., and Stoecker, B.J., Vanadium and ascorbate effects on 3-hydroxy-3-methylglutaryl coenzyme A reductase, cholesterol and tissue minerals in guinea pigs fed low-chromium diets, *Magnes. Trace Elem.*, 10, 327, 1991.
38. Nielsen, F.H., Vanadium in mammalian physiology and nutrition, *Metal Ions in Biological Systems*, Sigel, H. and Sigel, A. Eds., Marcel Dekker, Inc., New York, 1995, p. 543.
39. Nielsen, F.H., Nutritional requirements for boron, silicon, vanadium, nickel, and arsenic: current knowledge and speculation, *FASEB J.*, 5, 2661, 1991.
40. Vallee, B.L. and Auld, D.S., Zinc coordination, function, and structure of zinc enzymes and other proteins, *Biochem.*, 29, 5647, 1990.
41. Hambidge, K.M., Casey, C.E., and Krebs, N.F., Zinc, *Trace Elements in Human and Animal Nutrition*, Mertz, W., Ed., Academic Press, Inc., New York, 1986, p. 1.
42. O'Dell, B.L., Burpo, C.E., and Savage, J.E., Evaluation of zinc availability in foodstuffs of plant and animal origin, *J. Nutr.*, 102, 653, 1972.
43. Hankins, C.C., Veum, T.L., and Reeves, P.G., Effects of autoclaved spray dried egg white as the sole source of dietary protein on zinc requirement and performance of the baby pig, *Nutr. Rep. Int.*, 31, 1057, 1985.
44. Hankins, C.C., Veum, T.L., and Reeves, P.G., Zinc requirement of the baby pig when fed wet autoclaved spray dried egg albumen as the protein source, *J. Nutr.*, 115, 1600, 1985.
45. Reeves, P.G. and Stallard, L., Zinc deficiency reduces the activity of angiotensin-converting enzyme in testicular germ cells and sperm of adult rats, *J. Trace. Elem. Exp. Med.*, 7, 125, 1995.
46. Linder, M.C., *Biochemistry of Copper*, Plenum Press, New York, 1991.
47. Johnson, W.T., Dufault, S.N., and Thomas, A.C., Platelet cytochrome c oxidase activity is an indicator of copper status in rats, *Nutr. Res.*, 13, 1153, 1993.
48. Schuschke, L.A., Saari, J.T., Miller, F.N., and Schuschke, D.A., Hemostatic mechanisms in marginally copper-deficient rats, *J. Lab. Clin. Med.*, 125, 748, 1995.
49. Matz, J.M., Blake, M.J., Saari, J.T., and Bode, A.M., Dietary copper deficiency reduces heat shock protein expression in cardiovascular tissues, *FASEB J.*, 8, 97, 1994.
50. Reeves, P.G., Copper status of adult male rats is not affected by feeding an AIN-93G based diet containing high-zinc and marginal-copper contents, *J. Nutr. Biochem.*, 6, 7, 166, 1996.
51. Saari, J.T., Bode, A.M., and Dahlen, G.W., Defects of copper deficiency in rats are modified by dietary treatments that affect glycation, *J. Nutr.*, 125, 2925, 1995.
52. Matz, J.M., Saari, J.T., and Bode, A.M., Functional aspects of oxidative phosphorylation and electron transport in cardiac mitochondria of copper-deficient rats, *J. Nutr. Biochem.*, 6, 644, 1995.
53. Kramer, T.R., Johnson, W.T., and Briske-Anderson, M., Influence of iron and the sex of rats on hematological, biochemical and immunological changes during copper deficiency, *J. Nutr.*, 118, 214, 1988.

54. Lynch, S.M. and Klevay, L.M., Contrasting effects of a dietary copper deficiency in male and female mice, *Proc. Soc. Exp. Biol. Med.*, 205, 190, 1994.
55. Nederbragt, H. and Lagerwerf, A.J., Strain-related patterns of biliary excretion and hepatic distribution of copper in the rat, *Hepatology*, 6, 601, 1986.
56. Nederbragt, H., Strain- and sex-dependent differences in response to a single high dose of copper in the rat, *Comp. Biochem. Physiol.*, 81, 425, 1985.
57. Nielsen, F.H., Effects of sulfur amino acids and genetic makeup on the signs of copper deficiency in rats, *J. Trace. Elem. Exp. Med.*, 2, 225, 1989.
58. Wein, E.M. and Van Campen, D.R., Mucus and iron absorption regulation in rats fed various levels of dietary iron, *J. Nutr.*, 121, 92, 1991.
59. Baly, D.L., Curry, D.L., Keen, C.L., and Hurley, L.S., Effect of manganese deficiency on insulin secretion and carbohydrate homeostasis in rats. *J. Nutr.*, 114, 1438, 1984.
60. Hurley, L.S. and Keen, C.L., Manganese, *Trace Elements in Human and Animal Nutrition*, Mertz, W., Ed., Academic Press, Inc., New York, 1987, p. 185.
61. Davis, C.D., Ney, D.M., and Greger, J.L., Manganese, iron and lipid interactions in rats, *J. Nutr.*, 120, 507, 1990.
62. Lee, D.-Y., Korynta, E.D., and Johnson, P.E., Effects of sex and age on manganese metabolism in rats, *Nutr. Res.*, 10, 1005, 1990.
63. Lee, D.-Y. and Johnson, P.E., Factors affecting absorption and excretion of ^{54}Mn in rats, *J. Nutr.*, 118, 1509, 1988.
64. Vadhanavikit, S. and Ganther, H.E., Selenium requirements of rats for normal hepatic and thyroidal 5′-deiodinase (Type I) activities, *J. Nutr.*, 123, 1124, 1993.
65. van Middlesworth, L., T-2 mycotoxin intensifies iodine deficiency in mice fed low iodine diet, *Endocrinology*, 118, 583, 1986.
66. Smit, J.G.G., Van der Heide, D., Van Tintelen, G., and Beynen, A.C., Thyroid function in rats with iodine deficiency is not further impaired by concurrent, marginal zinc deficiency, *Br. J. Nutr.*, 70, 585, 1993.
67. Baker, D.H., Construction of assay diets for sulphur-containing amino acids, *Methods Enzymol.*, 143, 297, 1987.
68. Klevay, L.M., The biotin requirement of rats fed 20% egg white, *J. Nutr.*, 106, 1643, 1976.

CHAPTER 2

Basic Tissue Preparation for Electron Microscopy Assessment of Rodents

Robert E. C. Wildman
Department of Nutrient Dietetics
University of Delaware
Newark, Delaware

CONTENTS

0-8493-9611-5/97/$0.00+$.50

I. INTRODUCTION

The advent of the electron microscope has brought a new dimension to biological research. No longer is the visualization of tissue limited to the magnification of 1500X and the resolution of 0.2 μm associated with light microscopy. The transmission electron microscope (TEM) extended the magnification to 250,000X and the resolution to 0.5 nm or better. This relatively new investigative tool has been very kind to trace element researchers. It has allowed high magnification examination of the effects of trace element imbalances and/or factors that alter their metabolism within rodents and other mammals.

The goal of specimen preparation for electron microscopy is to fix and preserve cellular and extracellular components with minimal alteration. Chemical fixatives, such as paraformaldehyde, glutaraldehyde, and osmium tetroxide, are utilized to cross-link and stabilize cellular molecules. Specimens are ultimately infiltrated with an embedding resin, which when polymerized (hardened), will allow for ultrathin sectioning of specimens. Samples are then viewed at varying magnifications and photographed. Electron micrographs can be used to qualitatively and quantitatively assess cellular components and ultrastructure and to generate data to support other aspects of the investigator's research.

While the applications of electron microscopy are indeed broad, this chapter will be limited to basic tissue preparation for ultrastructural assessment. Thorough reviews of electron microscopy, theory, and practice can be found elsewhere.[1–4]

A. Materials for Tissue Preparation for Electron Microscopy

0.2 *M* Sorenson's phosphate buffer
25% EM grade glutaraldehyde
16% EM grade formaldehyde*
Spurr resin components**
  ERL 4206 (4-vinylcyclohexane dioxide)
  DER 736 (diglycidyl ether of polypropylene glycol)
  NSA (nonenyl succinic anhydride)
  DMAE (dimethylaminoethanol)
Osmium tetroxide
Acetone
Sucrose

pH meter
Vacuum oven
Razor blades
Syringes
Needles
Petri dish
Embedding molds
Tooth picks or wooden applicator sticks
Glass vials
Plastic pipettes

* Necessary if McDowell and Trump's fixative is to be used as the primary fixative.
** Can be purchased as a kit.

II. INITIAL TISSUE SECTIONING AND PRIMARY FIXATION

Chemical fixatives, when introduced to tissue, form cross-links between cellular components and themselves. They are provided within a buffer solution that often closely approximates cellular pH and tonicity. Most rodent tissue fixes well when the pH of the fixative-buffer solution is between 7.2 to 7.4. The primary fixative solution can be 2% glutaraldehyde in 0.1 *M* Sorenson's phosphate buffer in 0.1 *M* sucrose at pH 7.2 to 7.4.[6,7] An alternative primary fixative is McDowell and Trump's fixative,[8] which is a 4% formaldehyde and 1% glutaraldehyde in a phosphate buffer solution with a final pH of 7.2 to 7.4. McDowell and Trump's fixative may be more appropriate if tissue samples are also to be used for other electron microscopy techniques, such as immunocytochemistry, in conjunction with basic ultrastructural assessment. However, McDowell and Trump's fixative is relatively hypotonic (176 mOsm/liter).

Tissue that is to be visualized by electron microscopy is first perfused (100 to 140 mmHg) through a supplying artery. Instead of using a perfusion apparatus,[1] many investigators choose to cautiously perfuse the supplying artery using a hypodermic needle and a syringe.[5–7] Tissue should be first perfused with a buffer solution, such as 0.1 *M* Sorenson's phosphate buffer in 0.1 *M* sucrose, and then the primary fixative. Tissue is then excised and rapidly transferred to a small flat container, such as a petri dish, containing fresh primary fixative solution. While periodically pipetting fresh fixative onto the tissue, small sections (1 mm^3 or smaller) of the tissue are made. Thin razor blades (0.004 in.thick) or very fine scalpels work well and limit mechanical damage to tissue. Sections are made perpendicular to the cutting surface, not by a dragging motion. Tissue samples are then transferred to small glass vials containing a fresh fixative. Toothpicks or wooden applicator sticks, shaved to produce a thin reed-like end, can be used to delicately transfer tissue. Samples are maintained in the primary fixative for at least 2 h to allow adequate penetration. If necessary, tissue samples can be stored in the primary fixative at 4°C for several days to weeks with only minimal alterations in cellular ultrastructure. To reduce tissue exposure to air, vials should be capped during processing and storage.

A. Preparation of Solutions

0.2 *M* Sorenson's phosphate buffer (pH 7.2 to 7.4)
(stable at 4° for 2 to 3 months)

1. Prepare Solution A: 0.2 *M* of $NaH_2PO_4 \cdot H_2O$ (27.6 g/liter distilled H_2O) and Solution B: 0.2 *M* of Na_2HPO_4 anhydrous (28.4 g/liter distilled H_2O).
2. Mix 23 ml Solution A and 77 ml Solution B. This will yield a 0.2 *M* Sorenson's phosphate buffer.
3. Adjust the pH as necessary.

2% Glutaraldehyde in 0.1 *M* Sorenson's phosphate buffer in 0.1 *M* sucrose (pH 7.2 to 7.4) (stable at 4°C for 1 to 2 months)

1. Dilute an EM grade glutaraldehyde solution with 0.2 *M* sucrose solution to produce a 4% glutaraldehyde in 0.2 *M* sucrose.
2. Add an equivalent amount of 4% glutaraldehyde in 0.2 *M* sucrose solution to stock 0.2 *M* Sorenson's phosphate buffer to produce 2% glutaraldehyde in 0.1 *M* Sorenson's phosphate buffer in 0.1 *M* sucrose.
3. Adjust the pH as necessary.

McDowell and Trump's 4F:1G Fixative (pH 7.2 to 7.4) (stable at 4°C for 2 to 3 months)

1. Dilute glutaraldehyde with distilled water producing a 4% glutaraldehyde solution. Mix an equivalent amount of 16% formaldehyde with 4% glutaraldehyde producing a 8% formaldehyde-2% glutaraldehyde solution.
2. Dilute an equivalent amount of the 8% formaldehyde-2% glutaraldehyde solution with 0.2 *M* Sorenson's phosphate buffer creating a 4% formaldehyde-1% glutaraldehyde in 0.1 *M* Sorenson's phosphate buffer.
3. Adjust the pH as necessary.

III. SECONDARY FIXATION WITH OSMIUM TETROXIDE

Primary fixatives contain aldehydes that form cross-links between protein molecules in cells.[5] Secondary fixation with 1% osmium tetroxide in 0.1 *M* Sorenson's phosphate buffer (pH 7.2 to 7.4) allows for stabilization of lipid molecules, especially unsaturated double bonds, as well as functioning as a micrographic "stain." When osmium tetroxide interacts with lipids, it is reduced to an electron-dense osmium black, which will inhibit an electron beam at that site.[3] This results in an improved visualization of lipid regions of cells, such as membranes. A thorough description of the careful preparation of 1% osmium tetroxide in a phosphate buffer is found elsewhere.[3]

Since osmium tetroxide penetrates only about 0.25 mm, small sections are necessary to ensure proper penetration.[5] Prior to and after secondary fixation in osmium tetroxide, tissue samples are rinsed in 0.1 *M* Sorenson's phosphate buffer. Rinsing tissue prior to the introduction of osmium tetroxide will remove excessive primary fixative. This can be accomplished by extracting the primary fixative from the tissue sample vials using plastic pipettes. To avoid direct exposure to air, enough fixative is left in the vials to just cover the tissue sections. Using a clean pipette, gently introduce a couple of milliliters of buffer. After 10 min the fluid is removed, again leaving just enough fluid to cover the samples. Fresh

Sorenson's phosphate buffer is quickly pipetted into the vials. After 10 min, repeat this rinsing procedure one more time (Table 1).

After the last rinse, the fluid is pipetted out of the vial, again leaving only enough to cover the tissue. Then, 1% Osmium tetroxide in 0.1 Sorenson's phosphate buffer is pipetted into the vials. After 1 h, excessive osmium tetroxide is removed from the tissue samples with three rinses of 0.1 *M* Sorenson's phosphate buffer following the same procedure as prior to the introduction of osmium tetroxide (Table 1).

TABLE 1.

Fixation and Rinse Schedule

	Fixative/Rinse Buffer	Time (min)
1.	Primary fixative	120
2.	0.1 *M* Sorenson's phosphate buffer	10
3.	0.1 *M* Sorenson's phosphate buffer	10
4.	0.1 *M* Sorenson's phosphate buffer	10
5.	1% osmium tetroxide	60
6.	0.1 *M* Sorenson's phosphate buffer	10
7.	0.1 *M* Sorenson's phosphate buffer	10
8.	0.1 *M* Sorenson's phosphate buffer	10

Note: Vials should be capped and gently agitated periodically.

IV. DEHYDRATION

After secondary fixation with osmium tetroxide, tissue samples are dehydrated by an increasing concentration gradient of either ethanol or acetone. This step is necessary because most commonly used embedding resins are not miscible in water. Therefore, it is necessary to replace the water in tissue samples with an organic solvent. Since the described procedure employs Spurr resin as the embedding medium, acetone will be more appropriate. Rapid, yet thorough, dehydration is necessary to ensure minimal alterations in cellular components and maximal polymerization of the resin throughout the tissue sections (Table 2).

V. RESIN INFILTRATION

Polymerized plastic resins provide an appropriate medium for cutting ultrathin sections (50 to 100 nm) of tissue samples. The most common embedding resins can be classified into two general classes: epoxy resins (i.e., Spurr, Araldite, and Epon) and acrylics (i.e., LR White, LR Gold, Lowicryl, and glycol methacrylate).

Because of its relatively lower viscosity, Spurr resin can successfully infiltrate small samples in as little as 3 h. After the last rinse stage of dehydration (100% acetone), the fluid is pipetted out of the sample vials,

TABLE 2.

Dehydration Schedule for Acetone

	Acetone Concentration	Time (min)
1.	30% acetone	10
2.	50% acetone	10
3.	70% acetone	10
4.	80% acetone	10
5.	85% acetone	10
6.	90% acetone	10
7.	95% acetone	10
8.	100% acetone	10
9.	100% acetone	10

Note: Vials should be capped and gently agitated periodically.

again leaving just enough fluid to cover the samples. Spurr resin is first introduced to the tissue in a 1:1 Spurr to acetone mixture. This is followed by three subsequent changes of 100% Spurr resin (Table 3). The last change can be left overnight if desired. Flat embedding molds can be prepared during the last change of Spurr resin.

TABLE 3.

Schedule for Spurr Infiltration

	Spurr Concentration	Time (min)
1.	1:1 Spurr:acetone	60
2.	100% Spurr	60
3.	100% Spurr	60
4.	100% Spurr	60

Note: Vials should be capped and gently agitated periodically.

Small paper inserts containing identifying information (i.e., date, treatment, subject code) should be placed at the tail end of each embedding mold. Pipette fresh Spurr resin into the embedding mold cavities, filling them about halfway. Transfer one tissue sample to each cavity using a toothpick or shaved wooden applicator stick. It might be necessary to first pipette the samples out of the vial onto a glass plate or other clean surface. A wide stem plastic pipette, with its tip trimmed to yield a wider opening, works well in extracting the tissue samples. Adjust each sample so that it is close to the front end of the mold. Slowly add fresh resin until it slightly exceeds the rim of the cavity, producing a convex appearance.

When investigating specific types of tissue, there may be a desired visual perspective. For example, an investigator may be particularly interested in examining the longitudinal structure of muscle cells.[6,7] To do so, an investigator might be well served by creating orientation markers, such as cutting tissue into rectangles, when performing the initial tissue sectioning. The marker is then used to orientate the tissue samples in the embedding mold cavities. Some tissue provide biological markers, such as the epicardium of cardiac tissue, which when magnified can be used to orientate tissue samples in a specific manner.

Once the transfer of tissue samples to the embedding molds is complete, the molds are transferred to a vacuum oven and polymerized at 60°C. Polymerization should be complete in 12 to 24 h and can be assessed by the rigidity of your "blocks." It is a good idea to fill a couple of embedding mold cavities with only resin and to utilize these as blanks to test whether polymerization is complete.

VI. TRIMMING: THICK AND THIN SECTIONING

Once the blocks have completely polymerized and cooled, they are ready for trimming. Blocks are placed in an ultramicrotome chuck held in a block-trimming base. Using a single-edge razor, slices of resin are cut away until the face of the block (tissue end) has a trapezoidal design.[5] A microtome is then used to cut semi-thin sections of the block, which are mounted on slides and viewed by light microscopy. This step will provide the investigator a glimpse of the tissue at low magnification to determine its suitability for continuation. Semi-thin sectioning also prepares the face of your block for subsequent ultrathin sectioning.

Ultrathin sectioning (50 to 100 nm) is accomplished using a diamond knife. Sections of this thinness are necessary as a beam of electrons demonstrates limited specimen transmission when accelerated at practical voltages (40 to 100 kV). Ultrathin sections are captured onto metal grids, such as copper, and are subsequently stained with uranium and lead salts to enhance tissue contrast. Grids can then be placed into a numbered grid storage box in a random manner by an uninvolved individual who also completes a grid inventory. Thus, the investigator visualizing the samples with the electron microscope will do so without bias to treatment.

VII. ELECTRON MICROSCOPY AND MICROGRAPH ASSESSMENT

Once ultrathin sections are mounted on grids and stained with heavy metal salts, they are ready to be viewed with the electron microscope. Tissue samples can be photographed at various magnifications; however, choose the most appropriate magnifications based upon investigative intentions. Producing negatives at a magnification 2 to 2.5X smaller than final desired magnification, and then enlarging the image to print the desired magnification, will improve the contrast of the micrograph. Lower magnifications (3,000 to 15,000X) will allow for a broader perspective of the componentry of tissue while higher magnifications (>20,000X) will allow for greater visual access to cellular ultrastructure.

Electron micrographs not only allow an investigator to qualitatively assess tissue samples, but also to quantitatively assess various aspects of cells. Quantitative comparison of cellular component dimensions of experimental-treatment rodents to controls can be used to assess the treatment impact. Measurement of small cellular objects can be obtained using a magnifier incorporating a graticule, while larger objects can be measured by a ruler calibrated in 0.5 mm increments. Obtain a number of different measurements and determine the average size. To calculate the actual size of that component, the obtained measure must be reduced by the magnification of the print. For example, if the average measurement of a cell component on a print is 20 mm or 20,000,000 nm and the negative was obtained at 5000X and then enlarged 2X to print, then the actual size of that component is 2,000 nm or 2 μm. (20,000,000 nm ÷ 5000 = 4,000 nm ÷ 2 = 2,000 nm). Typical dimensions of cell components are listed in Table 4.

TABLE 4.

Cell Component Dimensions

Component	Approximate Dimension
Plasma membrane thickness	7 to 8 nm
Endoplasmic reticulum membrane thickness	8 nm
Ribosome diameter	15 to 23 nm
Centriole diameter	150 nm
Mitochondrion diameter (cross section)	1000 nm
Nuclear pore diameter	150 nm
Microtubule diameter	18 to 25 nm
Glycogen	15 to 40 nm
Glycogen rosettes	95 nm
Nuclear envelope thickness	25 to 40 nm
Erythrocyte diameter	8 μm
Cytoplasmic secretory granule diameter	100 nm
Actin microfilament diameter	6 nm

Lower magnification micrographs such as 10,000 to 20,000X can be used to quantitate the volume of cellular components. A transparent grid overlay can be used to estimate the volume densities of cell components by counting the number of grid intersects that contact a specified component relative to the total number of grid intersects.[6,7]

VIII. CONSIDERATIONS DURING TISSUE PREPARATION FOR ELECTRON MICROSCOPY

A number of the chemicals used in the preparation of tissue for electron microscopy are toxic and should be handled with care. An investigator should wear protective clothing and gloves and work under a hood. Spurr resin is considered a carcinogen.

Excessive resin, including that removed from tissue vials, should be collected in a disposable plastic beaker and left under a fume hood for several weeks or until it completely polymerizes. Do not put Spurr, or any resin, down a sink. Any unused osmium tetroxide, including any fluid pipetted out of your vials containing osmium tetroxide, should be stored in a labeled container under a fume hood until it completely oxidizes (turns black). It should then be collected by proper waste disposal personnel. Lastly, many resins, such as Spurr, are hygroscopic and must be maintained in sealed container. Also, sterile water must be used when producing solutions. Bacterial contamination of tissue will lead to alterations in micrograph quality.

REFERENCES

1. Hayat, M.A., *Principles and Techniques of Electron Microscopy: Biological Applications*. CRC Press, Boca Raton, FL, 1989.
2. Dykstra, M.J., *Biological Electron Microscopy: Theory, Techniques, and Troubleshooting*. Plenum Publishing, New York, NY, 1992.
3. Dykstra, M.J., *A Manual of Applied Techniques for Biological Electron Microscopy*. Plenum Publishing, New York, NY, 1993.
4. Beesley, J.E., *Colloidal Gold: A New Perspective for Cytochemical Marking*. Oxford University Press, London, 1989.
5. Dawes, C.J., *Biological Techniques for Transmission and Scanning Electron Microscopy*. Ladd Research Industries, Burlington, VT, 1979.
6. Wildman, R.E.C., Hopkins, R., Failla, M., and Medeiros, D.M., Marginal copper-restricted diets produce altered cardiac ultrastructure in the rat. *Proc. Soc. Exp. Biol. Med.* 1995, 210, 43.
7. Wildman, R.E.C., Medeiros, D.M., and McCoy, E., Cardiac changes with dietary copper, iron, or selenium restriction: Organelle and basal laminae aberrations, decreased ventricular function, and altered gross morphometry. *J. Trace Elem. Exp. Med.* 1995, 8, 11.
8. McDowell, E.M., and Trump, B.F., Histologic Fixatives Suitable for Diagnostic Light and Electron Microscopy. *Arch. Pathol. Lab. Med.* 1976, 100, 405.
9. Spurr, A.R., A Low-Viscosity Epoxy Resin Embedding Medium for Electron Microscopy. *J. Ultrastruct. Res.* 1969, 26, 31.

IRON

Chapter 3

Dietary Iron: Deficiency or Excess

Scott M. Smith
Henry C. Lukaski
Nutrition and Metabolism Laboratory
Biomedical Operations and Research Branch
NASA Johnson Space Center
Houston, Texas

CONTENTS

I. INTRODUCTION

Iron is an essential nutrient, the deficiency or excess of which has significant clinical and physiological implications. Iron deficiency results in multiple functional consequences, including impaired growth and energy metabolism,[1,2] altered thyroid hormone[3,4] and glucose metabolism,[1] and altered immune system function.[5] Some of these sequelae result from

0-8493-9611-5/97/$0.00+$.50

classic iron-deficiency anemia and subsequent reduced oxygen-carrying capacity of red blood cells, while others result from the depletion of iron from dependent enzymes and proteins.

The rat has proven to be a valuable model for the study of iron metabolism, and it has been extensively examined. While the absorption of iron in the rat differs from that in the human, similarities in iron metabolism and the functional consequences of iron deficiency between rats and humans are fairly well documented and have been reviewed.[6] Weanling rats are generally studied because iron deficiency is easier to induce before iron stores have become replete. The recommended amount of iron is 35 mg per kg diet, either in growing or adult rats.[7,8]

The mouse has been used for several studies of iron deficiency or excess, but not as frequently as the rat. Normal growth and hematopoiesis in mice is supported by diets containing between 25 and 100 mg iron per kg diet.[8] The National Research Council (NRC) report set the iron requirement at 35 mg per kg diet (the same as in the rat), based on the American Institute of Nutrition (AIN) 93 formulation.

Iron metabolism in hamsters and guinea pigs has not been as extensively studied as in the rat. The NRC[8] estimates that 50 mg iron per kg diet is adequate to support growth and hematology in these animals.

Careful planning of dietary regimens is extremely important. The amount of iron in the diet, as well as the interaction between iron and other dietary components, must be carefully considered before beginning an experiment to examine the impact of dietary iron deficiency or excess on physiological, behavioral, or nutritional outcomes.

The AIN-recommended diets have provided a solid guide for nutrition studies in rodents.[7,9,10] It is recommended that the current formulation, the AIN-93 diet,[7] be used. It is, however, necessary to slightly modify this diet to induce altered body iron balances (see below). Use of the growth or maintenance diet formulation depends on the age of the animals being studied.[7] The iron content of these two formulations is identical (35 mg iron per kg diet[7]).

II. IRON DEFICIENCY OR EXCESS

The degree of iron deficiency induced for research may vary from study to study, depending on the experimental hypotheses and objectives. It has been assumed (although never scientifically validated) that in animal studies, a severe iron deficiency for a short period of time (i.e., several weeks) is "equivalent" to a moderate deficiency over a longer period of time. Dietary iron concentrations of 3 to 8 mg iron per kg diet will produce a severe deficiency, typically resulting in hematocrits less than 0.25, and hemoglobin concentrations of less than 90 g/L. This is far below the values typically found in human iron-deficiency anemia, in which hematocrits rarely fall below 0.33. However, an extremely deficient animal model allows for the timely study of physiologic alterations without extensive durations (e.g., several months or years) of reduced iron intake, as is usually the case with human iron deficiency. In a study of varying degrees of iron deficiency in rats, Borel et al.[11] demonstrated physiological effects (e.g., altered cardiac catecholamine metabolism, hyperglycemia) at moderate to mild degrees of anemia (hemoglobin = 100 to 120 g/L). Thus, severe iron restriction is not required to study the biological functions of iron.

Iron deficiency in mice has been reported in diets ranging from 2 to 10 mg iron per kg diet.[8] Altered immunologic responses have been observed in mice fed 7 to 10 mg iron per kg diet.[12,13,14] These data demonstrate that a moderate iron deficiency will induce physiological consequences and that the use of a severe deficiency model may not be warranted, depending on the variable of interest.

Studies of iron excess warrant the same concern; that is, the amount of iron in the diet depends upon the desired degree of toxicity. One study in which exorbitant amounts of iron (1255 mg iron per kg diet, as ferric citrate) were used demonstrated a subsequent biological impact[15] in the form of increased hydroxyl radical formation. This experiment highlights an important concern with regard to study design: the need to provide additional dietary citrate to the control group at levels similar to those in the iron overload group. Dietary iron excess in mice has been induced by feeding diets containing 0.5 to 3.5% iron.[8] The majority of studies of iron overload use nondietary means to create this excess, the details of which are addressed elsewhere in this volume.

Experimental design merits careful consideration when one considers animal studies of iron deficiency or excess. While tedious and time consuming, food intake should be determined for studies of either iron deficiency or excess. Assessment of food intake is critical for determining the impact of decreased food intake, a common occurrence in iron-deficient animals,[16] on energy and nutrient intake. The use of pair feeding provides an added control for decreased food intake. This is, however, an imperfect control because the use and storage of energy are altered in a pair-fed (and thus meal-fed) animal compared to an *ad libitum*-fed animal.[17]

III. IRON-ADEQUATE CONTROL DIETS

Iron-sufficient control groups must be used for comparison with groups fed diets containing varied iron contents. Some studies have fed all animals the same iron-deficient diets, while providing the control animals iron in weekly iron dextran intraperitoneal injections.[18,19] While this ensures the provision of an identical (non-iron) dietary content among treatment groups, the metabolism of iron in the control groups is different than that with an iron sufficient diet. Because parenteral iron may actually create transient iron overload, the use of both iron-deficient and iron-sufficient diets is warranted.

The amount of iron missing from the diet formulation must be replaced to maintain similar nutrient densities in the iron deficient and iron adequate diets. Since the substitution with any component will alter that component's concentration in the diet, the use of carbohydrate, the largest diet component, provides the smallest relative change in nutrient concentration.

IV. IMPLICATIONS OF OTHER NUTRIENTS

Manipulation of dietary iron content to examine the physiological role of iron is relatively easy. However, it is important to consider the effects of interactions with other nutrients on the dependent variables under investigation.

The type and amount of carbohydrate in rodent diets has been the subject of discussion for several decades. The AIN-76 diet called for 100% of the carbohydrate source to be provided as sucrose.[9] For studies relating to feed efficiency, energy metabolism, or thermogenesis, this is a complicating factor because sucrose has a known thermogenic impact.[20] A provision to avoid this problem was given in the AIN-76A diet,[10] and many subsequent studies used cornstarch as the dietary carbohydrate source. The resulting diet matrix, however, is very soft and powdered. This presents technical concerns because the diet cannot be pelleted and measurements of food intake and pair-feeding become very difficult. The cornstarch diet alone, surprisingly, also has metabolic implications compared to other carbohydrate mixes,[16] with increased rates of thyroxine activation in 100% starch-fed animals.

The AIN-93 diet[7] suggests a mix of complex (400 g cornstarch per kg diet) and simple carbohydrates (100 g sucrose per kg diet), with dextrinized cornstarch (123 g per kg diet) included to increase the "pelletability." A similar diet formulation was used in iron-deficient rats.[16] This study found that animals fed the diet formulated with dextrins had similar rates of thyroid hormone metabolism as those fed a starch/sucrose mixture. The purpose of the AIN formulation is to provide a common diet formulation to allow cross-study comparisons. The available data regarding dietary carbohydrate and iron deficiency suggest that this is both rational and feasible.

For development of iron deficiency, selection of a fiber source is a concern because of its variable mineral, especially iron, content (Table 1). In some cases, fiber may be omitted from the diet, or if it is included, it must be accounted for in the iron content in the deficient diet formulation. The omission of fiber from either the iron-deficient, sufficient, or overloaded diet requires substitution with carbohydrate to maintain all other nutrients at the recommended levels.[11,16,21]

TABLE 1.

Mineral Content of Commercial Fiber Sources

	Fe (μg/g)	Cu (μg/g)	Zn (μg/g)	Ca (μg/g)	K (μg/g)	Na (μg/g)	Mg (μg/g)	Mn (μg/g)	P (μg/g)
Alphacel[A]	1.24	0.03	0.07	1.8	0.00	3.3	0.30	0.00	0.00
Celufil[B]	92.0	0.38	0.53	198.0	4.25	624.0	44.0	4.25	2.63
CF-11[C]	3.2	0.16	0.08	3.1	0.04	17.1	0.67	0.04	5.13
Celluflower[D]	100.00	0.57	0.30	107.0	1.60	657.0	26.0	1.6	3.57

[A] ICN Biochemicals, Irvine, CA.
[B] United States Biochemical, Cleveland, OH.
[C] Whatman Specialty Products, Maidstone, England.
[D] Harlan-Teklad, Madison, WI.

Courtesy of James Lindlauf, USDA Grand Forks Human Nutrition Research Center.

Dietary fat may also be a confounding variable in the diets, as interactions between type of fatty acid composition and iron absorption and metabolism have been documented (see reference 22 for a review). Dietary fatty acid patterns have also been shown to affect some of the physiological manifestations of iron deficiency, e.g., plasma thyroxine concentrations.[3] The AIN-93 recommendation of soybean oil, with the committee's consideration of P:S and *n*-6:*n*-3 ratios, is warranted.[7]

As with any nutrition study, knowledge and control of all dietary parameters is critical for success. The implications of nutrient-nutrient interactions could fill an entire volume in itself. Researchers must be aware of the diet that is used, and what affect each component may have. The use of a standardized diet formulation is the best way to allow for consistency in an area of research. Obviously, the standardization may need to be altered in certain cases, and this is where planning is required.

V. PREPARATION OF THE DIET

Care must be taken to clean all supplies and equipment with standard laboratory soap and/or a strong chelating agent (e.g., radiac) and water to reduce contaminants. While acid washing supplies are not a requirement for the preparation of iron deficient diets, thorough cleaning of all instruments and tools with soap and rinsing with distilled deionized water 2 to 3 times prior to diet preparation is critical for reducing iron contamination.

Contribution of iron from other individual dietary components must be minimized (e.g., casein, calcium carbonate). Analytical or ultra-pure grades of other mineral sources must be used to assure minimal contamination with iron (or other micronutrients). In our experience, with careful preparation, the iron content of the diet without any added iron is approximately 2 to 3 mg per kg diet. We recommend that the amount of iron be no less than 4 to 5 mg per kg diet to avoid extreme iron deficiency. However, depending on the degree of deficiency and the length of study planned, investigators may wish to titrate the amount of iron in the diet.[1,11]

Diets must be analyzed for iron contamination prior to initiation of the study to ensure use of the planned dietary iron content. Changes in suppliers or lots of any diet components may introduce contamination.

VI. SUMMARY

The further study of iron metabolism, including its deficiency and excess, will allow for a better understanding of the signs, symptoms, and treatments of iron deficiency and iron overload in affected populations. Thought must be given to the formulation of the experimental diet, ensuring that all nutrient requirements are met, while not biasing the results due to macro- or micronutrient interactions. Care must be taken in preparing the diet to ensure minimal iron contamination from other sources.

REFERENCES

1. Borel, M.J., Beard, J.L., and Farrell, P.A., Hepatic glucose production and insulin sensitivity and responsiveness in iron-deficient anemic rats. *Am. J. Physiol.*, 264, E380, 1993.
2. Masini, A., Salvioli, G., Cremones, P., Botti, B., Gallesi, D., and Ceccarelli, D., Dietary iron deficiency in the rat. I. Abnormalities in energy metabolism of the hepatic tissue. *Biochim. Biophys. Acta*, 1188, 46, 1994.
3. Smith, S.M., Johnson, P., and Lukaski, H.C., *In vitro* hepatic thyroid hormone deiodination in iron-deficient rats: effect of dietary fat. *Life Sci.*, 53, 603, 1993.
4. Smith, S.M., Finley, J., Johnson, L.K., and Lukaski, H.C., Indices of *in vivo* and *in vitro* thyroid hormone metabolism in iron-deficient rats. *Nutr. Res.*, 14, 729, 1994.
5. Dallman, P.R., Iron deficiency and the immune response. *Am. J. Clin. Nutr.*, 46, 329, 1987.
6. Dallman, P.R., Biochemical basis for the manifestations of iron deficiency. *Annu. Rev. Nutr.*, 6, 13, 1986.
7. Reeves, P.G., Nielsen, F.H., and Fahey, G.C., AIN-93 purified diets for laboratory rodents: final report of the American Institute of Nutrition Ad Hoc Writing Committee on the reformulation of the AIN-76A rodent diet. *J. Nutr.*, 123, 1939, 1993.
8. National Research Council Subcommittee on Laboratory Animal Nutrition, Nutrient Requirements of Laboratory Animals, Fourth revised edition. National Academy Press, Washington, D.C., 1995.
9. American Institute of Nutrition, Report of the American Institute of Nutrition Ad Hoc Committee on Standards for Nutritional Studies. *J. Nutr.*, 107, 1340, 1977.
10. American Institute of Nutrition, Second Report of the Ad Hoc Committee on Standards for Nutritional Studies. *J. Nutr.*, 110, 1726, 1980.
11. Borel, M.J., Smith, S.H., Brigham, D.E., and Beard, J.L., The impact of varying degrees of iron nutriture on several functional consequences of iron deficiency in rats. *J. Nutr.*, 121, 729, 1991.
12. Omara, F.O. and Blakely, B.R., The IgM and IgG antibody responses in iron-deficient and iron-loaded mice. *Biol. Trace Element Res.*, 46, 155, 1994.
13. Omara, F.O. and Blakely, B.R., The effects of iron deficiency and iron overload on cell-mediated immunity in the mouse. *Br. J. Nutr.*, 72, 899, 1994.
14. Kuvibidila, S., Baliga, B.S., and Murthy, K.K., Impaired protein kinase C activation as one of the possible mechanisms of reduced lymphocyte proliferation in iron deficiency in mice. *Am. J. Clin. Nutr.*, 54, 944, 1991.
15. Kadiiska, M.B., Burkitt, M.J., Xiang, Q-H, and Mason, R.P., Iron supplementation generates hydroxyl radical *in vivo*: An ESR spin-trapping investigating. *J. Clin. Invest.*, 96, 1653, 1995.
16. Smith, S.M. and Lukaski, H.C., Type of dietary carbohydrate affects thyroid hormone deiodination in iron-deficient rats. *J. Nutr.*, 122, 1174, 1992.
17. Cohn, C. and Joseph, D., Role of rate of ingestion of diet on regulation of intermediary metabolism ("meal eating" vs. "nibbling"). *Metab. Clin. Exp.*, 9, 492, 1960.
18. Beard, J. and Tobin, B., Feed efficiency and norepinephrine turnover in iron deficiency. *Proc. Soc. Exp. Biol. Med.*, 184, 337, 1987.
19. Beard, J., Tobin, B., and Smith, S.M., Norepinephrine turnover in iron deficiency at three environmental temperatures. *Am. J. Physiol.*, 255, R90, 1988.

20. Granneman, J.G. and Wade, G.N., Effect of sucrose overfeeding on brown adipose tissue lipogenesis and lipoprotein lipase activity in rats. *Metabolism*, 32, 202, 1983.
21. Smith, S.M. and Lukaski, H.C., Estrous cycle and cold stress in iron-deficient rats. *J. Nutr. Biochem.*, 3, 23, 1992.
22. Lukaski, H.C. and Johnson, P.E., Dietary fatty acids and minerals, in C.K. Chow, Ed., *Fatty Acids in Foods and Their Health Implications*. New York, Marcel Dekker, 1992, 501.

CHAPTER 4

The Use of Iron-Dextran to Produce Iron Overload in Rodents

Phillip Carthew
A. G. Smith
MRC Toxicology Unit
University of Leicester
Leicester, England

CONTENTS

I. INTRODUCTION

Iron in the mammalian body is highly conserved, and there appears to be no controlled route of excretion. Thus, there is only a very small

0-8493-9611-5/97/$0.00+$.50

daily excretion of iron, via intestinal desquamation and sweat, unless significant blood loss occurs, as in women through menstruation.[1] In contrast to excretion, the uptake of iron from food by the digestive tract is more regulated.[2] One exception is the case of primary genetic hemochromatosis, where the normal feedback mechanism which controls iron absorption from the intestine does not operate as normal.[3] This leads to a continuing uptake of iron even when there is no iron deficiency, and eventually results in a high level of tissue iron systemically, in the form of hemosiderin. If the condition is not diagnosed, tissue levels of iron in the liver and heart in particular rise to a level where cell death occurs, and hepatic and cardiac fibrosis ensue. This is referred to as hemochromatosis. Fibrosis of the liver in hemochromatosis progresses to eventual hepatoma and hepatocellular carcinoma, while severe cardiac fibrosis can affect the conductance system of the heart to an extent that can cause death.[3]

Another genetically inherited disease of hemoglobin regulation, β-thalassemia, requires that blood transfusions be given regularly from the first year of life to supplement an ineffectual red blood cell production.[4] Over long periods, this leads to iron overload, and the tissue burden of excess iron causes a similar pattern of cardiac pathology which can be life threatening by the second decade unless prophylactic iron chelation therapy is used to control iron accumulation.[5]

There are also other diseases, such as alcoholic cirrhosis of the liver, where iron deposition is commonly found associated with the predominant pathology.[5] Increased iron body stores presumed to be from elevated iron consumption have been positively correlated to an increased risk of cancer in man.[6,7] There are also many experimental studies which have shown a relationship between excess iron stores and the more rapid development of cancer for liver, kidney, colon, and mesothelioma. There is even a relationship between altered tissue iron stores and the development of mammary cancer.[8–13]

Because of the understanding of the redox chemistry of iron in biological systems and the generation of the hydroxyl free radical through Fenton chemistry, many experimentalists use iron overload as one of the best methods of inducing and studying the pathologies involving the hydroxyl free radical.[14] In addition, it is becoming increasingly apparent that iron plays regulatory roles in many crucial aspects of gene expression and the effects of increased iron status on these functions have barely been explored.[15]

Administration of excess iron in the diet may appear to be the best way of studying experimentally the mechanisms and consequences of iron overload. However, in practice, because of the tight regulatory control in normal animals, it can be difficult to produce high levels of tissue iron loading and there are often undesirable effects on growth because of the large amounts of iron required to be added to diets.[16] For these reasons, the parenteral administration of iron complexes has been found to be extremely useful in rapidly increasing tissue iron levels and in studying processes associated with iron overload.[8,17,18] Indeed, the most common complex, iron-dextran, was first produced for administration to patients (usually by intramuscular injection) as a treatment for anemia.[19] Here, we describe using iron-dextran solutions for mechanistic studies and in models of human iron overload.

II. IRON-DEXTRAN SOLUTIONS

Iron-dextran colloidal solutions can be obtained from a number of sources. Our original experience was with the pharmaceutical product Imferon, a sterile, isotonic solution of an iron-dextran complex containing 50 mg/ml of iron and 200 mg/ml of dextrans prepared by Fisons plc (Loughborough, UK). It was formed by neutralization of ferric chloride in the presence of an alkali-modified dextran 5000 to 7000 mol wt component, so that the dextran colloid stabilized colloidal ferric oxyhydroxide into 2 to 3 nm particles of mol wt 200,000 (data supplied by Fisons).

This product is no longer available and in recent years we have been using an iron-dextran complex marketed by Sigma Chemical Co. Full physical and chemical details are not available but it consists of ferric hydroxide complexed to dextrans at concentrations of 100 mg Fe/ml and 99 to 130 mg of dextrans/ml (average mol wt 5000). The solution also contains 0.5% phenol as a preservative and is highly stable. Our experience with this preparation is that it has given essentially the same results as we obtained previously with Imferon.[8,9] It is likely that other products of a related nature such as iron polysaccharate, polysorbitol, and polymaltose[21,22] will also behave in a similar manner. In experiments, a control dextran is required for the iron-dextran. Because of the polymerized nature of these iron complexes, it is impossible to use the exact non-iron dextran equivalent of iron-dextran. Instead, we employed originally the alkali-treated dextran used to manufacture Imferon, but more recently a dextran from Fluka of molecular weight 6000 has given satisfactory results.

III. METHODS FOR THE PARENTERAL ADMINISTRATION OF IRON DEXTRAN COMPLEX

A. Administration

The two most common methods of administering iron-dextran solution experimentally are by the intraperitoneal and subcutaneous routes. Intraperitoneal dosing has the advantage of more rapid tissue equilibration, due to the large surface area of the peritoneal cavity and the rapid uptake of the iron complex by peritoneal macrophages. However, this can result in acute toxicity at high doses. Subcutaneous dosing of rodents is normally carried out through the loose skin at the back of the neck, so that a depot of complex is established further down the dorsal side of the animal. This depot is gradually absorbed and iron-dextran is less toxic when administered in this way, although occasionally there can be problems with scabbing at the site of injection, especially if the dosing regime involves repeated injections at the same site. In our regimes, we have not observed injection site sarcomas.[23] To produce a low level of systemic iron overload primarily affecting the mononuclear phagocytic cells (referred to as reticuloendothelial cells in older literature), iron-dextran (Sigma) can be injected subcutaneously at a dose of 10 ml/kg body weight (1 g iron/kg) using a 23G × 1″ (0.6 × 25 mm) needle. To avoid leakage of the viscous solution from the skin after withdrawing the needle, the injection puncture should be gently closed as the needle is withdrawn by holding the fold of skin between the forefinger and thumb for 10 sec. Using these methods, rodents can be quickly and efficiently loaded with large amounts of iron.

B. Levels Attained

Parenchymal cells of the liver and cardiac myocytes do not rapidly develop significant amounts of Perl's stainable hemosiderin after a single dose, but iron will redistribute to hepatocytes within days of dosing.[17,18,24] The liver nonheme iron levels that are found in rodents after a single dose of iron-dextran are 250, 150, and 300 nmol/mg dry weight for gerbils, mice, and rats, respectively. Only relatively low levels of nonheme iron are found in the rodent heart after a single dose of iron-dextran. Very little, if any, hemosiderin will be found in cardiac myocytes at any time period after a single dose. The cardiac nonheme iron levels are 15, 10, and 20 nmol/mg dry weight for gerbil, mice, and rats, respectively.

To achieve a higher degree of iron loading in the liver and heart, the iron-dextran administration can be repeated on a weekly basis. Rats, mice, and gerbils will tolerate a repetition of dosing of up to 8 weeks if the dose given at weekly intervals is reduced to a half of that given as a single dose (5 ml/kg body weight, iron-dextran [Sigma]). Three months after the final dose, this will increase the nonheme iron tissue levels in the liver to around twice the levels achieved with a single dose, and comparable to those seen in cases of human hemochromatosis. These iron levels are approximately 550 nmol/mg dry weight of liver tissue in the gerbil, 600 nmol/mg dry weight in the rat liver, and 350 nmol/mg dry weight in the mouse liver.[18] The cardiac nonheme iron tissue levels are highest in gerbils at around 90 nmol/mg dry weight, while mice are very similar at around 80 nmol/mg dry weight. Rats are considerably lower at around 50 nmol/mg dry weight.[18] The cardiac iron levels for rats are consistent with the observation that there is very little Perl's stainable hemosiderin in rat cardiac myocytes.[18]

Studies in mice have shown that the $t_{1/2}$ for iron in the liver after a single dose of 600 mg Fe/kg (as Imferon) is approximately 40 weeks. The levels attained in the spleens of mice do not appear to reach the values for the liver, but have at least as long a half life.[20]

IV. CELLULAR DISTRIBUTION OF IRON

Iron given parenterally at first localizes in the cells of the mononuclear phagocytic system throughout the body. There is very little difference in the initial localization and tissue distribution of iron for the rodent species commonly used experimentally. However, as iron accumulates in organs, either with time as it redistributes from the site of injection, or because it is readministered, parenchymal cells accumulate hemosiderin that can be demonstrated by the Perl's reaction for iron.

The liver, which is a target organ of interest experimentally because it is severely affected in human hemochromatosis, shows differences in iron distribution between species with time and dose administered. With increasing amounts of iron in the rat liver, there is often a lining up of hemosiderin along the bile canaliculi in a characteristic fashion (Figure 1). The mouse tends to distribute hemosiderin in a heterogeneous manner between hepatocytes (Figure 2), while the Mongolian gerbil shows a more homogeneous distribution of hemosiderin in hepatocytes and Kupffer cells (Figure 3).[18] The levels of iron in isolated hepatocytes from mice, rats, and gerbils given iron-dextran have been compared (Table 1).

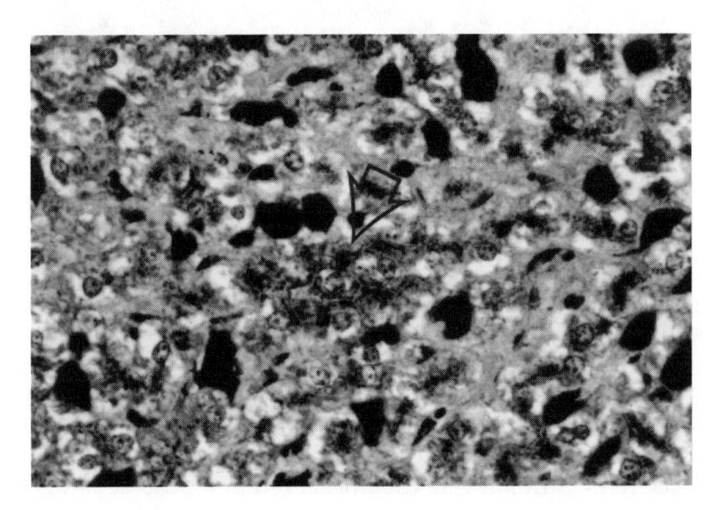

Figure 1
Hemosiderosis of the rat liver showing the characteristic "lining up" of Perl's positive hemosiderin granules along the bile canaliculi (arrow). Perl's reaction for iron × 330.

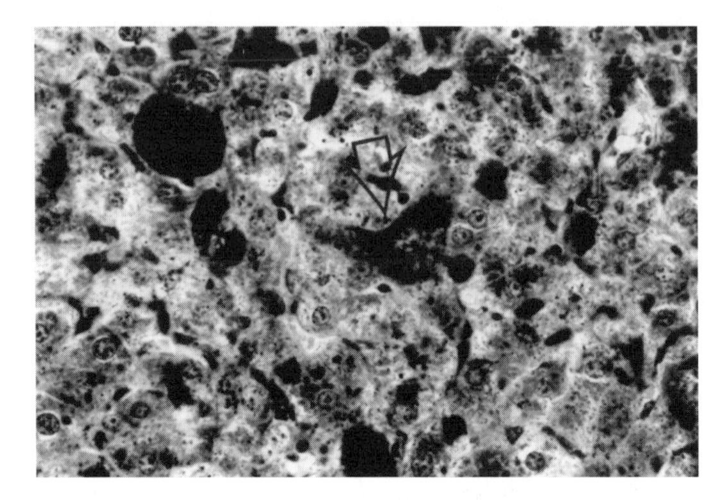

Figure 2
Hemosiderosis of the mouse liver showing the more heterogeneous distribution of stainable hemosiderin in particular hepatocytes (arrow). Perl's reaction for iron × 330.

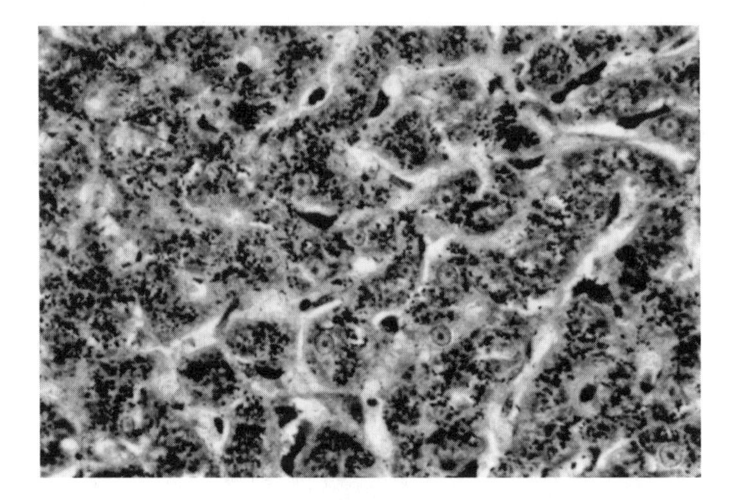

Figure 3
Hemosiderosis of the gerbil liver showing the more homogeneous distribution of stainable hemosiderin in hepatocytes and Kupffer cells. Perl's reaction for iron × 330.

Increases over controls are not as great as might be presumed from total iron levels of the liver. This is due to the presence of giant cells derived from fused iron-laden Kupffer cells in whole livers. However, appreciable rises in iron levels were observed in mouse hepatocytes after only 1 week and in a dose-dependent manner.[25] Loading of mice with iron-dextran may prove to be extremely useful in studying the influence of iron on nuclear function due to the chronic localization in hepatocyte nuclei.[20,25]

TABLE 1.

Iron Content of Isolated Hepatocytes From Iron-Dextran Treated Rodents

Species	Iron	Iron Content ($\mu g/10^6$ cells)
Mouse	–	2.6 ± 0.5
	+	28.2 ± 3.6
Rat	–	0.9 ± 0.1
	+	22.3 ± 6.7
Gerbil	–	0.1 ± 0.01
	+	9.9 ± 2.7

Note: Male C57BL/10ScSn mice, female F344 rats, and male Mongolian gerbils received a single dose of iron-dextran (600mg Fe/kg) or dextran as a control by i.p. injection, 2 weeks before isolation of hepatocytes. Results are means ± SD of 3 or 4 determinations. Hepatocytes were obtained by perfusion of livers with collagenase.

Data adapted from Madra, S., et al. *Carcinogenesis*, 16, 719, 1995. With permission.

The heart, which is the other major organ affected in hemochromatosis, initially only shows the presence of stainable hemosiderin in macrophages throughout the myocardium. As the amount of iron in the heart increases, myocytes of the mouse myocardium can be demonstrated to contain large amounts of hemosiderin. The rat appears to accumulate relatively little hemosiderin in myocytes, while the Mongolian gerbil also has considerable amounts of hemosiderin present in myocytes (Figure 4).[18]

The repeated administration of iron dextran to rodents will also lead to hemosiderin deposition in parenchymal cells of other organs. The tubular cells of the kidney and the acinar cells of the pancreas show increasing deposits of hemosiderin with time, after repeated administration of parenteral iron dextran.

V. DIFFERENCES IN THE RESPONSE OF SPECIES

There is a very marked difference in the response of some species to the iron overload induced by iron-dextran, which is also tissue variable.

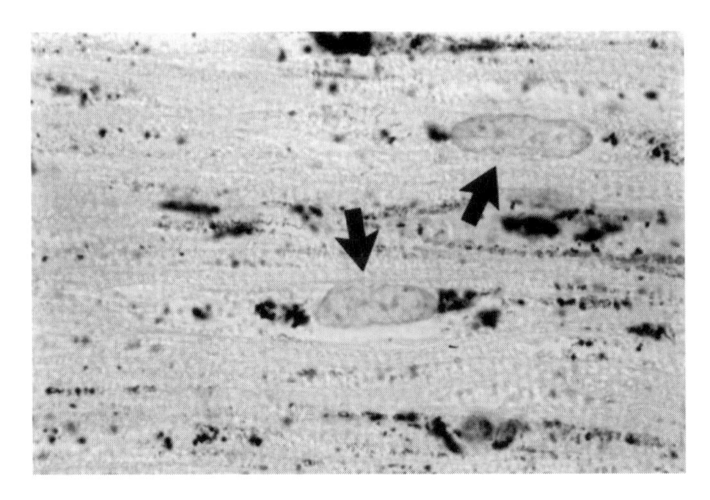

Figure 4
Hemosiderin in the myocytes of the gerbil myocardium. Note the characteristic perinuclear location of hemosiderin (arrows), as commonly seen in man. Perl's reaction for iron × 825.

This becomes particularly important when parenteral iron-dextran is being used to achieve an iron overload state to study the mechanism of the pathology of human hemochromatosis.[18] Even when high levels of tissue nonheme iron are achieved by repeated administration of iron dextran, the response of rats and mice does not model that seen in human hemochromatosis. There is no evidence of the severe degeneration of hepatocytes or cardiac myocytes in either rats or mice despite achieving the levels of iron which can be of the same order as seen in patients with this disease. The pattern of repair seen in patients with hemochromatosis, namely hepatic and cardiac fibrosis, just does not occur in rats and mice.[17,18]

In contrast, the Mongolian gerbil develops a severe hepatic and myocardial fibrosis with the same tissue levels of nonheme iron commonly seen in human hemochromatosis.[18,26] This is particularly of value in the examination of the effectiveness of iron chelators for use in the treatment of the hemosiderosis associated with repeated blood transfusions necessary for sufferers from β-thalassemia.[27] Thus, administration of iron-dextran to the gerbil, unlike either the rat or mouse, at least expresses the same pathological response to chronic iron overload as that seen in man,[18,28] and incidentally, as originally observed in gerbils suffering from endotoxin induced hemorrhage.[29]

The particular susceptibility of the gerbil to the development of the major pathologies associated with hemochromatosis in man is not due to a greater tissue concentration of iron, but rather to an intrinsic species susceptibility to iron overload.[18] This may be mediated through a lack of ability to control the levels of oxidative stress that are induced in gerbil hepatocytes and myocytes.[18]

Although the iron-loaded gerbil appears to be the best animal model so far for secondary hemochromatosis in humans, iron-dextran treatment of other rodents is still an important tool in the elucidation of fundamental toxicological and pathological processes as well as studying the action of iron chelating drugs.[8,9,30,31]

VI. LIMITATIONS

Of course, one limitation of this system is the considerable loading of liver macrophages as well as of hepatocytes. If the intention is to use it as a model for primary genetic hemochromatosis, in which the hepatocytes are primarily loaded at first, the Kupffer cell involvement must be taken into account. In addition, injection of iron-dextran via the subcutaneous route will stain the whole of the body muscle and skin of the animal.

REFERENCES

1. Brock, J.H., The biology of iron, in *Iron in Immunity, Cancer and Inflammation,* de Sousa, M. and Brock, J.H. Eds., John Wiley and Sons, Chichester, 1989, 35.
2. Conrad, M.E. and Umbreit, J.N., Iron absorption — the mucin-mobiliferrin-integrin pathway. A competitive pathway for metal absorption. *Am. J. Hematol.,* 42, 67, 1993.
3. Bothwell, T.H., Charlton, R.W., and Motulsky, A.G., Hemochromatosis, in *The Metabolic and Molecular Bases of Inherited Disease,* Scriver, C.R., Beaudet, A.L., Sly, W.S., and Valle, D., Eds., McGraw-Hill, 1995, 2237.
4. Weatherall, D.J., Clegg, J.B., Higgs, D.R., and Wood, W. G., The hemoglobinopathies, in *The Metabolic and Molecular Bases of Inherited Disease,* Scriver, C.R., Beaudet, A.L., Sly, W.S., and Valle, D., Eds., McGraw-Hill, 1995, 3417.
5. Porter, J.B., Huehns, E.R., and Hider, R.C., The development of iron chelating drugs, *Balliere's Clin. Haematol.,* 2, 257, 1989.
6. Stevens, R.G., Graubard, B.I., Micozzi, M.S., Neriiski, K., and Blumberg, B.S., Moderate elevation of body iron level and increased risk of cancer occurrence and death. *Int. J. Cancer,* 56, 364, 1994.
7. Weinberg, E., Roles of iron in neoplasia: Promotion, prevention and therapy. *Biol. Trace Element Res.,* 34, 123, 1992.
8. Smith, A.G., Cabral, J.R.P., Carthew, P., Francis, J.E., and Manson, M.M., Carcinogenicity of iron in conjunction with a chlorinated environmental chemical, hexachlorobenzene, in C57BL/10SCSn mice, *Int. J. Cancer,* 43, 492, 1989.
9. Smith, A.G, Francis, J.E., and Carthew, P., Iron as a synergist for hepatocellular carcinoma induced by polychlorinated biphenyls in Ah-responsive C57BL/10ScSn mice, *Carcinogenesis,* 11, 437, 1990.
10. Ebina, Y., Okada, S., Hamazaski, S., Ogino, F., Li, J., and Midorikawa, O., Nephrotoxicity and renal cell carcinoma after use of iron- and aluminum-nitrilotriacetate complexes in rats, *JNCI,* 76, 107, 1986.
11. Siegers, C.P., Bumann, D., Baretton, G., and Younes, M., Dietary iron enhances the tumor rate in dimethylhydrazine-induced colon carcinogenesis in mice, *Cancer Letters,* 41, 252, 1988.
12. Okada, S., Hamazaki, S., Toyokuni, S., and Idorikawa, O., Induction of mesothelioma by intraperitoneal injections of ferric saccharate in male Wistar rats, *Brit. J. Cancer,* 60, 708, 1989.

13. Thompson, H.J., Kennedy, K., Witt, M., and Juzefyk, J., Effect of dietary iron deficiency or excess on the induction of mammary carcinogenesis by 1 methyl-1-nitrosourea, *Carcinogenesis,* 12, 111, 1991.
14. Aisen, P., Cohen, G., and King, J.O., Iron Toxicosis, *Int. Rev. Exp. Pathol.*, 31, 1, 1991.
15. Theil, E.C., Iron regulatory elements (IRES): a family of mRNA non-coding sequences. *Biochem. J.*, 304, 1–11, 1994.
16. Park, C.H., Bacon, B.R., Brittenham, G.M., and Tavill, A.S., Pathology of dietary carbonyl iron overload in rats, *Lab. Invest.*, 57, 555, 1987.
17. Carthew, P., Edwards, R.E., Smith, A.G., Dorman, B., and Francis, J.E., Rapid induction of hepatic fibrosis in the gerbil after the parenteral administration of iron-dextran complex, *Hepatology,* 13, 534, 1991.
18. Carthew, P., Dorman, B.M., Edwards, R.E., Francis, J.E., and Smith, A.G., A unique rodent model for both the cardiotoxic and hepatotoxic effects of prolonged iron overload, *Lab. Invest.*, 69, 217, 1993.
19. Capell, D.F., Hutchinson, H.E., Hendry, E.B. and Conway, H., A new carbohydrate-iron haematinic for intramuscular use, *Brit. Med. J.*, 2, 1255, 1954.
20. Smith, A.G., Carthew, P., Francis, J.E., Edwards, R.E., and Dinsdale, D., Characterization and accumulation of ferritin in hepatocyte nuclei of mice with iron overload, *Hepatology,* 12, 1399, 1990.
21. Moore, R.D., Muman, V.R., and Schoenberg, M.D., The transport and distribution of colloidal iron and its relation to the ultrastructure of the cell, *J. Ultrastructure Res.*, 5, 244, 1961.
22. Iancu, T.C., Rabinowitz, M., Brissot, P., Guillouzo, A, Deugnier, Y., and Bourel, M., Iron overload of the liver in the baboon, An ultrastructural study. *J. Hepatol.*, 1, 261, 1985.
23. Haddow, A. and Horning, E.S., On the carcinogenicity of an iron-dextran complex, *JNCI,* 24, 109, 1960.
24. Pechet, G.S., Parenteral iron overload. Organ and all distribution in rats, *Lab. Invest.*, 20, 119, 1969.
25. Madra, S., Styles, J., and Smith, A.G., Perturbation of hepatocyte nuclear populations induced by iron and polychlorinated biphenyls in C57BL/10ScSn mice during carcinogenesis, *Carcinogenesis,* 16, 719, 1995.
26. Carthew, P., Edwards, R.E., Smith, A.G. and Cox, T.M., Excessive ferritin accumulation in fibroblast cells is associated with hepatic fibrosis due to iron overload in man, in *Molecular and Cell Biology of Liver Fibrogenesis*, Gressner, A.M. and Ramadori, G., Eds., Kluwer Academic Publishers, Dordrecht, Boston, London, 1992, 316.
27. Carthew, P., Smith, A.G., Hider, R.C., Dorman, B.M., Edwards, R.E., and Francis, J.E., Potentiation of iron accumulation in cardiac myocytes during the treatment of iron overload in gerbils with the hydroxypyridinone iron chelator CP94, *Biol. Metals.*, 7, 267, 1994.
28. Pietrangelo, A., Gualdi, R., Casalgrandi, G., Montosi, G., and Ventura, E., Molecular and cellular aspects of iron-induced hepatic cirrhosis in rodents. *J. Clin. Invest.*, 95, 1824, 1995.
29. Carthew, P., Edwards, R.E., and Dorman, B.M., Hepatic fibrosis and iron accumulation due to endotoxin induced haemorrhage in the gerbil, *J. Comp. Pathol.*, 104, 303, 1991.
30. Urquhart, A.J., Elder, G.H., Roberts, A.G., Lambrecht, R.W., Sinclair, P.R., Bement, W.J., Gorman, N., and Sinclair, J., Uroporphyria produced in mice by 20-methylcholanthrene and 5-aminolaevulimic acid, *Biochem. J.*, 253, 357, 1988.
31. Porter, J.B., Morgan, J., Hoyes, K.P., Burke, L.C., Huehns, E.R., and Hider, R.C., Relative oral efficacy and acute toxicity of hydroxy-pyridin-4-one iron chelators in mice, *Blood,* 76, 2389, 1990.

Chapter 5

EXTRACTION AND ANALYSIS OF IRON SPECIES IN DIET AND GUT

R. J. Simpson
C. Ogunkoya
T. J. Peters
College School of Medicine
London, England

CONTENTS

0-8493-9611-5/97/$0.00+$.50

I. INTRODUCTION

The first step in intestinal absorption is the digestion of diet and presentation of soluble nutrients to mucosal transporters. Bioavailability of trace metals is critically dependent on the speciation of these metals in the gastrointestinal lumen during the absorptive process. Digestion of food serves to solubilize and release iron along with a complex, diet-dependent mixture of potential iron ligands. The iron species formed during this process diffuse across the mucous layer and interact with the brush border membrane of the duodenal enterocytes. The absorption of iron is affected by the quantity and chemical forms which arrive at the brush border surface.

Differing degrees of solubilization of iron from foods, together with the chemical species which result, are both determinants of iron bioavailability. Methods capable of distinguishing between different iron species are required for analyzing foods and digesta from the intestinal lumen and *in vitro* digestion products. This approach aims to develop chemical methods which can assay the bioavailable iron content of foods and thereby predict food iron bioavailability in humans.[1]

Several different approaches to this problem have been tried. A sophisticated multi-step analysis for iron was developed by Lee and Clydesdale[2] and applied to several foods. The determination of dialysability has also been useful in iron speciation analysis.[3] Other work has concentrated on the valency of iron and possible reducing agents in foods and the intestinal lumen.[4-8] No suitable noninvasive method for investigating solid material (such as food or digesta) is available; therefore, the application of these methods requires cautious validation and interpretation.

Recently we have applied a novel approach to trace element speciation to the process of intestinal iron absorption in the rat.[9] We showed that the method of sequential extraction[10] can be applied to the contents of rat gastrointestinal tract and to *in vitro* simulated digestions. This method can demonstrate changes in speciation of iron on passage down the gastrointestinal tract and during *in vitro* simulated digestion. The method is not restricted to iron and is suitable for simultaneous analysis of many trace metals in solid and liquid samples. This method has the potential to be useful for the investigation of the speciation of trace metals in the gastrointestinal lumen and for the further investigation of *in vitro* simulation of digestion.

The technique can give useful information on changes in iron speciation in both solid, semi-solid and liquid material, and therefore provides a useful approach to metal ion speciation in foods and digesta. The method yields five fractions, termed (after Tessier et al.[10]) exchangeable, carbonate- , oxide- , organic-bound, and residual fractions. The rodent diets being studied contain little haem iron and the organic-bound fraction was found to be small and relatively unchanging in our initial study.[9] It should be noted that these fraction names are operational and that their selectivity and specificity requires experimental validation for a given substrate.[11,12] Model iron complexes have been identified with each of the first three fractions;[9] however, empirical validation, such as has been applied to other *in vitro* speciation techniques,[1] has not yet been attempted.

The principle problems with the extraction technique are caused by incomplete extraction of a fraction, cross contamination of fractions, and difficulties of interpretation of fractions as specific iron species. Three approaches to interpretation of the speciation technique are possible: (1) incorporation of known species into the material to be analyzed, (2) varying the order of extractions and repeating extraction steps, and (3) investigation of the absorption of iron from individual fractions. It is hoped that foods, being less variable in overall composition than soils or sediments, will not require individual validations except where large changes in composition occur. In this chapter, we describe the application of the above sequential extraction technique to the analysis of a rodent diet. However, the same technique can be applied to gastrointestinal contents as described by Simpson et al.[9] We further describe some approaches to validation and interpretation of the speciation method based on (1) and (2) above.

II. SEQUENTIAL EXTRACTION METHOD

Sequential extractions were performed with the method of Tessier et al.,[10] as applied to rodent diet and gastrointestinal contents by Simpson et al.[9] Five fractions are obtained and analyzed for iron. Volumes given are for 1g samples. Each fraction was the supernatant obtained after 30 min centrifugation at 10,000 to 12,000 x *g* combined with a wash obtained by mixing the pellet with 8 ml H_2O and recentrifuging. The remaining pellet was subjected to the next extraction step (Figure 1).

Exchangeable fraction: incubate for 1 h with 8 ml 1*M* $MgCl_2$ with continuous agitation at room temperature.

Carbonate-bound: incubate for 4 h with 8 ml 1*M* sodium acetate (pH 5.0, adjusted with acetic acid) with continuous agitation at room temperature.

Oxide-bound: incubate with 20 ml 0.04*M* $NH_2OH.HCl$ in 25% (v/v) acetic acid at 95°C for 16 h.

Bound to organic matter: incubate with 3 ml 0.02*M* HNO_3 and 5 ml 30% H_2O_2 (pH 2, adjusted with HNO_3) for 2 h at 85°C with occasional agitation, add a further 3 ml of 30% H_2O_2 (pH 2) and heat for a further 3 h at 85°C. The mixture was cooled, 5 ml of 3.2 *M* NH_4 acetate in 20% (v/v) HNO_3 added, then diluted to 20 ml and agitated continuously for 30 min.

Residual fraction: the final pellet.

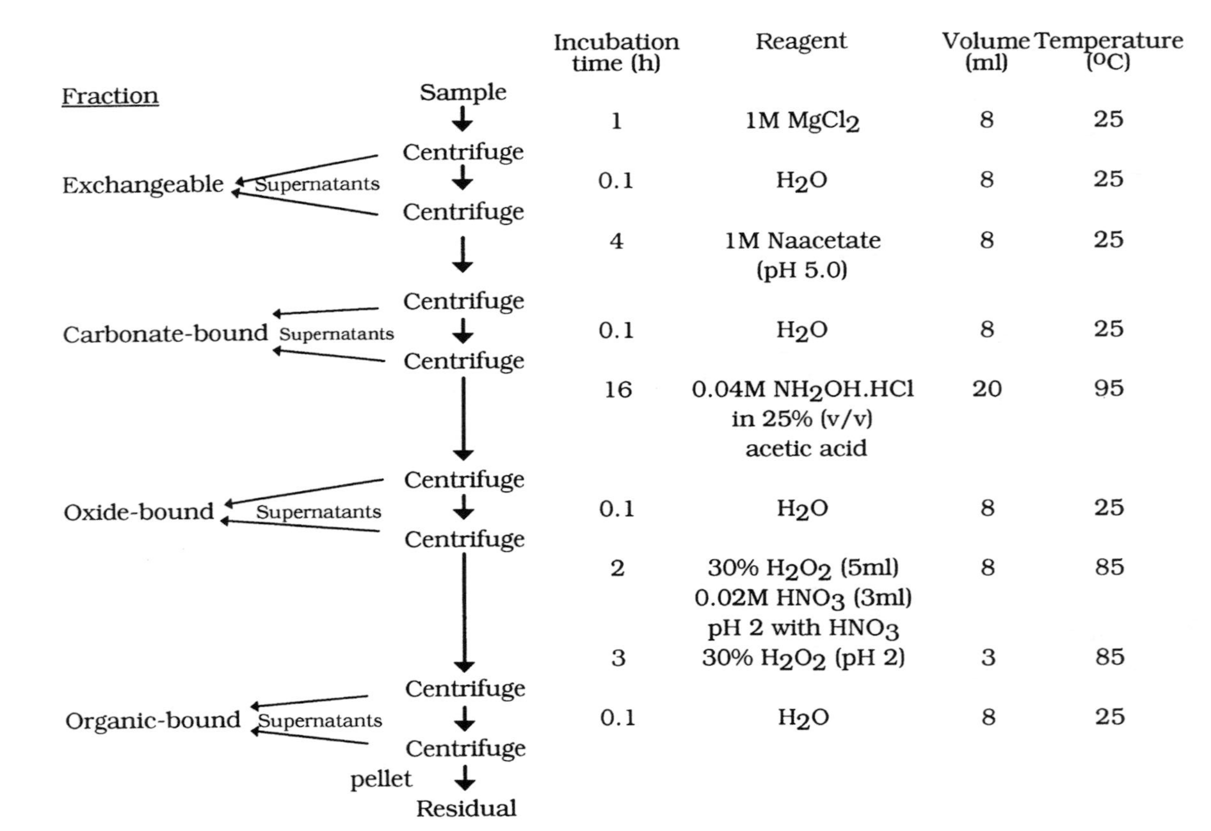

Figure 1
Flow chart for sequential extraction procedure. Each centrifugation was for 30 min at 10,000to 12,000 x *g*.

The total Fe in all five fractions was summed and the content of each fraction expressed as a percentage of the total.

The exchangeable fraction includes soluble Fe complexes and iron displaced from insoluble ligands by excess Mg^{2+}. The carbonate-bound fraction represents Fe bound to insoluble ligands with a pK of 5 or more. The oxide-bound fraction includes Fe bound in a form susceptible to reduction and/or bound to ligands with a pK of 2 or more. The organic-bound fraction includes insoluble haem Fe as well as Fe bound to other ligands degraded by oxidation. The residual fraction represents Fe bound to ligands not degraded by the above procedures. The names used to describe the fractions are taken from the soil analysis method[10] and therefore are somewhat arbitrary in the context of the application to diet or intestinal contents. These names are retained here for brevity and to avoid confusion.

III. IRON ASSAY

Fractions from sequential extractions were assayed according the modified method of Foy et al.,[13] as described by Simpson et al.[6] Aliquots of fractions, 100 µl, were extracted three times by boiling for 10 min with 200 µl of 12.5% trichloroacetic acid/ 2% sodium pyrophosphate and centrifuging for 5 min at 10,000 x *g*. The supernatants were combined and a 200 µl aliquot was mixed with 200 µl sodium ascorbate, 590 µl of sodium acetate (pH 4.8), and 10 µl 0.1*M* ferrozine. Absorbance of the mixture was read at 562 nm. Internal and external standards were prepared from stock 10 mM $FeCl_3{\cdot}6H_2O$ in 10 m*M* HCl. Recovery of internal standards was in the range 87 to 108% for the five fractions.

IV. ANIMALS, DIETS AND MODEL COMPOUNDS

A. Animals and Diet

Diet was obtained from SDS (Witham, Essex, UK; diet RM1), and is based on a soya bean concentrate as protein source. The diet contains 3.6% crude fiber, 5.3% ash, 2.4% crude oil, 17.9% crude protein, and 45% starch, and includes 106 mg Fe/kg (manufacturer's data). Assayed iron content of the diet was variable from batch to batch, with values in the range 167 to 212 mg/kg diet. Mice (Balb/c strain) were fed this diet *ad libitum* from weaning until they weighed 25 to 30 g. Mice were fasted for 12 h (grid-bottom cage) then offered ^{59}Fe-labeled diet or diet remains and left for 1 to 4 h. They were killed by cervical dislocation and the entire gastrointestinal tract was removed. The stomach and intestinal tract were rapidly removed, without allowing leakage of contents, and the duodenum (first

35 mm of small intestine) and proximal jejunum (next 65 mm) taken and the contents flushed into vials with 5 ml each of 0.5 M NaCl. The stomach, including contents, ileum including contents, duodenum, proximal jejunum, duodenal, and proximal jejunal washings were separately counted for radioactivity on a Beckman gamma-7000 (LKB 1282 Compugamma CS). The carcass was counted for radioactivity in a large volume, high-resolution gamma counter.[14] Appropriate standards were employed to relate counts from the two machines. Intestinal uptake was calculated as the sum of duodenal and carcass radioactivity.

B. Radiolabeled Diets and Model Compounds

$^{59}FeCl_3$ was prepared by mixing 2 μl of 200 μCi/ml $^{59}FeCl_3$ ($^{59}FeCl_3$, 25μCi/g, Amersham International, Bucks, UK) with 100 μl of 10 m*M* $FeCl_3$ in 10 m*M* HCl, then adding 100 μl H_2O. ^{59}Fe/ascorbate (100 μ*M* Fe, 2 m*M* sodium ascorbate) was prepared within 5 min of use by mixing 2 μl of 200 μCi/ml $^{59}FeCl_3$ in HCl with 100 μl of 10 m*M* $FeCl_3$ in 10 m*M* HCl and 200 μl fresh 0.1 *M* sodium ascorbate. This mixture was brought to the required volume with 0.04 *M* NaHEPES (pH 7.4), 0.3 *M* NaCl. ^{59}FeNTA and ^{59}FeEDTA solutions were prepared as above except that 20 μl 0.1 *M* $NTA(Na)_3$ or 0.1 *M* EDTA (pH 7.4) were substituted for the ascorbate solution.

Fe/bicarbonate was prepared by mixing 2 μl of $^{59}FeCl_3$ stock with 100 μl of 10 m*M* $FeCl_3$ in 10 m*M* HCl, then adding 100 μl 0.1 *M* $NaHCO_3$. Fe/phosphate was prepared by mixing 2 μl of $^{59}FeCl_3$ stock with 100 μl of 10 m*M* $FeCl_3$ in 10 mM HCl, then adding 100 μl 0.1 *M* sodium phosphate (pH 7.0) followed by 100 μl of 40 m*M* HEPES-NaOH/0.2 *M* NaCl (pH 7.4). Fe/oxide was prepared by mixing 2 μl of $^{59}FeCl_3$ stock with 100 μl of 10 m*M* $FeCl_3$ in 10 mM HCl, then adding 100 μl 10 m*M* Na HEPES/0.15 *M* NaCl (pH 7.4) followed by 10 μl 1 *M* NaOH. These mixtures were analyzed by sequential analysis alone or after mixing with 100 mg of powdered RM1 diet. ^{59}Fe-labeled diet was prepared by mixing 3 g of powdered RM1 diet with $^{59}FeCl_3$ (2.5 μCi), prepared as described above, and drying the mixture.

V. IRON SPECIATION IN DIETS AND MODEL COMPOUNDS

A. Endogenous Iron Speciation in RM1 Diet

Figure 2 shows the speciation of endogenous iron in the diet. This speciation pattern differed from that seen with a distinct rodent diet;[9] in particular, the RM1 diet has much less iron in the residual fraction than the diet studied previously. Note that the previously studied diet was a cereal-based diet while the present diet is based on soy bean concentrate.

This suggests that the method is indeed able to detect dietary differences in iron speciation. It was therefore necessary to perform experiments to determine whether such differences are attributable to differences in chemical speciation of iron within the diet.

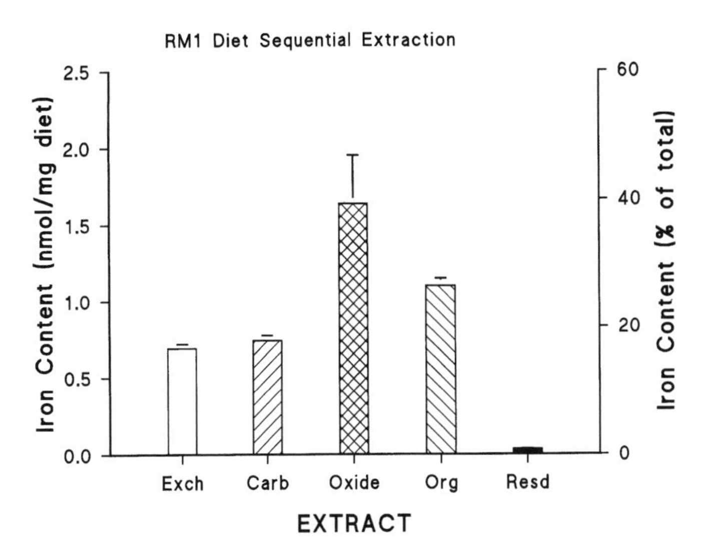

Figure 2
Sequential extraction analysis of RM1 diet. The five fractions analyzed are termed exchangeable iron, iron bound to carbonates, oxides or organic material, and residual iron. For details of analysis, see materials and methods.

B. Effect of Diet on Speciation of Model Compounds

Various model compounds were subjected to sequential extraction in the absence and presence of diet. Figure 3 shows the effect of diet on the extraction of various soluble iron compounds. It can be seen that iron complexed to chelators such as EDTA or NTA was relatively unaffected by the presence of diet (Figure 3A). However, iron chloride or iron ascorbate gave different patterns of extraction if diet was present (Figure 3B), with iron appearing in carbonate-, oxide-, and even organic-bound fractions. This latter observation is surprising as the diet contains little haem and suggests that interpretation of the organic-bound fraction as haem is incorrect. The tendency of iron chloride to give different speciation in the presence of diet may be attributed to the high pH of the diet (6.15), which probably promoted oxidation of Fe^{2+} and formation of hydroxides. NTA and EDTA, on the other hand, formed iron complexes which were stable at neutral pH values even in the presence of diet. The iron/ascorbate model solution had already been neutralized; therefore, the change in speciation in the presence of diet indicates the presence of ligands of various affinities which can bind the iron. The change in speciation seen with iron chloride or iron ascorbate probably reflects a real change in iron speciation.

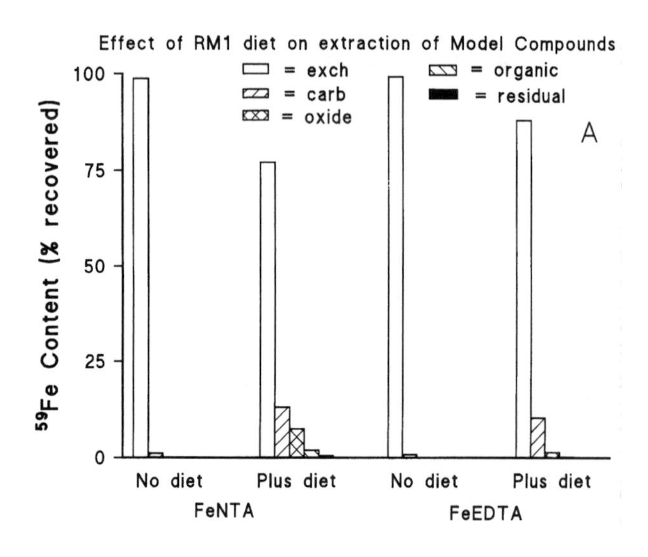

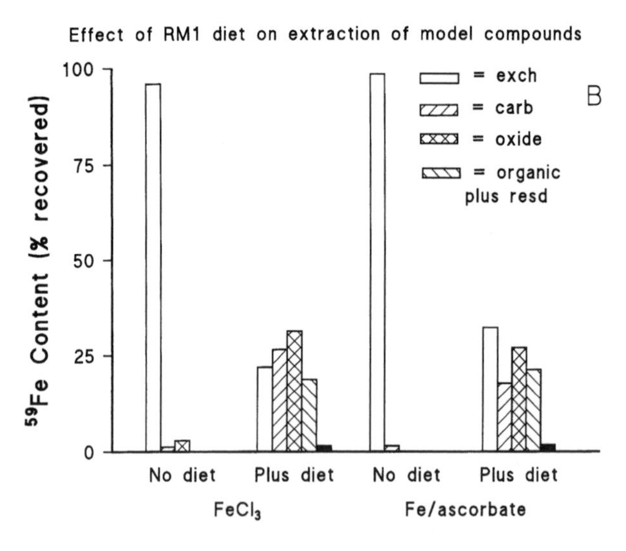

Figure 3
Effect of RM1 diet on extraction of model soluble iron compounds. Iron compounds were prepared as described in the materials and methods section. (A) FeNTA, FeEDTA; (B) $FeCl_3$, Fe/ascorbate.

Figure 4 shows the effect of diet on the extraction of a model for the carbonate-bound fraction. It can be seen that the presence of diet caused this compound to be extracted in the oxide-bound fraction. This effect cannot be attributed to the pH as the iron/bicarbonate compound had already been neutralized. It may be that diet has binding sites which affect the extraction of the carbonate-bound iron. Figure 5 shows the effect of diet on the extraction of a model for the oxide-bound fraction. It can be seen that a large amount of iron is redistributed to the organic-bound fraction. As before, the pH of the model compound seems unlikely to be responsible for this effect.

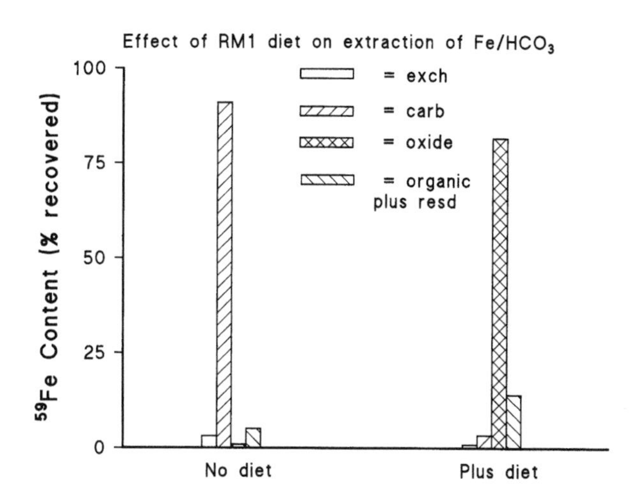

Figure 4
Effect of RM1 diet on extraction of Fe/bicarbonate. Fe/bicarbonate was prepared as described in the materials and methods section.

The data clearly show that the presence of diet causes iron to be extracted from different fractions than those seen in the absence of diet. The sequential extraction technique is therefore not a truly selective technique, and it appears that the extraction process itself may modify the extraction of subsequent fractions. This latter conclusion may be deduced from the fact that the model compounds for both carbonate- and oxide-bound fractions are shifted by diet to the next fraction in the sequential extraction. They do not both appear in the organic-bound fraction.

VI. RELATIONSHIP OF IRON SPECIES TO IRON ABSORPTION

In order to try to identify which iron fractions may be available for absorption, the RM1 diet was mixed with $^{59}FeCl_3$ in a manner similar to that used for Figure 3B. This can be seen to label the first four fractions (exchangeable, carbonate-, oxide-, and organic-bound) fairly equally. We then subjected portions of the labeled diet separately to each of the four extraction steps and dried the remains. Note that the extractions were not

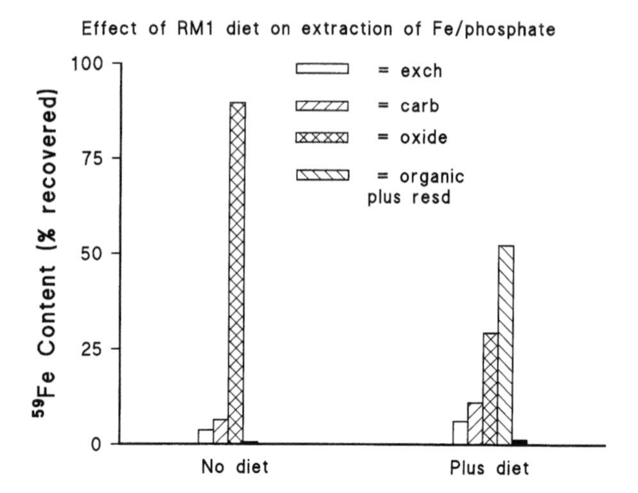

Figure 5
Effect of RM1 diet on extraction of Fe/phosphate. Fe/phosphate was prepared as described in the materials and methods section.

performed sequentially on any sample (Figure 6A). This was because the remains from some of the sequential extractions were unsuitable for feeding to mice. In addition, a water extraction was performed on a separate portion of diet and a sixth portion of diet was left untreated as a control. Figure 6B shows the proportions of radioiron removed by each of the five extraction methods. Note that water extraction removed little radioiron while the other four extractions removed broadly similar amounts of radioiron.

When the dried remains from these extractions were fed to mice, the proportion of the iron dose taken up by duodenum was determined and is shown in Figure 7. It can be seen that water extraction had little effect on iron uptake but all of the other four extraction techniques tended to increase the absorption of the radioiron in the remains. The experiment was performed twice with similar results to those shown.

The result of this experiment could be consistent with the major bioavailable fraction being the residual fraction, as this is the only common fraction present in all the remains. The reason for the higher apparent absorption seen with the remains, compared with the whole diet, could be that nonabsorbable radioiron has been removed by the extractions. This conclusion is consistent with the previous finding that the residual fraction is mobilized by the digestion process, especially in the duodenum, which is the main site of iron absorption.[9] On the other hand, it is surprising that the most chemically resistant iron fraction should be bioavailable. Furthermore the extraction experiment shown in Figure 3B would suggest that there is little radioiron in the residual fraction, thus ruling out the above interpretation. An alternative interpretation would be that

A

Sample:	a	b	c	d	e
Extraction Procedures:					
Incubation time (h):	1	4	16	2 plus 3	1
Reagent:	1M $MgCl_2$	1M Naacetate (pH 5.0)	0.04M $NH_2OH.HCl$ in 25% (v/v) acetic acid	30% H_2O_2 (2.5ml) 0.02M HNO_3 (1.5ml) pH 2 with HNO_3 30% H_2O_2 (pH 2)	H_2O
Volume (ml)	4	4	10	4 plus 1.5	4
Temperature (^oC)	25	25	95	85	25

B

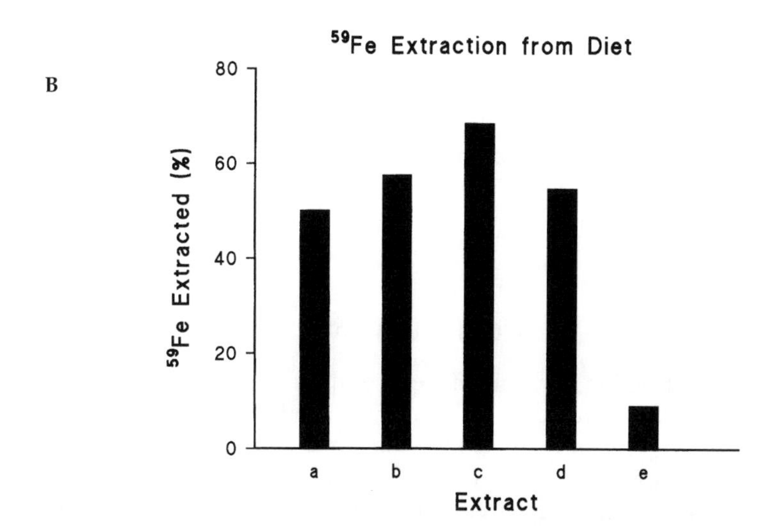

Figure 6
Single-step extractions of radiolabeled diet. (A) RM1 diet was labeled with $^{59}FeCl_3$ and divided into equal portions of 500 mg. Five of these (a to e) were separately extracted either with 4ml water (e) or with the procedure for exchangeable (a), carbonate-bound (b), oxide-bound (c) or organic-bound (d) iron to give five remainders. The remainders were each washed once with 4ml of water and the washings combined with the extracts. (B) The remainders were freeze-dried and counted for radioiron. Aliquots of each extract were also counted for radioiron and the percentage extraction calculated.

the extraction steps affect the bioavailability of the remains. This latter possibility agrees with the conclusions from the model extraction experiments described above and suggests that it is not possible to draw any conclusions about the bioavailability of any of the iron species identified

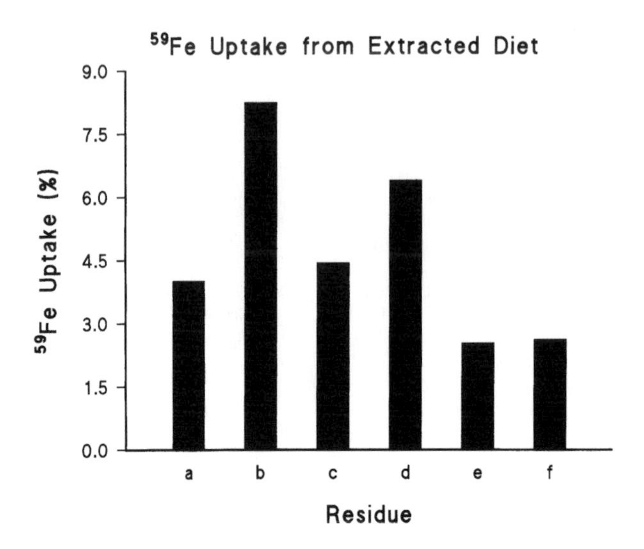

Figure 7
In vivo intestinal uptake of extracted diet remainders. Remainders from the five extractions described in Figure 6, or untreated radiolabeled diet (f), were fed to mice and uptake determined as described in the materials and methods section.

by this particular extraction technique, using *in vivo* absorption studies of the remains from extractions. Future work will examine the absorption of labeled model compounds in the presence of diet.

VII. CONCLUSION

Sequential extraction analysis of iron compounds in food gives speciation patterns which are specific for a given food. The extraction technique, however, significantly affects the intestinal absorption and speciation behavior of iron in the residues from the extraction steps. The technique is therefore not a selective extraction technique and the fractions identified in food cannot yet be identified with model iron compounds. It is not yet possible to specify the bioavailability of the fractions.

ACKNOWLEDGMENTS

This is a contribution from the King's College Centre for the Study of Metals in Biology and Medicine.

REFERENCES

1. Forbes, A.L., Adams, C.E., Arnaud, M.J., Chichester, C.O., Cook, J.D., Harrison, B.N., Hurrell, R.F., Kahn, S.G., Morris, E.R., Tanner, J.T., and Whittaker, P., Comparison of *in vitro*, animal and clinical determinations of iron bioavailability: international nutritional anemia consultative group task force report on iron bioavailability. *Am. J. Clin. Nutr.* 49, 225, 1989.
2. Lee, K. and Clydesdale, F.M., Quantitative determination of the elemental, ferrous, ferric, soluble, and complexed iron in foods. *J. Food Sci.* 44, 549, 1979.
3. Miller, D.D. and Bernier, L.A., Is solubility *in vitro* a reliable predictor of iron bioavailability? *Biol. Trace El. Res.* 4, 11, 1989.
4. Nojeim, S.J., Clydesdale, F.M., and Zajicek, O.T., Effect of redox potential on iron valence in model systems and foods. *J. Food Sci.* 46, 1265, 1981.
5. Reddy, M.B., Chidambaram, M.V., Fonesca, J., and Bates, G., Potential role of *in vitro* bioavailability studies in combating iron deficiency: a study of the effects of phosvitin on iron mobilization from pinto beans. *Clin. Physiol. Biochem.* 4, 78, 1986.
6. Simpson, R.J. and Peters, T.J., Forms of soluble iron in mouse stomach and duodenal lumen: significance for mucosal uptake. *Br. J. Nutr.* 63, 79, 1990.
7. Kapsokefalou, M. and Miller, D.D., Effects of meat and selected food components on the valence of nonheme iron during *in vitro* digestion. *J. Food. Sci.* 56, 352, 1991.
8. Wien, E.M. and Van Campen, D.R., Ferric iron absorption in rats: relationship to iron status, endogenous sulfhydryl and other redox components in the intestinal lumen. *J. Nutr.* 121, 825, 1991.
9. Simpson, R.J., Sidhar, S., and Peters, T.J., Application of selective extraction to the study of iron species present in diet and rat gastrointestinal tract contents. *Br. J. Nutr.* 67, 437, 1992.
10. Tessier, A., Campbell, P.G.C., and Bisson, M., Sequential extraction procedure for the speciation of particulate trace metals. *Anal. Chem.* 51, 844, 1979.
11. Kheboian, C. and Bauer, C.F., Accuracy of selective extraction procedures for metal speciation in model aquatic sediments. *Anal. Chem.* 49, 1417, 1987.
12. Nirel, P.M.V. and Morel, F.M.M., Pitfalls of sequential extractions. *Water Res.* 24, 1055, 1990.
13. Foy, A.L., Williams, H.L., Cortell, S., and Conrad, M.E., A modified procedure for the determination of non heme iron in tissue. *Anal. Biochem.* 18, 559, 1967.
14. Cronquist, G., Mackenzie, J., and Smith, T., A high resolution bulk sample counter with variable geometry. *Int. J. Appl. Radiat. Isotop.* 26, 86, 1975.

MANGANESE

CHAPTER 6

Manganese Deficiency and Excess in Rodents

John W. Finley
Phyllis E. Johnson
U.S. Department of Agriculture/ARS
Grand Forks Human Nutrition Research Center
Grand Forks, North Dakota

CONTENTS

0-8493-9611-5/97/$0.00+$.50

I. INTRODUCTION

Manganese is an essential trace element because of its role in a number of enzyme systems,[1,2] and perhaps, because of its role as an intracellular metabolic regulator.[3] The involvement of manganese in enzyme systems can be differentiated between manganese-containing enzymes and manganese-activated enzymes. The former group is comprised of enzymes that need manganese for catalytic activity and retain manganese in a relatively nondissociated form; arginase, manganese superoxide dismutase (MNSOD), and pyruvate carboxylase are the only known examples of this group. Manganese-activated enzymes are numerous and include hydrolases, kinases, decarboxylases, and transferases. Many of these enzymes may also be activated by other metal ions, such as Mg^{++}, and in some cases, only *in vitro* activation has been demonstrated.[2]

II. MANGANESE METABOLISM

The efficiency of manganese absorption in rats decreases with age.[4] In adult rats, only a small percentage of manganese is absorbed, apparently along the entire intestinal tract. Absorption of manganese is influenced most notably by iron; these two elements may utilize a common absorptive pathway.[5] Manganese enters the portal system bound to a ligand, and recent findings indicate that the ligand may be albumin,[6] although other findings suggest transferrin.[7] Manganese is rapidly taken up into the hepatocyte and secreted into the bile; the latter process occurs against a strong concentration gradient.[8] Maximum biliary excretion of intravenously injected manganese in rats occurs 15 to 60 min after injection.[9] Manganese escaping first pass liver clearance enters the systemic circulation bound primarily to transferrin as Mn^{3+}.[7] Tissue distribution and retention have been extensively reviewed.[1]

III. EFFECTS OF DIETARY MANGANESE DEFICIENCY

A. Manganese and Skeletal Abnormalities

Primary problems encountered in manganese-deficient animals include skeletal abnormalities. The primary causative factor apparently is decreased glycosyltransferase activity that leads to a subsequent decrease in proteoglycan formation.[10] Birds develop a condition known as "slipped tendon," which is a result of a defective attachment of the gastrocnemius tendon to the bone.[11] Gross skeletal abnormalities, such as shortened and misshaped bones, brittleness, and joint problems, have been observed in cattle, swine, goats, and sheep.[12]

Hurley and co-workers did much of the initial characterization of the effect of manganese deficiency on skeletal development in the rat. They reported severe skeletal abnormalities in offspring of manganese-deficient females, including shortened and misshaped long bones, and other abnormalities in skeletal development, particularly the skull.[13,14] Defective development of the otolith of the inner ear, caused by manganese deficiency, results in an irreversible congenital ataxia in rats, mice, and guinea pigs.[15–17] The *pallid* mutant mouse also develops a congenital ataxia caused by defective otolith development, which can be overcome with very high amounts of manganese.[17,18]

Induction of skeletal abnormalities is best done by producing second generation manganese-deficient animals. The procedure followed by Hurley et al.[15] was to feed weanling females a manganese-deficient diet (less than 1 μg/g manganese) consisting of whole market milk fortified with vitamins and minerals (Diet 1, Table 1). Females were maintained on this diet until they were of breeding age, and then were mated with males fed diets of standard rat chow. Females were fed the manganese-deficient diet throughout gestation until weaning, and the second generation animals were fed the same diet. This protocol resulted in a high percentage of animals with severe skeletal abnormalities and ataxia.[19–21]

Strause and co-workers[22] fed rats a casein/corn syrup diet deficient in copper and manganese (less than 1mg of Mn/kg diet; Diet 2, Table 1). Their diet resulted in radiographic differences in long bones and in changes in osteoinduction.[23] Animals were fed the diet for up to 1 year; animal growth and diet consumption were not different than that of animals consuming diets containing adequate manganese and copper.

Erway et al. produced second generation manganese-deficient mice with a dextrose/casein diet containing either 1 or 3 mg Mn/kg diet (Diet 3, Table 1).[24] Mice used for reproductive purposes were fed 3 mg of Mn/kg diet, and some of their offspring continued on this diet. Using this approach, manganese-deficient mice were produced through six generations.

Other factors that influence development of skeletal abnormalities in rodents fed manganese-deficient diets include the timing of the supplemental manganese[20,25] and the genetic strain of the animal.[17] Hurley and

TABLE 1.

Diets Used for Study of Skeletal Abnormalities in Rats and Mice Fed Manganese Deficient Diets for More Than One Generation

	Diet 1	Diet 2	Diet 3
Investigator	Hurley et al.[15]	Strause et al.[22]	Erway et al.[24]
Ingredient (%)	Whole milk fortified	High fructose	
	with Vitamin D	Corn syrup (41.9%)	Dextrose (54.5%)
	Corn oil (0.3%)	Corn oil (6.5%)	Corn oil (8.0)
	Pyridoxin (0.1%)	Water (13.1%)	
		Casein (24.5%)	Casein (30%)
		Non-nutritive bulk (10%)	
		Choline (0.1%)	
		Methionine[3] (0.2%)	
	Mineral mix[a]	Mineral mix[c] (3.2%)	Mineral mix[e] (6.0)
	Vitamin mix[b]	Vitamin mix[d] (0.5%)	Vitamin mix[f] (1.5%)

[a] Mineral mix composition (mg/100 ml milk): Cu as $CuSO_4$, 1.3 mg; Fe as $FeSO_4$, 1.3 mg; I as KI (given as a supplement in cod liver oil three times weekly; mg/d).

[b] Vitamin mix composition: Ca pantothenate 500, ρ-aminobenzoic acid and riboflavin, each 100; thiamine·HCl, pyridoxine and nicotinic acid, each 300; menadione, 250; folic acid, 6; biotin, 2.5; vitamin B_{12}, 0.3; and choline, 10 mg; inositol 5 mg; α-tocopherol, 1.1 mg; ascorbic acid, 1 mg; vitamin A and vitamin D, 15 U.S.P. units each.

[c] Mineral mix composition (mg/kg diet): $CaCO_3$, 5383; $MgSO_4$ anhydrous, 431; KH_2PO_4, 5436; NaCl, 659; $Na_2SeO_3 \cdot 5H_2O$, 0.32; $NaMoO_4 \cdot 2H_2O$, 1.08; NH_4VO3, 0.11; NaF, 3.23; KI, 0.16; H_3BO_3, 0.11; $Na_2SiO_3 \cdot 9H_2O$, 48; $MnSO_4 \cdot H_2O$, 66; $NiS \cdot 6H_2O$, 0.13; $ZnSO_4 \cdot 7H_2O$, 26; $Cr(NO_3)_3 \cdot 9H_2O$, 0.66; $SnCl_2 \cdot 2H_2O$, 0.13; $CuSO_4 \cdot 5H_2O$, 5; and $FeSO_4 \cdot 7H_2O$, 250.

[d] Vitamin mix composition (mg/kg diet): retinol palmitate, 54; ergocalciferol, 0.06 (2400 IU/kg diet); α-D-tocopherol acetate, 123; *myo*-inositol, 100; menadione, 5; nicotinic acid, 90; riboflavin, 20; pyridoxine · HCl, 20; thiamin · HCl, 20; calcium pantothenate, 60; biotin, 0.4; folic acid, 1.8; and cyafocobalamin, 0.00003.

[e] Mineral mix composition (g/kg): $CaCO_3$, 300.0; K_2HPO_4, 321.0; NaCl, 168.0; $MgSO_4 \cdot 7H_2O$, 125.0; $CaHPO_4$, 60.0; $FeSO_4 \cdot 7H_2O$, 25.0; KI, 0.80; $ZnCO_3$, 0.25; $CuSO_4 \cdot 5H_2O$, 0.30; $MnSO_4 \cdot H_2O$, varied.

[f] Vitamin mix composition (g/kg): folic acid, 0.030; biotin, 0.125; vitamins A and D (each 325,000 IU/g), 0.230; ρ-amino benzoic acid, 0.500; riboflavin, 0.500; menadione, 1.250; nicotinic acid, 1.500; pyridoxine, 1.500; thiamin · HCl, 1.500; vitamin B_{12} (1 mg/g), 1.500; vitamin A (325,000 IU/g), 2.100; Ca pantothenic acid, 2.500; ascorbic acid, 5.000; vitamin E (125,000 IU/g), 21.400; inositol, 25.000; choline chloride, 50,000; dextrose to make up, 1,000.000.

co-workers found that supplementing deficient, gestating females with manganese for 24 h, on day 13 of gestation, reduced the incidence of many of the skeletal abnormalities.[25] Conversely, manganese supplementation later in gestation had no effect.[20]

B. Manganese and Abnormalities in Carbohydrate Metabolism

Abnormalities in carbohydrate metabolism in several species, as a consequence of manganese deficiency, have been described. An initial indication of this relationship came from the report of an insulin-resistant man whose blood glucose was lowered by administration of oral manganese supplements.[26]

Baly and associates[27] studied the effects of manganese deficiency on carbohydrate metabolism in second generation manganese-deficient rats. Pregestational female rats were fed a casein-based diet containing 1 mg of Mn/kg diet (Table 2, Diet 4; Diet cited in Keen et al.[28]). Dietary content of manganese was increased to 3 mg/kg of diet during gestation, and following birth, the dam, and subsequently, weanling animals were fed the diet containing 1 mg of Mn/kg of diet. During gestation, 3 mg of Mn/kg of diet resulted in more viable births. Although the pups were deficient in manganese, they did not exhibit many of the overt signs of deficiency, including skeletal malformations and ataxia.[29]

TABLE 2.

Diets Used for Studies of Effects of Manganese Deficiency on Carbohydrate Metabolism in Rats and Mice

	Diet 4	Diet 5	Diet 6
Investigator	Keen et al.[28]	Werner et al.[35]	Werner et al.[35]
Ingredient	Cerelose (54.5)		
	Casein (30.0)	Casein (20%)	Casein (65%)
	Corn Oil (8.0)	Corn Oil (5%)	Corn Oil (5%)
		Cornstarch (65%)	Cornstarch (20%)
		Cellulose (5%)	Cellulose (5%)
		DL-methionine (0.3%)	DL-methionine (0.3%)
		Choline bitartrate (0.2%)	Choline bitartrate (0.2%)
	Vitamin mix[a] (1.5)	Vitamin mix[c] (1%)	Vitamin mix[c] (1%)
	Salt mix[b] (6.0)	Salt mix[d] (3.5%)	Salt mix[d] (3.5%)

[a] Vitamin mix composition (mg/kg diet): inositol, 375.0; ascorbic acid, 75.0; Ca-pantothenate, 37.5; thiamin-HCl, 22.5; pyridoxine-HCl, 22.5; nicotinic acid, 22.5; menadione, 18.8; riboflavin, 7.5; P-aminobenzoic acid, 7.5; folic acid, 0.45; biotin, 0.19; vit E (Rovomix E-50), 178.5; vit A (Rovomix A-250), 41.0, vit D (Rovomix $AD_3$325/325), 3.4; vit B_{12} (Merck 12 + mannitol), 22.5; choline-Cl (70% solution), 1072.5.

[b] Salt mix composition (gm/kg diet): $CaCO_3$, 18.05; K_2HPO_4, 19.3; NaCl, 10.1; $MgSO_4$, 3.6; $CaHPO_4$, 3.6; $FeSO_4 \cdot 7H_2O$, 1.5; KI, 0.05; $ZnCO_3$, 0.19; $CuSO_4 \cdot 5H_2O$, 0.02; $CrK(SO_4)_2 \cdot 12H_2O$, 0.02; $Mn(SO_4)_2 \cdot H_2O$, 0.14.

[c] AIN-76 vitamin mix composition.[72]

[d] AIN-76 mineral mix composition.[72]

By using these manganese-deficient animals, Baly et al.[27,30] produced evidence that intracellular manganese concentrations may influence insulin synthesis and secretion, perhaps by acting as a metabolic regulator.[31] Other studies showed that the development and integrity of the pancreas may be impaired by manganese deficiency. Pancreata of manganese-deficient guinea pigs exhibited marked aplasia and hypoplasia, fewer islets, and decreased β-cell populations.[32] It also has been postulated that some of these effects may be mediated by free-radical damage or caused by a decrease in MnSOD activity in the β-cell.[29]

Manganese deficiency might also influence carbohydrate metabolism through interaction with carbohydrate-metabolizing enzyme systems, most notably pyruvate carboxylase (E.C. 6.4.1.1)[33] and phosphoenolpyruvate carboxykinase (E.C. 4.1.1.32).[34] Both enzymes are critical to gluconeogenesis, the former serving to carboxylate pyruvate and form oxaloacetate (OAA), and the latter to convert OAA to phosphoenolpyruvate, a step considered to be rate-limiting in gluconeogenesis.

Werner and co-workers[35] used diets high in carbohydrate (Table 2, Diet 5), protein (Table 2, Diet 6) or fat (Table 3, Diet 7) to investigate the influence of manganese deficiency on pancreatic metabolism. The basal diet contained less than 1 mg Mn/kg diet, and control animals were fed 40 mg Mn/kg of diet. These diets were formulated to be isocaloric with 67% of the energy coming from carbohydrate, protein, or fat, respectively. Feed consumption and animal weight were not altered by diet, but 6 weeks of consuming the low manganese diet resulted in decreased tissue manganese content, lower pancreas weight, greater amylase activity (for rats on Diets 5 and 6[35]), and increased amylase mRNA.[36] Manganese repletion restored the manganese content of the pancreas, but did not reverse elevated amylase activity.[37]

C. Manganese and Lipid Metabolism

The association between manganese status and lipid metabolism is well established and is extensively reviewed elsewhere.[1,38,39] Werner et al.[35] showed that manganese deficiency could alter lipase expression. Sprague-Dawley rats fed a manganese-deficient diet (Diet 7, Table 3) high in corn oil for 6 weeks had increased lipase activity, as compared to controls.

Manganese also may affect lipid metabolism through an involvement in cholesterol synthesis; however, animal studies have been inconclusive. Wistar rats and genetically-hypercholesterolemic RICO rats were fed egg albumin/glucose-based diets that supplied 0.12 mg of Mn/kg of diet (Diet 8, Table 3[40]). Manganese deficiency did not alter cholesterol metabolism in the Wistar rats fed this diet for 8 weeks. Consequently, Klimis-Tavantzis et al. concluded that manganese was probably not involved in lipid metabolism. Conversely, Davis and co-workers[41] found that Sprague-Dawley rats fed a manganese-deficient diet high in linoleic or oleic acid safflower oil (Table 3, Diet 9) for 8 weeks exhibited changes in cholesterol and lipid metabolism, probably as a result of lowered cholesterol synthesis. Discrepancies between the above studies may be partially the result of the use of different strains of rat.

TABLE 3.

Diets Used for Studies of Effects of Manganese Deficiency on Lipid Metabolism in Rats and Mice

	Diet 7	Diet 8	Diet 9
Investigator	Werner et al.[35]	Klimis-Tavantzis et al.[40]	Davis et al.[41]
Ingredient		Egg albumin (20%)	
	Casein (20%)		Casein (20%)
	Corn Oil (28.9%)	Corn Oil (6%)	
			Safflower oil (12%)
		Cerelose (68.6%)	
			Sucrose (50%)
	Cornstarch (11.3%)		Cornstarch (7.98%)
	Cellulose (34.8%)		Cellulose (5%)
	DL-methionine (0.3%)	DL-methionine (0.4%)	DL-methionine (0.3%)
		Biotin (0.0002%)	
		$CaCO_3$ (0.85%)	
		KH_2PO_4 (1.25%)	
		$CaHPO_4 \cdot 2H_2O$ (0.67%)	
		NaCl (0.26%)	
	Choline bitar. (0.2%)		Choline bitar. (0.2%)
			BHT (0.02)
	Vitamin mix[a] (1%)	Vitamin mix[c] (1%)	Vitamin mix[a] (1.0%)
	Salt mix[b] (3.5)	Trace mineral mix[d] (1.0%)	Mineral mix[b] (3.5)

[a] AIN-76 vitamin mix composition.[72]

[b] AIN-76 mineral mix composition.[72]

[c] Vitamin mix composition (g/kg diet): Vit A and D powder (500,000 units of vit A acetate/g), 20,000 IU; dry vit E acetate (500 units/g), 0.20; menadione, 0.005; inositol, 0.1; niacin, 0.04; D-calcium pantothenate, 0.04; riboflavin, 0.008; thiamin · HCl, 0.005; pyridoxine · HCl, 0.005; folic acid, 0.002; biotin, 0.0004; vit B-12 (0.1% trituration in mannitol), 0.03; choline dihydrogencitrate (41% choline), 4.87; ρ-aminobenzoic acid, 0.1; dextrose diluent.

[d] Mineral mix composition (g/kg diet): $FeCl_3$ (prepared from Fe powder low in manganese and magnesium), dissolved in ultrapure HCl (Ultrex), 0.194; $MgSo_4$, 2.50; KI, 0.0026; $CuSO_4 \cdot 5H_2O$, 0.0167; $ZnCO_3$, 0.115; $CoCl_2 \cdot 6H_2O$, 0.0017; $NaMoO_2 \cdot 2H_2O$, 0.0083; $NaSeO_3 \cdot 5H_2O$, 0.0033; $CrK(SO_4)_2 \cdot 12H_2O$, 0.020; glucose diluent.

IV. DIETARY INTERACTIONS MAY AFFECT THE ABSORPTION/BIOAVAILABILITY OF MANGANESE

There are several dietary interactions that have been proposed or demonstrated to affect manganese bioavailability. These interactions have been the subject of several exhaustive reviews.[42–44] Potential interactions should be kept in mind when developing a diet specifically for the study of manganese metabolism.

A. Iron

The interaction between iron and manganese has been noted in numerous studies and reviewed extensively.[45] In humans, differences in manganese absorption and retention between males and females have been correlated to iron status.[46] Competition for absorption occurs between the two metals,[44] perhaps because of competition for a common transport mechanism.[47]

Davis and co-workers [48] fed weanling rats diets either 0.9, 48, or 188 mg Mn/kg diet (Table 4, Diet 10); dietary iron (supplied as ferric citrate) was 19 or 276 mg of Fe/kg diet. After 45 days of the dietary regimen, weight gain and food consumption were not significantly affected. High iron intake inhibited manganese absorption and decreased tissue manganese concentrations and MnSOD activity. Endogenous losses of manganese were not affected by diet, and it seemed that the greatest effect was a decrease in the uptake of manganese by the mucosal cell.

TABLE 4.

Diets Used to Study the Interaction Between Manganese and Other Dietary Components

	Diet 10	Diet 11
Interaction studied	Iron	Protein source
Investigator	Davis et al.[48]	Lee and Johnson[60]
Ingredient		Soy protein (20%)
	Lactalbumin (20%)	
	Sucrose (50%)	Sucrose (65%)
	Corn oil (5%)	Corn Oil (5%)
	Cornstarch (15%)	
	Cellulose (5%)	Cellulose (5%)
	DL-methionine (0.3%)	DL-methionine (0.3%)
	Choline-bitartrate (0.2%)	Choline chloride (0.2%)
	Vitamin mix (1%)	Vitamin mix (10%)[a]
	Trace mineral mix (3.5%)[b]	Trace mineral mix (3.5%)[c]

[a] AIN-76 vitamin mix.[72]
[b] AIN-76 trace mineral mix with out manganese or iron.[72]
[c] AIN-76 trace mineral mix without manganese.

B. Other Trace Elements

In addition to iron, several other minerals have been shown to interact with manganese, especially during absorption. Phosphorous, but not calcium, impaired manganese absorption in chicks,[49,50] whereas calcium affected manganese absorption in humans[51] and mice.[52] Interactions between manganese and nickel have been noted in rats[53] and in cultured colon cells.[54] Pigs fed manganese-deficient diets have impaired selenium status.[55] Copper alters manganese distribution in cultured glial cells[56] and affects manganese absorption and retention in rats,[57] and manganese alters copper distribution in rats.[58]

C. Other Dietary Interactions

Johnson and Korynta[59] showed that dietary protein source could affect manganese bioavailability in the rat. Manganese absorption was greatest with tuna and lowest with beef as the protein source. The biologic half-life of manganese was longest with beef and shortest with soy as the protein source. Lee and Johnson[60] fed casein-based (Diet 5, Table 2) or soybean protein-based diets (Diet 11, Table 4), formulated to contain 2.5 or 52 mg of Mn/kg of diet, to weanling Long-Evans rats for 4 weeks. Liver manganese concentrations and MnSOD activity were greater when soy, rather than casein, was the protein source. It should be noted that the concentration of manganese in soybean protein is higher than in casein, and may be sufficiently high to prevent formulation of a soy-based manganese-deficient diet.

Lee and Johnson[61] also reported that several other factors could affect manganese homeostasis in rats. Rats were fed Diet 10 (Table 4), formulated to contain between 0.4 and 49 mg of Mn/kg of diet. Rats fed 1.4 mg of Mn/kg of diet grew as well as animals fed the control diet of 49 mg of Mn/kg of diet, but increased dietary manganese reduced manganese absorption. Manganese absorption was enhanced by fasting the rats prior to oral administration of tracer manganese and by sucrose (diets composed of 65% sucrose, and no cornstarch, as compared to diets containing 65% cornstarch).[61] Another study that used similar diets fed to male Sprague Dawley rats showed that older rats accumulated more manganese, although they exhibited decreased efficiency of manganese absorption.[62]

Manganese retention in rats was impaired by phytate[63] added to the diet as the soluble sodium salt, and the impairment was similar to that observed for other trace metals. The degree of interaction of manganese with natural phytate in foodstuffs is not known, and the reports of phytate inhibition of manganese absorption in humans have been contradictory.[51,64] Consumption of reducing organic acids, such as ascorbate, should theoretically enhance manganese absorption because manganese is best absorbed in the Mn^{2+} state, but ascorbate did not alter manganese absorption in rats,[57] cultured cells,[54] or humans.[51,64] An earlier report, however, suggested that ascorbic acid supplementation improved manganese bioavailability in humans.[65] Lactose[66] and lipid content of the diet also have been demonstrated to alter manganese bioavailability.[67]

D. Chemical Forms of Manganese

Many different chemical forms of manganese have been used in formulating diets. The most common are chloride, sulfate, and carbonate.

There has been limited work comparing the bioavailability of these forms, but most researchers have noted that all forms are equally available.[42,68]

V. MANGANESE DEFICIENCY AND ISOLATED RODENT TISSUES AND CELLS

A. Isolated Tissues

A number of investigators have used *in vitro* techniques to study basic aspects of manganese metabolism. Korc[69] studied the influence of manganese deficiency on insulin production in isolated pancreatic acini. Isolated acini were perfused with Hepes-Ringers buffer, pH 7.4, with or without 7×10^{-4} *M* manganese, supplemented with minimal Eagle's medium (MEM) amino acid mixture, 11.2 m*M* glucose, and 0.01 soybean trypsin inhibitor. Addition of manganese to the medium enhanced protein synthesis in the acini from diabetic rats, which suggests a role for manganese as an intracellular regulator. Baly et al.[70] perfused pancreata *in vitro*, from rats fed a manganese-deficient diet, with a Krebs-Ringer's bicarbonate, 4% dextran buffer. Pancreata from manganese-deficient animals exhibited depressed synthesis and enhanced degradation of insulin, which further demonstrates the involvement of manganese in metabolic regulation of the pancreas.

B. Cultured Cells

Schramm and Brandt[71] utilized isolated rat hepatocytes to characterize a high affinity manganese-specific transport system. Nishida et al.[72] found that hepatocytes isolated from rats injected with manganese for 2 weeks prior to the experiment bound more manganese in the nucleus than did control rats.

Wedler et al.[73] studied the uptake and distribution of manganese by isolated rat astrocytes.[73] Cells from neonatal rat brains were cultured for one week in 12-well cell culture plates with Eagle's minimal essential medium (EMEM) and 10% horse serum. Results led them to the hypothesis that free cytoplasmic manganese may regulate glutamine synthetase activity. Kim and co-workers cultured rat pheochromocytoma (PC12) cells in Dulbecco's Modified Eagles Medium (DMEM) containing horse serum and either no added or 10 μ*M* added manganese. This study showed the existence of a manganese-stimulated Na/K/Cl co-transport pathway.

Our laboratory has studied the influence of dietary manganese on manganese uptake by isolated and cultured rat hepatocytes.[74] Hepatocytes were isolated from rats fed diet 10 (Table 4), which was formulated to contain 2 or 50 mg of Mn/kg diet, for 6 weeks. Isolated hepatocytes were cultured on collagen-coated semipermeable membranes for 24 h in

MEM with 14% fetal calf serum (FCS), which had been previously dialyzed to reduce the concentration of trace elements;[75] manganese (as manganese chloride) was added to give a final concentration of 30 or 300 nmol/L. Cells from animals fed the low manganese diet took up significantly less radioactive manganese than cells from animals fed the high manganese diet.[74]

VI. MODIFICATION OF EXISTING DIETS TO MEET AIN-93 DIETARY GUIDELINES

Most currently published diets that are used to study manganese metabolism in rodents follow the guidelines of the AIN-76 diet.[76,77] The publication of new guidelines as the AIN-93 diet,[78] however, necessitates several changes. The most notable is a decrease in the manganese recommendation from 50 to 10 mg of Mn/kg of diet. Another major change is the recommendation for a replacement of sucrose by cornstarch because of several complications caused by feeding high sucrose diets to rodents. The enhancement of manganese absorption by sucrose has been discussed previously. The protein of choice for the AIN-93 diet is casein. Casein has been used in many previous dietary manganese studies and seems to function well. Investigators are urged to carefully consider the recommendations and changes given in the AIN-93 diets.

VII. MANGANESE EXCESS BY DIETARY MANIPULATION

Manganese is apparently one of the least toxic trace elements for mammals. Toxicity resulting from high dietary manganese is uncommon in domestic animals such as cattle, sheep, pigs, and chickens; they can consume diets with 400 to 3000 mg Mn/kg diet without showing signs of toxicity.[1] Most toxicity in humans occurs because of inhalation of manganese in environments such as manganese mines; chronic inhalation of airborne manganese at levels > 5 mg/m^3 results in neurotoxicity with signs similar to schizophrenia and Parkinson's disease. There are limited reports of manganese toxicity in humans who consumed drinking water with more than 2 mg Mn/L over many years, but no reports of manganese toxicity resulting from food alone. Experimental manganese toxicity in primates has also been produced by oral dosing with manganese.

A. Manganese and Neurological Symptoms

Because of the severe neurological effects caused by manganese toxicity in humans, many studies with rats and mice have been performed. Although manganese toxicity has been produced in rodents via inhalation, the scope of this review is limited to toxicity produced by nonrespiratory intake of manganese. As in primates, chronic manganese toxicity in adult rodents results in low brain concentrations of dopamine, norepinephrine, and serotonin, while the opposite occurs in neonates.[79] In contrast, acute toxicity can increase the concentrations of these neurotransmitters.[79] Rats gavaged with high doses of manganese in the form of particulate Mn_3O_4 showed a high rate of manganese accumulation in tissues, especially the cerebrum, hypothalamus, and pituitary; neonatal rats were much more susceptible and had higher tissue concentrations than older animals.[80,81] It should be noted that preweanling rats and mice are unable to excrete manganese via the bile as adult animals do.[82] Third-generation mice given drinking water containing 200 mg Mn/L, and consuming a commercial pelleted diet, exhibited both severe growth retardation (30%) and moderate to severe effects on gait.[83] Mice given tap water containing 10,000 mg Mn/mL for 7 weeks accumulated manganese in the cerebral cortex, cerebellum, hypothalamus and hippocampus and had decreased monoamine oxidase activity in the brain. Neonates of the third generation were described as "dwarfed" and suffered from ataxia,[84] but by the fourth generation, susceptibility to manganese toxicity seemed to be decreased, although litter size was also reduced.[85] Locomotor activity was decreased in male mice fed diets containing 3550 mg Mn/kg,[86] and the excessive manganese intake resulted in impaired learning and memory consolidation in rats, in addition to increased locomotor activity.

B. Effects on Other Organ Systems

The most visible effects of manganese toxicity are on the central nervous system, but other organs are affected as well. Liver damage and cholestasis can result from acute manganese toxicity in rodents; this may be caused by a manganese-bilirubin complex binding to canalicular membranes or to the pericanalicular microfilament network.[87]

Other signs of toxicity may be gastrointestinal damage,[88] nephritis,[89] testicular damage,[90,91] goiter,[84] and pancreatitis.[89,92] Rats fed diets containing 14.4 g Mn/kg for 8 weeks had increased intestinal blood loss, probably the result of irritation caused by the $MnCl_2$ in the diet.[88] High concentrations of dietary manganese (>1050 mg Mn/kg diet) also causes reproductive dysfunction in both male and female rats and mice.[86,91,93] Mice are apparently more susceptible to adverse effects on the reproductive system than are rats.[86]

At moderately high intakes of dietary manganese (50 to 100 mg/kg), some biochemical changes are observed, principally in parameters associated with iron metabolism, although other trace elements can also be affected.[48,57]

In an early study, adolescent male rats consuming a diet with approximately 11,000 mg Mn/kg diet and receiving intraperitoneal (i.p.)

injections of $MnCl_2$ (25.5 mg total, given 3×/wk for 120d), developed a mild microcytic, hypochromic, anemia accompanied by a slight increase in reticulocytes.[94] Copper in plasma and brain decreased. The anemia also occurred when copper was injected along with the manganese. Acute widespread necrosis of the liver was also observed.

C. Interactions With Other Nutrients

As noted above, interactions between manganese and iron metabolism are well known. Postnatal growth of rats was not affected by dietary manganese levels as high as 1000 to 2000 mg/kg, as long as dietary iron was adequate, though diets with low iron (20 mg/kg) and high manganese (1100 mg/kg) depressed growth.[95] When rats were fed adequate iron and 3550 mg Mn/kg diet, their offspring exhibited retarded growth but normal survival; when the diet with 3550 mg Mn/kg was also made iron-deficient, both growth retardation and increased mortality were observed.[93,95] When rats were intubated with 25 μg manganese as particulate Mn_3O_4 in a 50% sucrose suspension, infant rats retained more of the dose at 24 h than did adolescent or adult rats. Compared to a high iron diet (240 mg Fe/kg diet), a low iron diet (20 mg Fe/kg diet) resulted in increased manganese retention by adolescent, but not adult rats.[81] In post-weanling rats, some growth retardation may also result from moderately high levels of dietary manganese. In the authors' laboratory, rats fed diets with 82 mg Mn/kg diet grew slightly less well than those fed 43 mg Mn/kg diet or less.[61]

In rodents, dietary manganese levels that are only moderately high (50 mg Mn/kg diet) depressed hemoglobin when dietary iron was deficient (10 mg Fe/kg diet); however, when dietary iron was 140 mg/kg diet, addition of manganese at 50 mg/kg diet increased hemoglobin in rats.[48,57] In a similar study[48] employing three concentrations of dietary manganese (0.9, 48, or 188 mg/kg diet), increasing dietary manganese decreased hematocrits when dietary iron was 19 mg/kg diet, but had no effect when dietary iron was 272 mg/kg diet. In the same study, high dietary manganese (188 mg/kg diet) depressed intestinal mucosal iron concentrations significantly, but did not affect spleen or kidney iron. In pre-weanling rats intubated with particulate Mn_3O_4, there was a manganese dose-related acceleration of liver iron depletion, and decreases in red blood cell count, hematocrits, hemoglobin, body weight, and survival.[96]

Diez-Ewald et al.[88] found increased iron absorption and decreased hematocrits and liver iron concentrations in rats fed high concentrations of manganese for long periods (Tables 5 and 6). Iron absorption and serum iron concentrations were highest in rats fed 320 mg Fe and 14.4 g Mn/kg diet, but liver iron concentrations were markedly depressed. This suggests that although iron was being absorbed, it was not being utilized. A possible explanation for this suggestion is that heme synthesis is inhibited through

TABLE 5.

Diets Producing Symptoms of Manganese Toxicity

Diet ingredients	Other conditions	Effects	Ref.
"Standard rat diet" circa 1968; (General Biochemicals, Chagrin Falls, OH)	+ 33 mg $MnCl_2$/g diet	Gastrointestinal irritation	88
	+ 320 or 800 μg Fe/g diet	Impaired Fe utilization	
Casein 20% Sucrose 64% Lard 11% Wintrobe salt mix 5% Vitamins	+ 4% $MnCl_2 \cdot 4H_2O$ and/or 0.1% $CuSO_4$ + i.p. injections of 4.5 or 25.5 mg $MnCl_2$ in divided doses and 4.5 04 25.5 mg "Cupralene" in divided doses	Anemia, changes in tissue Cu distribution	57
Casein, high prot., 210 g/kg DL-met, 1 Sucrose, 576 Cornstarch, 100 Corn oil, 50 Teklad vitamin mix, 10 #40060 $CaCO_3$, 20 $NaH_2PO_4 \cdot H_2O$, 20 NaCl, 5 KCl, 5 $MgSO_4$, 1.993 $ZnSO_4 \cdot 7H_2O$, 0.5307 $MnSO_4 \cdot H_2O$, 0.1548	20 or 240 mg Fe/kg + 0,350, 1050 or 3550 mg Mn/kg as particulate Mn_3O_4 Basal diet was 50 mg Mn/kg	20Fe + 1100Mn — growth retardation 240Fe + 3550Mn — growth retardation 20Fe + 3550Mn — growth retardation and increased mortality; reproductive effects	86,92,94

$CuSO_4 \cdot 5H_2O$, 0.0198 KIO_3, 0.0017 $Fe(NH_4)_2(SO_4)_2 \cdot 6H_2O$, 0.1404			
Casein, vitamin-free, 20% Sucrose, 50 Cornstarch, 15 Corn oil, 5 Cellulose, 5 AIN-76 mineral mix, 3.5 (Cu, Mn, Fe-free) AIN-76 vitamin mix, 1.0 D-methionine, 0.3 Choline chloride, 0.2	Fe added to provide 10, 35, 140 μg/g diet; Mn to provide 1 to 80 μg/g diet; Cu to provide <0.5 or 5 μg/g diet.	10Fe+50Mn — depressed Hb	48,95
Lactalbumin, 20% Sucrose, 50 Cornstarch, 14.84 to 15.0 Corn oil, 5 Cellulose, 5 AIN-76 mineral mix, 3.5 (Fe, Mn-free) AIN-76 vitamin mix, 1.0 DL-methionine, 0.3 Choline bitartrate, 0.2	Mn added as $MnCO_3$ to provide 0.9 to 188 μg/g diet; Fe citrate added to provide 19 or 276 μg/g diet.		93

TABLE 6.

Drinking Water or Gavage Treatments Producing Manganese Toxicity

Diet	Water	Gavage Dose	Effects	Ref.
"Regular pelleted diet" (composition unspecified)	200 mg Mn/L in tap water		Growth retardation, ataxia	83
"Solid food," type MF, Oriental Yeast Co, Tokyo	10,000 mg Mn/Ml for 7 weeks		Increased brain Mn; decreased brain MAO activity; F_2 neonates dwarfed, ataxic.	84, 85
Not specified	55 mg/Mn/L		Decreased brain RAN and protein synthesis	100
10% or 21 and casein with 1 mg/mL starch, sucrose, oil, supplemented with vitamins, minerals. Basal Mn and Cu in diet not given; Cu added at 0 or 250 mg/kg.			Impaired learning, memory consolidation. Increased locomotor activity. More vulnerable with 10% casein diet.	99
Not specified		Particulate Mn_3O_4Mn in 50% sucrose; 1 μl/g bw of Mn-sucrose soln w/21, 71, or 214 μg Mn/μl.	Dose-related liver Fe depletion decreased RBC, Hct, Hb, body weight, survival.	61
10% or 21% casein + starch. sucrose, ground nut oil, vitamins and minerals.		3 mg Mn/Ml as $MnCl_2 \cdot H_2O$	Delayed development of reflexes, lowered seizure threshold, increased seizure duration. Symptoms worse with low protein diet.	97, 98

inhibition of 5-aminolevulinate synthase activity by high amounts of dietary manganese.[97] Very high amounts of dietary iron (800 mg/kg diet) partially counteracted the deleterious effects of the high dietary manganese.

It has also been suggested that the toxicity of dietary manganese can be exacerbated by dietary protein deficiency.[98–100] Young rats given drinking water containing 2 mg Mn/mL exhibited developmental delays, compared to rats given water without added manganese; the effect was increased by feeding a diet containing 10% casein instead of 21% casein.[98,99] A 10% casein diet also increased vulnerability to the combined neurotoxic effects of manganese and copper, compared to the same treatments with a 21% casein diet.[101]

VIII. MANGANESE TOXICITY VIA DRINKING WATER

Manganese toxicity is more easily produced via high concentrations in the drinking water than in the diet. In young rats, only 1 mg Mn/mL of drinking water produced impaired learning ability and memory consolidation.[100] A much higher concentration of manganese in drinking water, 55 μg/mL, produced reductions in brain RNA and protein synthesis in young rats after only 3 weeks of consumption,[101] but both protein synthesis and RNA content returned to control levels in the fourth week; this indicates an ability to adapt to the high manganese intake.

IX. MANGANESE EXCESS PRODUCED BY GAVAGE

Although toxic doses of manganese administered via oral gavage are not quite comparable to toxic intakes through feeding, this technique has been frequently used in studies of manganese toxicity because much inhaled manganese dust ends up in the gastrointestinal tract. To model the intake of manganese via dust, usually particulate Mn_3O_4 (which is soluble in hydrochloric acid, but not water), is used. The doses used in such rodent studies have been proportionally much larger than those achieved by humans consuming dietary supplements. In these studies, when insoluble Mn_3O_4 was administered via gavage, it was generally in an aqueous suspension with 50% sucrose.

X. SUMMARY

Manganese is nutritionally important because of both its essential and toxic actions; laboratory rodents have been extensively utilized to study both aspects. Manganese deficiency is characterized by skeletal abnormalities, altered carbohydrate metabolism, and perhaps, altered lipid metabolism. Deficiency signs are apparent when rodents are fed diets containing 3 mg Mn/kg diet, but marked deficiency consequences can be best demonstrated by feeding diets containing only 1 mg Mn/kg diet. The latter dietary intake of manganese may impair reproduction. Feeding manganese deficient diets through the second generation results in animals with extreme deficiencies.

Manganese is apparently one of the least toxic of the trace elements; however, toxicity does occur, especially in human miners who breathe in manganese-laden dust. Manganese toxicity causes neurological problems and can also affect many organ systems. Manganese toxicity has been induced in laboratory rodents by feeding diets or providing drinking water high in manganese, as well as by gavage.

Many dietary ingredients affect the bioavailability of manganese; most notable is iron, but other trace elements, phytate, and protein source, may all also affect manganese bioavailability. Thus, the severity of manganese deficiency and toxicity may be affected by dietary composition.

REFERENCES

1. Hurley, L. and Keen, C., Manganese, in *Trace Elements in Human and Animal Nutrition*, 5th ed., Mertz, W., Ed., Academic Press, San Diego, 1987, 185–224.
2. Wedler, F., Biochemical and nutritional role of manganese: An overview, in *Manganese in Health and Disease,* Klimis-Tavantzis, D., Ed., CRC Press, Boca Raton, FL, 1994, 1–38.
3. Williams, R., Free Mn(II) and Fe(II) cations can act as intracellular controls, *FEBS Lett.*, 140, 3, 1982.
4. Keen, C., Bell, J., and Lönnerdal, B., The effect of age on manganese uptake and retention from milk and infant formulas in rats, *J. Nutr.*, 116, 395, 1986.
5. Thomson, A., Olatunbosun, D., Valberg, L., and Ludwig, J., Interrelation of intestinal transport system for manganese and iron, *J. Lab. Clin. Med.*, 78, 642, 1971.
6. Davis, C., Zech, L., and Greger, J., Manganese metabolism in rats: An improved methodology for assessing gut endogenous losses, *Proc. Soc. Exp. Biol. Med.*, 202, 103, 1993.
7. Davidsson, L., Lönnerdal, B., Sandstrom, B., Kunz, C., and Keen, C., Identification of transferrin as the major plasma carrier protein for manganese introduced orally or intravenously or after *in vitro* addition in the rat, *J. Nutr.*, 119, 1461, 1989.
8. Brandt, M. and Schramm, V., Mammalian manganese metabolism and manganese uptake and distribution rat hepatocytes, in *Manganese in Metabolism and Enzyme Function,* Schramm, V. and Wedler, F., Eds., Academic Press, Orlando, 1986, 3–16.
9. Klassen, C., Biliary excretion of manganese in rats rabbits and dogs, *Toxicol. Appl. Pharma.*, 29, 458, 1974.
10. Leach, R., Mn(II) and glycosyltransferases essential for skeletal development, in *Manganese in Metabolism and Enzyme Function,* Schramm, V. and Wedler, F., Eds., Academic Press, Orlando, 1986, 81–89.
11. Wilgus, H., Norris, L., and Heuser, G., The role of manganese and certain other trace elements in the prevention of perosis, *J. Nutr.*, 14, 155, 1937.
12. Hidiroglou, M., Zinc, copper and manganese deficiencies and the ruminant skeleton: A review, *Can. J. An. Sci.*, 60, 579, 1980.

13. Hurley, L., Wooten, E., and Everson, G., Disproportionate growth in offspring of manganese-deficient rats: II, Skull, brain and cerebrospinal fluid pressure, *J. Nutr.*, 74, 282, 1961.
14. Hurley, L., Everson, G., Wooten, E., and Asling, W., Disproportionate growth in the offspring of manganese-deficient rats: I, The long bones, *J. Nutr.*, 74, 274, 1961.
15. Hurley, L., Wooten, E., Everson, G., and Asling, W., Anomalous development of ossification in the inner ear of offspring of manganese-deficient rats, *J. Nutr.*, 71, 15, 1960.
16. Shrader, R. and Everson, G., Anomalous development of otoliths associated with postural defects in manganese-deficient guinea pigs, *J. Nutr.*, 91, 453, 1967.
17. Hurley, L. and Bell, L., Genetic influence on response to dietary manganese deficiency in mice, *J. Nutr.*, 104, 133, 1974.
18. Erway, L., Fraser, A., and Hurley, L., Prevention of congenital otolith defect in pallid mutant mice by manganese supplementation, *Genetics*, 67, 97, 1971.
19. Hurley, L. and Everson, G., Delayed development of righting reflexes in manganese-deficient rats, *Proc. Soc. Exp. Biol. Med.*, 102, 360, 1959.
20. Hurley, L. and Asling, W., Localized epiphyseal dysplasia in offspring of manganese-deficient rats, *Anat. Rec.*, 145, 25, 1963.
21. Hurley, L., Everson, G., and Geiger, J., Manganese deficiency in rats: Congenital nature of Ataxia, *J. Nutr.*, 66, 309, 1958.
22. Strause, L., Hegenauer, J., Saltman, P., Cone, R., and Resnick, D., Effects of long-term dietary manganese and copper deficiency on rat skeleton, *J. Nutr.*, 116, 135, 1986.
23. Strause, L., Saltman, P., and Glowacki, J., The effect of deficiencies of manganese and copper on osteoinduction and on resorption of bone particles in rats, *Calcif. Tissue Int.*, 41, 145, 1987.
24. Erway, L., Hurley, L., and Fraser, A., Congenital ataxia and otolith defects due to manganese deficiency in mice, *J. Nutr.*, 100, 643, 1970.
25. Hurley, L. and Everson, G., Influence of timing of short-term supplementation during gestation on congenital abnormalities of manganese-deficient rats, *J. Nutr.*, 79, 23, 1963.
26. Rubenstein, A., Levin, M., and Elliott, G., Manganese induced hypoglycemia, *Lancet*, 2, 1348, 1962.
27. Baly, D., Curry, D., Keen, C., and Hurley, L., Effect of manganese deficiency on insulin secretion and carbohydrate homeostasis in rats, *J. Nutr.*, 114, 1438, 1984.
28. Keen, C., Mark-Savage, P., Lönnerdal, B., and Hurley, L., Teratogenic effect of D-penicillamine in rats: Relation to copper deficiency, *Drug-Nut. Inter.*, 2, 17, 1983.
29. Hurley, L., Keen, C., and Baly, D., Manganese deficiency and toxicity: Effect on carbohydrate metabolism in the rat, *NeuroToxicology*, 5, 97, 1984.
30. Baly, D., Curry, D., Keen, C., and Hurley, L., Dynamics of insulin and glucagon release in rats: Influence of dietary manganese, *Endocrinology*, 116, 1734, 1985.
31. Baly, D., Walter, R., and Keen, C., Manganese metabolism and diabetes, in: *Manganese in Health and Disease*, Klimis-Tavantzis, D., Ed., CRC Press, Boca Raton, FL, 1994, 101.
32. Shrader, R. and Everson, G., Pancreatic pathology in manganese-deficient guinea pigs, *J. Nutr.*, 94, 269, 1968.
33. Scrutton, M., Manganese and pyruvate carboxylase, in *Manganese in Metabolism and Enzyme Function*, Schramm, V., Wedler, F., Eds., Academic Press, Orlando, 1986, 147.
34. Nowak, T., Manganese and phosphoenolpyruvate carboxykinase, in *Manganese in Metabolism and Enzyme Function*, Schramm, V. and Wedler, F., Eds., Academic Press, Orlando, 1986, 165.
35. Werner, L., Korc, M., and Brannon, P., Effects of manganese deficiency and dietary composition on rat pancreatic enzyme content, *J. Nutr.*, 117, 2079, 1987.
36. Chang, S., Brannon, P.M., and Korc, M., Effects of dietary manganese deficiency on rat pancreatic amylase mRNA levels, *J. Nutr.*, 120, 1228, 1990.
37. Brannon, P., Collins, V., and Korc, M., Alterations of pancreatic digestive enzyme content in the manganese-deficient rat, *J. Nutr.*, 117, 305, 1987.

38. Freeland-Graves, J. and Llane, C., Models to study manganese deficiency, in *Manganese in Health and Disease,* Klimis-Tavantzis, D., Ed., CRC Press, Boca Raton, FL, 1994, 59.
39. Klimis-Tavantzis, D., Taylor, P., and Wolinsky, I., Manganese, lipid metabolism and atherosclerosis, in *Manganese in Health and Disease,* Klimis-Tavantzis, D., Ed., CRC Press, Boca Raton, FL, 1994, 87.
40. Klimis-Tavantzis, D., Leach Jr., R., and Kris-Etherton, P., The effect of dietary manganese deficiency on cholesterol and lipid metabolism in the Wistar rat and the genetically hypercholesterolemic RICO rat, *J. Nutr.,* 113, 328, 1983.
41. Davis, C.D., Ney, D.M., and Greger, J.L., Manganese, iron and lipid interactions in rats, *J. Nutr.,* 120, 507, 1990.
42. Wapnir, R., Nutritional status and the effect of protein and other foodstuffs on the absorption of manganese, molybdenum and vanadium. In: *Protein Nutrition and Mineral Absorption,* CRC Press, Boca Raton, FL, 1990, 243.
43. Kies, C., Bioavailability of manganese, in *Manganese in Health and Disease,* Klimis-Tavantzis, D., Ed., CRC Press, Boca Raton, FL, 1994, 39.
44. Mills, C., Dietary interactions involving the trace elements, *Ann. Rev. Nutr.,* 5, 173, 1985.
45. Johnson, P., Manganese and iron metabolism, *Manganese in Health and Disease,* Klimis-Tavantzis, D., Ed., CRC Press, Boca Raton, FL, 1994, 159.
46. Finley, J.W., Johnson, P.E., and Johnson, L.K., A compartmental model of manganese metabolism in rats, *FASEB J.,* 6, A1947, 1992. (abst)
47. Thomson, A.B.R., Olatunbosun, D., Valberg, L.S., and Ludwig, J., Interrelation of intestinal transport system for manganese and iron, *J. Lab. Clin. Med.,* 78, 642, 1971.
48. Davis, C., Wolf, T., and Greger, J., Varying levels of Manganese and iron affect absorption and gut endogenous losses of manganese by rats, *J. Nutr.,* 122, 1300, 1992.
49. Anonymous, Excess phosphorous impairs manganese utilization in chicks, *Nutr. Rev.,* 49, 125, 1991.
50. Wedekind, K., Titgemeyer, E., Twardock, A., and Baker, D., Phosphorus, but not calcium, affects manganese absorption and turnover in chicks, *J. Nutr.,* 121, 1776, 1991.
51. Davidsson, L., Cederblad, A., Lönnerdal, B., and Sandstrom, B., The effect of individual dietary components on manganese absorption in humans, *Am. J. Clin. Nutr.,* 54, 1065, 1991.
52. Barneveld, A. and Van Den Hammer, C., The influence of calcium and magnesium on manganese transport and utilization mice, *Biol. Tr. Elem. Res.,* 6, 489, 1984.
53. Nielsen, F., Poellot, R., and Uthus, E., Manganese deprivation affects response to nickel deprivation, *J. Tr. Elem. Exp. Med.,* 7, 167, 1995.
54. Finley, J. and Monroe, P., Mn absorption: The use of CACO-2 cells as a model of the intestinal epithelium, *FASEB J.,* 8, A429, 1994.
55. Burch, R., Williams, R., Hahn, H., Jetton, M., and Sullivan, J., Tissue trace element and enzyme content in pigs fed a low manganese diet. I. A relationship between manganese and selenium, *J. Lab. Clin. Med.,* 86, 132, 1975.
56. Wedler, F.C. and Brenda, W.L., Cu(II) and Zn(II) ions alter the dynamics and distribution of Mn(II) in cultured chick glial cells, *Neurochem. Res.,* 15, 1221, 1990.
57. Johnson, P. and Korynta, E., Effects of copper, iron, and ascorbic acid on manganese availability to rats, *Proc. Soc. Exp. Biol. Med.,* 199, 470, 1992.
58. Scheuhammer, A. and Cherian, M., The influence of manganese on the distribution of essential trace elements, *Toxicol. Appl. Pharma.,* 61, 227, 1981.
59. Johnson, P. and Korynta, E., The effect of dietary protein source on manganese bioavailability to the rat, *Proc. Soc. Exp. Biol. Med.,* 195, 230, 1990.
60. Lee, D. and Johnson, P., 54Mn absorption and excretion in rats fed soy protein and casein diets (42852), *Proc. Soc. Exp. Biol. Med.,* 190, 211, 1989.
61. Lee, D. and Johnson, P., Factors affecting absorption and excretion of 54Mn in rats, *J. Nutr.,* 118, 1509, 1988.
62. Lee, D., Korynta, E., and Johnson, P., Effects of sex and age on Manganese metabolism in rats, *Nutr. Rev.,* 10, 1005, 1990.

63. Davies, N. and Nightingale, R., The effects of phytate on intestinal absorption and secretion of zinc, and whole-body retention of Zn, copper, iron and manganese in rats, *Br. J. Nutr.*, 34, 243, 1975.
64. Davidsson, L., Almgren, A., Juillerat, M., and Hurrell, R., Manganese absorption in humans: the effect of phytic acid and ascorbic acid in soy formula, *Am. J. Clin. Nutr.*, 62, 984, 1995.
65. Kies, C., Aldrich, K., Johnson, J., Creps, C., Kolwalski, C., and Wang, R., Manganese availability for humans: Effects of selected dietary factors, in *Nutritional Bioavailability of Manganese*, Kies, C., Ed., American Chemical Society, Washington, 1995, 221.
66. King, B., Lassiter, J., Neathery, M., Miller, W., and Gentry, R., Effect of lactose, copper and iron on manganese retention and tissue distribution in rats fed dextrose-casein diets, *J. Anim. Sci.*, 50, 452, 1980.
67. Kies, C.V., Mineral utilization of vegetarians: impact of variation in fat intake, *Am. J. Clin. Nutr.*, 48, 884, 1988.
68. King, B.D., Lassiter, J.W., Neathery, M.W., Miller, W.J., and Gentry, R.P., Manganese retention in rats fed different diets and chemical forms of manganese, *J. Anim. Sci.*, 49, 1235, 1979.
69. Korc, M., Manganese action on protein synthesis in diabetic rat pancreas: Evidence for a possible physiological role, *J. Nutr.*, 114, 2119, 1984.
70. Baly, D., Curry, D., Keen, C., and Hurley, L., Dynamics of insulin and glucagon release in rats: Influence of dietary manganese, *Endocrinology*, 116, 1734, 1985.
71. Schramm, V. and Brandt, M., The manganese(II) economy of rat hepatocytes, *Fed. Proc.*, 45, 2817, 1986.
72. Nishida, M., Ogata, K., Sakurai, H., Morimoto, A., Yamashita, K., and Kwada, J., A binding profile of manganese to the nucleus of rat liver cells, and manganese induced aberrations in thyroid hormone content and RNA synthesis in the nucleus, *Biochem. Inter.*, 27, 209, 1992.
73. Wedler, F., Vichnin, M., Ley, B., Tholey, G., Ledig, M., and Copin, J., Effects of Ca(II) ions on Mn(II) dynamics in chick glial cells and rat astrocytes: Potential regulation of glutamine synthetase, *Neurochem. Res.*, 19, 145, 1994.
74. Finley, J., Briske-Anderson, M., and Gregoire, B., Metabolism of manganese by isolated rat hepatocytes and by the Hep-G2 cell line, *FASEB J.*, 10: A819, 1996.
75. Finley, J., Reeves, P., Briske-Anderson, M., and Johnson, L., Zinc uptake and transcellular movement by CACO-2 cells: Studies with media containing fetal bovine serum, *J. Nutr. Biochem.*, 6, 137, 1995.
76. American Institute of Nutrition, Second report of the ad hoc committee on standards for nutritional studies, *J. Nutr.*, 110, 1726, 1980.
77. American Institute of Nutrition, Report of the American Institute of Nutrition ad hoc committee on standards for nutritional studies, *J. Nutr.*, 107, 1340, 1977.
78. Reeves, P., Nielsen, F., and Fahey, G., AIN-93 diets purified diets for laboratory rodents: Final report of the American Institute of Nutrition ad hoc committee on the reformulation of the AIN-76 rodent diet, *J. Nutr.*, 123, 1939, 1993.
79. Seth, P.K. and Chandra, S.V., Neurotransmitters and neurotransmitter receptors in developing and adult rats during manganese poisoning, *Neurotoxicology*, 5, 67, 1984.
80. Rehnberg, G.L., Hein, J.F., Carter, S.D., Link, R.S., and Laskey, J.W., Chronic ingestion of Mn_3O_4 by young rats: tissue accumulation, distribution, and depletion, *J. Toxicol. Environ. Health*, 7, 263, 1981.
81. Cahill, D.F., Bercegeay, M.S., Haggerty, R.C., Gerding, J.E., and Gray, L.E., Age-related retention and distribution of ingested Mn_3O_4 in the rat, *Toxicol. Appl. Pharmacol.*, 53, 83, 1980.
82. Miller, S.T., Cotzias, G.C., and Evert, H.A., Control of tissue manganese: initial absence and sudden emergence of excretion in the neonatal mouse, *Am. J. Physiol.*, 229, 1080, 1975.

83. Ishizuka, H., Nishida, M., and Kawada, J., Changes in stainability observed by light microscopy in the brains of ataxial mice subjected to three generations of manganese administration, *Biochem. Intl.*, 25, 677, 1991.
84. Nishida, M., Kawada, J., Ishizuka, I., and Katsura, S., Goitrogenic action of manganese on female mice thyroid through three generations, *Zool. Sci.*, 5, 1043, 1988.
85. Nishida, M., Yakashij, H., Kawada, J., Sakura, H., and Takada, J., Changes in mouse brain monoamine oxidase activity in the first, second, and fourth generations after manganese administration, *Biochem. Intl.*, 23, 307, 1991.
86. Gray, L.E. Jr. and Laskey, J.W., Multivariant analysis of the effects of manganese on the reproductive physiology and behavior of the male house mouse, *J. Toxicol. Environ. Health*, 6, 861, 1980.
87. Dahlström-King, L., Couture, J., and Plaa, G.L., Functional changes of the biliary tree associated with experimentally induced cholestasis: sulfobromophthalein on manganese-bilirubin combinations, *Toxicol. Appl. Pharmacol.*, 108, 559, 1991.
88. Diez-Ewald, M., Weintraub, L.R., and Crosby, W.H., Interrelationship of iron and manganese metabolism, *Proc. Soc. Exp. Biol. Med.*, 129, 448, 1968.
89. Keen, C.L. and Zidenberg-Cherr, S., Manganese toxicity in humans and experimental animals, in *Manganese in Health and Disease*, Klimis-Tavanzis, D.J., Ed., CRC Press, Boca Raton, FL, 1994, pp. 193–205.
90. Singh, J., Husain, R., Tandon, S.K., Seth, P.K., and Chandra, S.V., Biochemical and histopathological alterations in early manganese toxicity in rats, *Environ. Physiol. Biochem.*, 4, 16, 1974.
91. Gray, L.E. Jr., Kutzman, M., Laskey, J., and Reiter, L., The effects of manganese (Mn_3O_4) administration on the spontaneous behavior of male and female rats, *Toxicol. Appl. Pharmacol.*, 45, 356, 1978.
92. Scheuhammer, A.M. and Cherian, M.G., The influence of manganese on the distribution of essential trace elements, II, The tissue distribution of manganese, magnesium, zinc, iron, and copper in rats after chronic manganese exposure, *J. Toxicol. Environ. Health*, 12, 361, 1983.
93. Laskey, J.W., Rehnberg, G.L., Hein, J.F., and Carter, S.D., Effects of chronic manganese (Mn_3O_4) exposure on selected reproductive parameters in rats, *J. Toxicol. Environ. Health*, 9, 677, 1982.
94. Gubler, C.J., Taylor, D.S., Eichwald, E.J., Cartwright, G.E., and Wintrobe, M.M., Copper metabolism, XII., Influence of manganese on metabolism of copper, *Proc. Soc. Exp. Biol. Med.*, 86, 223, 1954.
95. Rehnberg, G.L., Hein, J.F., Carter, S.D., Linko, R.J., and Laskey, J.W., Chronic ingestion of Mn_3O_4 by rats: tissue accumulation and distribution of manganese in two generations, *J. Toxicol. Environ. Health*, 9, 175, 1982.
96. Rehnberg, G.L., Hein, J.F., Carter, S.D., and Laskey, J.W., Chronic manganese oxide administration to preweanling rats: manganese accumulation and distribution, *J. Toxicol. Environ. Health*, 6, 217, 1980.
97. Maines, M.D., Regional distribution of the enzymes of haem biosynthesis and the inhibition of 5-aminolevulinate synthase by manganese in the rat brain, *Biochem. J.*, 190, 315, 1980.
98. Ali, M.M., Murthy, R.C., Saxena, D.K., and Chandra, S.V., Effect of low protein diet on manganese neurotoxicity, II., Brain GABA and seizure susceptibility, *Neurobehav. Toxicol. Teratol.*, 5, 385, 1983.
99. Ali, M.M., Murthy, R.C., Saxena, D.K., Srivastava, R.S., and Chandra, S.V., Effect of low protein diet on manganese neurotoxicity: I. Developmental and biochemical changes, *Neurobehav. Toxicol. Teratol.*, 5, 377, 1983.
100. Murthy, R.C., Lal, S., Saxena, D.K., Shukla, G.S., Mohd, M., and Chandra, V., Effects of manganese and copper interaction on behavior and biogenic amines in rats fed a 10% casein diet, *Chem. Biol. Interact.*, 37, 299, 1981.
101. Magour, S., Mäser, H., and Steffen, I., Effect of daily oral intake of manganese on free polysomal synthesis of rat brain, *Acta Pharmacol. Toxicol.*, 53, 88, 1983.

Chapter 7

Manganese Uptake in Tissues *In Vitro*: Tissue Slices as Model

George D. V. van Rossum
Tommaso Galeotti
Department of Pharmacology
Temple University School of Medicine
Philadelphia, Pennsylvania

CONTENTS

0-8493-9611-5/97/$0.00+$.50

I. INTRODUCTION

Manganese is accumulated in many types of cells to concentrations much in excess of the blood level of 0.05 to 0.1 μM.[1] The liver is an important site of accumulation where one of its functions is activation of the mitochondrial Mn-dependent superoxide dismutase (Mn-SOD), which contributes to defense against reactive oxygen radicals. The liver is also the major excretory organ of Mn, the concentration of which in the parenchymal cells is several times the biliary level.[2] Nevertheless, very little is known of the transport mechanism(s) by which Mn crosses the plasma membrane of any type of cell. Manganese is able to substitute for Ca in the Na/Ca exchange mechanism which transports Ca out of erythrocytes and other cells, with half-maximal effects at 5 to 10 μM,[3] and it can traverse receptor-operated Ca channels of hepatocytes, although this has only been studied at rather high concentrations.[4,5] The mechanism of transport is also of interest in view of the low content of Mn found in a number of tumors, compared to their tissue of origin.[1] In a limited number of hepatomas with varying growth rates, the Mn content parallels a decrease of Mn-SOD levels as the growth rate and degree of cellular dedifferentiation increase.[6] It was this last point that led to our own studies of Mn uptake by tissue slices of rat liver and hepatoma, the results of which have been published elsewhere.[7,8]

For this work, our choice of tissue model was to a considerable extent dictated by the difficulty of preparing satisfactory isolated cells from the hepatomas. This led us to use tissue slices, a preparation with which we have had extensive experience in transport, metabolic, and morphological studies of liver, hepatoma, and other tissues.[9–12] Mn uptake was measured in two ways, namely by atomic absorption spectrometry of total tissue Mn and by the radioactive tracer, ^{54}Mn.

II. METHODS

A. Tissue Preparation

1. General Characteristics

Tissue slices are currently less extensively used than cells in isolation or culture and it is therefore useful to describe some details of their use and preparation as we have experienced them. Probably the most crucial point is that the slices should be sufficiently thin, and agitated sufficiently rapidly during incubation, for there to be an adequate exchange of O_2 between tissue and medium to ensure oxygenation of the central cells of the slices by diffusion.[13,14] The estimated maximal thickness required to achieve this is 0.3 to 0.5 mm. An additional factor of importance here is the incubation temperature, for while slices of most tissues we have studied are adequately oxygenated at 37 to 38°C, slices of renal cortex, although respiring rapidly, show reduced levels of ATP and transport activity (e.g., K^+ uptake) at this temperature; they are therefore best incubated at 25°C.[11]

2. *Method of Preparation*

Although machines are available for tissue slicing, we have always preferred slices cut manually. We place a single lobe of liver (or conveniently sized piece of other tissues) on a moistened filter paper held on a Petri dish containing cracked ice. The organ is steadied with a horizontally-held microscope slide, exerting only very light pressure, while cutting slices vertically with a moistened, one-sided razor blade or Stadie-Riggs blade, using a gentle sawing motion; the edge of the slide also acts as a partial guide for the blade. The aim is to produce thin slices rather than uniform or complete cross-sections of the tissue and to work as quickly as possible. Appropriately thin slices appear somewhat translucent when suspended in fluid medium and viewed against the light. A large lobe of rat liver can be completely sliced within 5 to 10 min of excision of the liver. The average thickness of sample slices may be estimated by tracing the outline onto mm graph paper to estimate the area and then weighing the slice after gentle blotting to determine the volume, but this is time-consuming and not necessary as a routine. In any case, each slice cut will be different. Slices of hepatoma, lung, and renal cortex (after removal of the red-colored medulla with fine-pointed scissors) are readily prepared similarly.

B. Incubation of Tissue Slices

1. *General Considerations*

During preparation and the initial stages of incubation at 37°C, liver slices lose some K^+ and swell, gaining water, Na^+, Cl^-, and Ca^{2+}. In order to standardize the initial condition of slices for experiments from different animals, we usually preincubate them at 1°C for 30 to 45 min, a period in which they approach maximal changes of water and ions. They are then allowed to recover for a standard period (45 min) in an oxygenated medium at 37°C, during which the changes are largely reversed, before starting our experiments by addition of Mn. Measuring slice K^+ on samples taken just before (i.e., at 1°C) and after the 45 min at 37°C gives the net uptake of K^+ during this period, a convenient indicator of the metabolic and transporting condition of the slices (Table 1). These preincubation conditions also allow a convenient period for equilibration of the slices with any desired addition such as a marker for the extracellular water (see below), inhibitors, or substrates.

TABLE 1.

Change in Endogenous Composition of Liver Slices During the Standard Experimental Procedure Described in the Text

Substance	"Fresh" tissue	45 min at 1°C	+ min at 37°C: 45	55	65
K^+ mmol/kg dry wt	259 ± 5	64 ± 8	152 ± 5	170 ± 15	204 ± 15
Mn μmol/kg dry wt	165 ± 7	215 ± 11	248 ± 22	—	—
Total water kg/kg dry wt	2.26	3.39 ± 0.10	2.93 ± 0.06	2.99 ± 0.07	3.09 ± 0.03
Intracellular water	—	2.34	2.02	2.06	2.13

Note: "Fresh" tissue samples were taken from the liver immediately *post mortem*. Other samples of liver slices were taken after incubation for 45 min at 1°C followed by increasing times at 37°C. Manganese (20 μ*M*) was added to the medium after 45 min at 37°C; subsequent Mn contents therefore no longer represented endogenous Mn alone and are not shown. Intracellular water contents were calculated from the fractional volume of distribution of inulin in the slice water (mean value 0.31), as described in the text. No intracellular value is given for fresh tissue water. Values are mean ± SEM.

2. *Incubation Media*

We have found incubation media with various buffers (HCO_3^-, phosphate, "TRIS," "HEPES") and substrates to be suitable for transport studies. For our experiments with Mn uptake by liver slices, the medium composition was (mM): NaCl 140, KCl 5.0, $MgCl_2$ 1.0, $CaCl_2$ 1.25, Tris.HCl 10.0, Na phosphate 2.0, glucose 20.0 at pH 7.4. However, glucose was added only because it was needed for the hepatoma slices in the same work, liver slices containing sufficient endogenous substrates to maintain respiration and transport at 37°C for at least 90 to 120 min.

3. *Incubation and Sample Collection*

Immediately after preparation, the slices are first incubated in 2 to 3 changes (about 5 min each) of medium, with occasional stirring, to remove blood and tissue debris and are then transferred in lots of 50 to 200 mg (4 to 6 slices) to Erlenmeyer flasks (25 ml capacity) containing 3 ml medium. The flasks stand in an ice-water bath, which ensures a medium temperature of 0.5 to 1.0°C, for 45 min (see above). At 5 to 10 min before the end of the 45 min, the vessels are gassed with 100% O_2 via fine plastic tubes dipping below the liquid surface. They are then stoppered and transferred to a shaking water bath at 37°C; shaking at a rate of approximately 100 rpm, together with the large fluid surface area arising from the suggested combination of medium volume and flask capacity, ensures a good exchange of O_2 between gas and liquid phases. Samples for analysis are collected at intervals (e.g., as in Table 1) by tipping the entire contents of a flask onto hardened filter paper (Whatman no. 54) held under suction (house vacuum line or water-driven filter pump) on a sintered glass filter. After a single wash on the filter with a cold medium, the filter paper bearing the slices is removed and very gently blotted; the slices are then picked off with forceps and transferred to tared weighing bottles which are then stoppered.

4. *Assay Procedures*

The bottles are weighed to obtain the wet weight of the slices and are then dried in an air oven at 110°C to obtain the dry weight and, by difference, water content. Drying is complete in approximately 3 h,[15] although for convenience the samples can be left in the oven for 2 to 3 days or overnight. Ions are extracted from the dried tissue with 0.1 N HNO_3 for a minimum of 3 h, as shown previously for Cl^-, Na^+, K^+,[15–17] Ca^{2+} and Mg^{2+}.[18] The cations are assayed by standard techniques of emission flame photometry (Na, K) or atomic absorption spectrometry (Ca, Mg).

5. *Expression of Results*

Slice water contents can vary considerably during the course of experiments, especially in response to changes of temperature (see Table 1) or in the presence of inhibitors of metabolism and ion transport[9–12] and some metabolic substrates.[19] Expressing ion contents in terms of tissue wet weight is therefore unsatisfactory and we express them per unit dry weight, for the tissue solids remain practically unchanged during experiments of this type. Further, drying the tissue gives a measure of tissue water so that the apparent concentration of ions in the slices can be calculated if desired.

C. MANGANESE UPTAKE

1. *Incubation*

Our work was intended to compare the ability of liver and hepatoma cells to bring about a net uptake of Mn and to determine the kinetic characteristics of the uptake system(s). At relatively high medium Mn concentrations (≥ 5 μ*M*), atomic absorption reliably measured of the difference of slice Mn content before and after addition of Mn to the medium. In these experiments, liver and hepatoma slices were preincubated at 1°C followed by 45 min at 37°C, as described above. Analysis of the endogenous Mn content of the slices at various stages of this procedure showed that it differed little from the initial content of Mn sampled immediately *post mortem* (Table 2). After the 45 min period of warm preincubation, a small volume (≤ 20 μl) of a concentrated solution (1 to 10 mM) of $MnCl_2$ was then added to bring the medium in each flask to the desired concentration of Mn. Sample flasks were subsequently taken at intervals for collection and analysis of the slices, as described above. Subtraction of the endogenous Mn content from the total Mn after incubation gives an estimate of the net uptake.

The above approach achieved some success with studies in the range 1 to 3 μ*M*. However, at these concentrations, the net uptake of Mn is small compared to the endogenous content so that it is difficult to obtain statistically significant measures of the uptake. It was therefore preferable at these concentrations, and essential at concentrations less than 1 μ*M*, to use the radioisotope, ^{54}Mn, as a tracer for the uptake. As well as being highly sensitive, this obviated the need for an endogenous tissue sample. Conduct of such experiments was identical to the previous type except that the various quantities of Mn added were labeled with the isotope. Calculation of the quantity of total Mn taken up required a knowledge of the specific radioactivity of the Mn in the medium, for which purpose a medium sample was taken from each incubation flask. In a series of experiments carried out at 5 and 10 μ*M* total medium Mn, labeled with ^{54}Mn, the net uptake determined simultaneously by both methods in the same slices was very close[8] (see Table 2).

2. *Assay of Mn*

Procedures followed the general methods described above for K^+ and other ions. Both total Mn and ^{54}Mn were satisfactorily extracted from the dried tissue with 0.1 *N* HNO_3. The former was assayed by atomic absorption spectrometry and the latter by counting γ-radiation.

3. *Extracellular Water and Intracellular Mn Content*

Tissue slices inevitably comprise both interstitial (extracellular) and intracellular water compartments of the tissue with the consequences: (1) that net entry of Mn added acutely, as in our experiments, must diffuse through differing lengths of extracellular spaces before entering cells at different levels in the slice, thus causing a delay in overall uptake, and (2) that the measured Mn content of the slice is not wholly intracellular. Knowledge of the extracellular water content of the slices and the assumption that the concentration of Mn in the medium rapidly equilibrates with this water permits an allowance to be made for these factors.

Many compounds have been used as markers for the extracellular water, the ideal requirement being that a marker enters the whole of the extracellular water without penetrating into the cells, either directly through the plasma membrane itself or by endocytosis, or becoming bound to extracellular structures. The volume of distribution of the marker in the slice can then be equated with the extracellular water volume and the extracellular Mn content per unit tissue estimated by assuming an extracellular concentration equal to that of the medium. In practice, it is impossible to be sure that the requirements are met and it is found that different markers have different volumes of distribution.[20] For example,

TABLE 2.

Comparison of the Increase of Liver Slice Mn Content as Determined by Atomic Absorption Spectrometry and by Influx of ^{54}Mn

t after Mn addition (min)	Mn in medium: 5 μ*M*			10 μ*M*		
	Total Mn (μmol/kg dry wt)	^{54}Mn (μmol/kg dry wt)	Concentration in cell water (μM)	Total Mn (μmol/kg dry wt)	^{54}Mn (μmol/kg dry wt)	Concentration in cell water (μM)
2	59.4 ± 9.9	46.5 ± 11.2	20.8	88.1 ± 19.3	69.2 ± 7.6	26.4
5	69.4 ± 11.2	65.0 ± 9.6	27.7	109.4 ± 9.2	109.3 ± 15.2	54.9
10	107.5 ± 65.8	85.0 ± 14.2	39.6	188.3 ± 50.8	181.1 ± 28.3	77.4

Note: Slices were preincubated for 30 min at 1°C and 45 min at 37°C, at which point Mn labeled with ^{54}Mn was added to the medium to give final concentrations of 5 and 10 μM. Samples were then taken after a further 2, 5 and 10 min. Total cellular Mn was determined by atomic absorption spectrometry, with correction for the calculated extracellular content, and the increase after addition of Mn was determined by subtraction of endogenous Mn assayed in slices from parallel incubation vessels. Uptake based on the influx of ^{54}Mn was calculated from the specific radioactivity of the isotope in the medium. The apparent concentration of Mn taken into the slices was determined from the calculated intracellular water content of the same slices.

in two sets of experiments comparing different substances frequently used as extracellular markers, we obtained quite different volumes of distribution (referred to as α) in liver slices, at least partly associated with the molecular weight and/or charge. The values obtained in the two series were: (1) polyglucose (mol weight 22,000) α 0.32 ± 0.01, inulin (approximately 5,000) α 0.34 ± 0.02, SO_4^{2-} (96) α 1.40 ± 0.03[21]; (2) polydextran (mol weight 90,000) α 0.10 ± 0.01, inulin α 0.30 ± 0.03, sucrose (342) α 0.66 ± 0.02.[22] It is therefore necessary to choose a marker that gives an intrinsically reasonable value for α in the tissue studied. For example, in early work it was found that the extracellular volume in fresh (unincubated) liver tissue, estimated from the volume of distribution of Cl (corrected for the influence of membrane potential) was similar to that obtained from inulin distribution in liver slices *in vitro*,[23] and that inulin and polyglucose gave a similar value for α despite a 4 to 5 fold difference of mol weight. On these grounds, inulin was chosen as the routine marker of extracellular water for liver slices as well as other tissues *in vitro*.[9,11,24]

For these measurements with liver slices, inulin is added to all media (see below for concentration), allowing it to equilibrate in the slices during the preincubation periods, for which approximately 75 min is required.[21] Slices are collected from the incubation flask as above but, after blotting, slices from each flask are divided into two weighing bottles. Slices in one bottle are dried and assayed for water and ions. Inulin is not readily extracted from the dried tissue and the slices in the second weighing bottle, after weighing, are therefore directly extracted overnight with 10% (w/v) trichloroacetic acid (TCA). A medium sample from each incubation flask is also taken and treated with TCA. Samples of both tissue and medium are assayed for inulin.

Inulin can be used radiolabeled with ^3H or ^{14}C, in which case it is added to the medium with nonlabeled "carrier" inulin to a total concentration of 0.1% (w/v). Alternatively, as a cheaper method which is just as satisfactory, unlabeled inulin can be used at 0.5% (w/v) in the medium and assayed by the colorimetric method for ketohexoses and ketopentoses.[25] In this latter case, it is necessary to subtract a tissue-blank value which is obtained from a set of slices incubated in inulin-free medium.

The volume of distribution of the marker expressed, as a fraction of the total tissue water, is the estimate of the fraction of total water which is extracellular (i.e., α):

$$\alpha = \frac{(\text{Inulin in slices} - \text{tissue blank})/(\text{water content of slices})}{\text{Inulin concentration in medium}}$$

Then, when tissue water is expressed as kg water/kg dry weight,

$$\text{Extracellular water} = \alpha\ (\text{total tissue water})$$

and,

$$\text{Intracellular water} = \text{total water} - \text{extracellular water.}$$

In the case of problem (2), mentioned above, the quantity of Mn in the extracellular water can be estimated by assuming its presence at a concentration equal to that in the medium and the Mn entering the cells is then determined by difference from the total Mn content of the slices. With regard to problem (1), the effect of time of penetration to the cells can to some extent be assessed by finding the time at which the total Mn in the tissue comes to exceed the quantity calculated to equilibrate with the extracellular water. In experiments at 5 and 10 μM choline, the tissue content already exceeded the supposed extracellular content tenfold at the first observation time (2 min after Mn addition). Thus, the penetration time appeared to have a minimal effect on the rate of uptake by the cells themselves.[8]

III. DISCUSSION

A. Choice of Tissue Preparation

Our intention to compare Mn uptake by hepatocytes and hepatoma cells, and the unsatisfactory nature of cells isolated from the hepatomas used, dictated our use of the tissue-slice preparation. Some general factors for the successful use of tissue slices in studies of cell physiology have been noted above. For precise studies of cellular transport mechanisms, the extracellular fluid compartment of tissue slices poses two problems. First, the exchange of transported substance between medium and cells is delayed by diffusion through the interstitial space. In the case of the studies of Mn uptake by liver, this effect appeared to be relatively slight, as the total slice content exceeded the quantity calculated to be in the extracellular water well within 2 min. Second, the accuracy of the estimates of cellular content is limited by the estimation of extracellular volume and the assumption that, at equilibrium, the concentration of Mn in the interstitial fluid is equal to that in the incubation medium. Our experience is that, irrespective of the marker substance used or the analytical method (e.g., radioactive or colorimetric), estimates of extracellular volume in tissue slices even in the same experiment are rather variable, and to obtain reliable results it is necessary to make replicate estimates in each experiment, preferably in samples from each incubation vessel. There is also the possibility that Mn may become bound to charged elements of the cell exterior or to extracellular structures.

Manganese uptake results obtained from tissue slices are the average of uptake by all the cells present. In the case of liver slices, approximately 90% of total cell volume (although 65% of cell number) consists of hepatic parenchymal cells with only a minor contribution (5 to 10% total volume) from Kupffer and other sinusoidal-lining cells.[26] However, there are indications of at least some metabolic differences between periportal and centrilobular hepatocytes.[27] Also, some peripheral cells of the slices become damaged during the slicing procedure. Histological and electron-microscopic observations indicate that the damaged zone represents no more than 10% of the slice volume.[22] Cell homogeneity is much less in, for example, renal cortex and lung, where slices contain many different types of cell.

Despite these drawbacks, tissue slices have a number of advantages over the currently more widely used alternative of isolated cells in suspension or culture. Tissue slices are rapidly prepared and incubation of slices from 1 or 2 lobes of rat liver can be underway within 5 to 10 min of the death of the animal. The structural environment of the cells, such as intercellular geometry and junctions with neighboring cells, remains relatively unchanged and the plasma membranes have not been exposed to the Ca-free media or proteolytic enzymes required to separate cells. On the other hand, isolated cells of a particular type, in suspension or primary culture, can be made relatively free of contaminating cells from the same tissue. For transport studies, they show rather fewer complications arising from the presence of extracellular fluid and limited access to the actual site of transport at the plasma membrane. Potentially a big advantage is that they can be studied with ion-sensitive, fluorescent dyes, allowing continuous recording to be made of ionic composition over a lengthy period (see below).

B. Measurement of Mn

In principle, the cellular content of an element such as Mn is maintained in a steady state by the relative rates of its influx and efflux across the plasma membrane, with binding or chelation by intracellular molecules and possible transport by organellar membranes modifying its intracellular distribution. Measurement of the change in total cellular content of Mn after its addition to the medium, such as in our experiments, therefore represents the net effect of these factors. The initial rate of entry of a tracer not initially present in the cells, such as the radioactive isotope, ^{54}Mn, can be used to estimate the inward transport activity and, conversely, efflux of a tracer previously loaded into the cells into an unlabeled medium measures the unidirectional efflux. In our experiments, we determined that, using media containing 5 or 10 μM total Mn, the initial rate of ^{54}Mn influx and the net increase of total Mn in the slices of liver and hepatomas were closely similar for at least 10 min.[8] The maximal rate of each was observed during the first 2 min after addition of Mn. This was therefore adopted as a suitable standard time at which to estimate the initial rate of entry and, therefore, at which to study the kinetics of the uptake mechanism(s). The equality between the uptake of total and labeled Mn shows that it is permissible to use either technique.

The limiting factor in using total Mn uptake, measured by atomic absorption spectrometry, is the ability to detect small changes of slice Mn when uptake is from a low to medium concentration into the relatively high endogenous Mn content of liver. We have found it possible to measure statistically significant uptake into liver slices from a medium containing 1 μM Mn, but not less. Accordingly, for further work at the more interesting concentration range embracing the plasma level, only ^{54}Mn influx was a feasible approach. With this method, statistically significant cellular uptake of Mn was readily detected at 0.05 μM medium Mn, after correction for extracellular Mn.[8]

The rate of efflux from the slices can be estimated by preloading the tissue with radioactive Mn for approximately 30 min at 37°C, at which time equilibration is approached (Figure 1 of Reference 8). The slices are then rinsed briefly in ice-cold, unlabeled Ringer's, to remove superficial labeled medium, and are resuspended in unlabeled medium from which samples are taken at intervals to determine the emergence of radioactivity from the cells. Eventually, back flux of ^{54}Mn into the slices will become significant as the medium concentration of label builds up. It is therefore desirable either to transfer the slices at intervals to vessels containing fresh medium or to arrange a system for the continuous perifusion of slices with medium, the outflowing perifusate being collected for counting.

A sensitive and apparently noninvasive method currently in wide use for the study of ion movements is that of ion-selective fluorescent probes loaded into cells.[28,29] This method is suitable for use with isolated cells in suspension, using a standard fluorimeter cuvette, or cells in culture, using a cuvette modified to take a small culture plate or cover slip orientated at 45° to the incident and emission light paths. It is also used for microfluorimetry of single cells. In principle, this allows continuous measurements of the *changing* element contents to be made. Accurate calibration for determining *absolute* levels is less certain. To date, there is no dye available with satisfactory selectivity for Mn. The Ca-sensitive dye, FURA-2, responds to increasing Mn with a decrease of fluorescence and has been used for studies at rather high concentrations of Mn (40 to 500 μM).[4,5] Clearly, correction for any Ca changes must be taken into account. In the past, the fluorescence of endogenous electron carriers in tissue slices has been used for studying metabolic responses, notably nicotinamide nucleotides and flavoprotein co-enzymes;[30,31] whether the ion-sensitive fluorescent probes could also be used in a similar manner after preloading into slices has not been considered.

C. Interpretations and Further Experiments

The experiments so far published were intended primarily to compare the kinetic characteristics of Mn uptake by liver with those of hepatomas. The work therefore exclusively studied the concentration-dependence of the uptake and was analyzed by fitting to Michaelis-Menten-type kinetic systems. The analysis of the data was carried out by computer curve-fitting. The results suggested a complex situation in which the slices studied had up to three apparently saturable components of uptake in the medium range of 0.05 to 100 μM. At the presumably physiologically relevant range of 0.05 to 2.0 μM, liver slices had two systems (apparent K_m 0.075 and 2.0 μM, respectively) while a hepatoma had only a single one (K_m 0.34 μM). The presence of several systems is not unusual for trace metals, as shown by uptake of Zn and Pb.[32,33]

These experiments give little information on the more precise nature of the uptake systems. Saturability suggests that the mechanisms may be dependent on carriers for Mn. Also, calculation of the apparent concentration achieved in the intracellular water by Mn taken up during the experiment (i.e., excluding the endogenous Mn) suggests a movement against a concentration gradient of several fold. This last is, however, not necessarily evidence for an active transport mechanism, for in the case of an ion the demonstration of active transport requires that a net movement take place against the electrochemical gradient, rather than against the concentration gradient alone. Moreover, the calculated intracellular concentration (i.e., total intracellular Mn/total intracellular water) is only an apparent concentration, for it takes no account either of the likely binding or chelation of Mn to intracellular molecules or of its uptake by organelles such as the mitochondria.[6] Further experiments are required to characterize the various systems fully and, in the case of liver, are planned for isolated hepatocytes. Preliminary experiments have shown a statistically significant uptake of total Mn by the isolated cells, measured by atomic absorption spectrometry, at a medium concentration of 2 μM. The time-course and amount of uptake was similar to that recorded previously in the slices, illustrating the feasibility of working with this preparation (Özdener, F., and van Rossum, G.D.V., unpublished observations). Important questions to be answered include whether the uptake system(s) require metabolic energy, as indicated by effects of inhibitors of oxidative phosphorylation, electron transport, or glycolysis, and whether they are coupled to the movement of other ions such as Na and Cl or are dependent on a pH gradient.

REFERENCES

1. Ling, G.N., Kolebic, T., and Damadian, R., Low paramagnetic-ion content in cancer cells: its significance in cancer detection by magnetic resonance imaging. *Physiol. Chem. Phys. Med. NMR*, 22, 1, 1990.
2. Papavasiliou, P.S., Miller, S.T., and Cotzias, G.C., Role of liver in regulating distribution and excretion of manganese. *Am. J. Physiol.*, 211, 211, 1966.
3. Milanick, M.A. and Frame, D.S., Kinetic models of Na-Ca exchange in ferret red blood cells. Interaction of intracellular Na, extracellular Ca,Cd, and Mn. *Ann. N.Y. Acad. Sci.*, 639, 604, 1991.
4. Renard-Rooney, D.C., Hajnoczky, G., Seitz, M.B., Schneider, T.G., and Thomas, A.P., Imaging of inositol 1,4,5-triphosphate-induced Ca^{2+} fluxes in single permeabilized hepatocytes. Demonstration of both quantal and nonquantal patterns of Ca^{2+} release. *J. Biol. Chem.*, 268, 23601, 1993.
5. Kass, G.E.N., Webb D.-L., Chow, S.C., Llopis, J., and Berggren, P.-O., Receptor-mediated Mn^{2+} influx in rat hepatocytes: comparison of cells loaded with Fura-2 ester and cells microinjected with Fura-2 salt. *Biochem. J.*, 392, 5, 1994.
6. Borrello, S., De Leo, M.E., Wohlrab, H., and Galeotti, T., Manganese deficiency and transcriptional regulation of mitochondrial superoxide dismutase in hepatomas. *FEBS Letters*, 310, 249, 1992.
7. Galeotti, T., Palombini, G., Borrello, S., and van Rossum, G.D.V., High affinity manganese transport by liver and hepatomas. *FASEB J.*, 9, A447, 1995.
8. Galeotti, T., Palombini, G., and van Rossum, G.D.V., Manganese content and high-affinity transport in liver and hepatoma. *Arch. Biochem. Biophys.*, 322, 453, 1995.
9. Russo, M.A., Galeotti, T., and van Rossum, G.D.V., Metabolism-dependent maintenance of cell volume and ultrastructure in slices of Morris hepatoma 3924A. *Cancer Res.*, 36, 4160, 1976.
10. Russo, M.A., van Rossum, G.D.V., and Galeotti, T., Observations on the regulation of cell volume and metabolic control *in vitro*: changes in the composition and ultrastructure of liver slices under conditions of varying metabolic and transporting activity. *J. Membrane Biol.*, 31, 267, 1977.
11. van Rossum, G.D.V., and Ernst, S.A., The effects of ethacrynic acid on ion transport and energy metabolism in slices of avian salt gland and of mammalian liver and kidney cortex. *J. Membrane Biol.*, 43, 251, 1978.
12. Mariani, M.F., Thomas, L., Russo, M.A., and van Rossum, G.D.V., Regulation of cellular water and ionic content in lungs of fetal and adult rat. *Exp. Physiol.*, 76, 745, 1991.
13. Longmuir, I.S. and Bourke, A., The measurement of the diffusion of oxygen through respiring tissue. *Biochem. J.*, 76, 225, 1960.
14. Figueroa, E., Vallejos, R., and Pfeiffer, A., Effect of oxygen pressure on glycogen synthesis by rat-liver slices. *Biochem. J.*, 98, 253, 1966.
15. Little, J.R., Determination of water and electrolytes in tissue slices. *Anal. Biochem.*, 7, 87, 1964.
16. Whittam, R., A convenient method for the estimation of tissue chloride. *J. Physiol., London*, 128, 65P, 1955.
17. van Rossum, G.D.V., Some observations on tissue swelling *in vitro*. D.Phil. Thesis, University of Oxford, U.K., 1960.
18. van Rossum, G.D.V., Net movements of calcium and magnesium in slices of rat liver. *J. Gen. Physiol.*, 55, 18, 1970.
19. Haussinger, D., Gerok, W., and Lang, F., Regulation of cell function by the cellular hydration state. *Am. J. Physiol.*, 267, E343, 1993.
20. Law, R.O., Techniques and applications of extracellular space determinations in mammalian tissues. *Experientia*, 38, 411, 1982.

21. Parsons, D.S. and van Rossum, G.D.V., On the determination of the extracellular water compartment in swollen slices of rat liver. *Biochim. Biophys. Acta*, 57, 495, 1962.
22. Mariani, M.F. DeFeo, B., Thomas, L., Schisselbauer, J.C., and van Rossum, G.D.V., Effects of chlorinated hydrocarbons on cellular volume regulation in slices of rat liver. *Toxicol. in vitro*, 5, 311, 1991.
23. Parsons, D.S. and van Rossum, G.D.V., Observations on the size of the fluid compartments of rat liver slices *in vitro*. *Quart. J. Exp. Physiol.*, 164, 116, 1962.
24. Russo, M.A., Ernst, S.A., Kapoor, S.C., and van Rossum, G.D.V., Morphological and physiological studies of rat kidney cortex slices undergoing isosmotic swelling and its reversal: a possible mechanism for ouabain-resistant control of cell volume. *J. Membrane Biol.*, 85, 1, 1985.
25. Kulka, R.G., Colorimetric estimation of ketopentoses and ketohexoses. *Biochem. J.*, 63, 542, 1956.
26. Oesch, F. and Steinberg, P., A comparative study of drug-metabolizing enzymes present in isolated rat liver parenchymal, Kupffer and endothelial cells. *Biochem. Soc. Trans.*, 15, 372, 1987.
27. Jungermann, K. and Katz, N., Functional specialization of different hepatocyte populations. *Physiol. Rev.*, 69, 708, 1989.
28. Tsien, R.Y., Fluorescent indicators of ion concentrations. *Meth. Cell Biol.*, 30, 127, 1989.
29. Wadsworth, S.J. and van Rossum, G.D.V., Role of vacuolar adenosine triphosphatase in the regulation of cytosolic pH in hepatocytes. *J. Membrane Biol.*, 142, 21, 1994.
30. Chance, B., Cohen, P., Jöbsis, F., and Schoener, B., Intracellular oxidation-reduction states *in vivo*. *Science*, 137, 499, 1962.
31. van Rossum, G.D.V., Simultaneous measurements of ^{24}Na-efflux and pyridine nucleotides in slices of rat liver. *Biochim. Biophys. Acta*, 122, 312, 1966.
32. Bobilya, D.J., Briske-Anderson, M., and Reeves, P.G., Zinc transport into endothelial cells is a facilitated process. *J. Cell Physiol.*, 151, 1, 1992.
33. Kapoor, S.C., van Rossum, G.D.V., O'Neill, K.J., and Mercorella, I., Uptake of inorganic lead *in vitro* by isolated mitochondria and tissue slices of rat renal cortex. *Biochem. Pharmacol.*, 34, 1439, 1985.

SELENIUM

Chapter 8

Selenium Diets: Deficiency and Excess

Merrill J. Christensen
Department of Food Science and Nutrition
Brigham Young University
Provo, Utah

CONTENTS

0-8493-9611-5/97/$0.00+$.50

I. INTRODUCTION

Selenium (Se) was first suspected to be an essential nutrient in the diets of rodents in 1957 when Klaus Schwarz identified this element as the active component of "Factor 3," which protected rats fed yeast-based diets from liver necrosis.[1] In 1973, its first biochemical role, as a component of the antioxidant enzyme glutathione peroxidase (EC 1.11.1.9) (GPX) was defined.[2] Since that time, additional Se-containing proteins have been discovered and new roles for the element identified.[3,4,5]

Diets deficient in Se have been used for many years to study the functions and metabolism of this nutrient in various rodent species. In other studies, diets containing high levels of this element have been fed to determine the toxicity of various organic and inorganic forms of Se and to examine the cancer chemopreventive effects of those forms.[6] In most cases, the induction of Se deficiency or toxicity is a comparatively straightforward proposition.

II. SELENIUM-DEFICIENT DIETS

A. Diet Formulation

The production of Se deficiency in the rat has been described in detail by Burk.[7] The basal Se-deficient diet used in our studies of Se regulation of gene expression[8,9,10,11] is based on that described by Burk and is shown in Table 1.

1. Protein

Almost all Se present in the diets of man and animals is supplied by protein. Hence, the choice of protein in the formulation of a Se-deficient diet is critical. We use *Torula* yeast, which has been the standard protein source in Se-deficient rodent diets for many years. We have purchased *Torula* yeast as a dried powder from U.S. Biochemical and from Rhinelander Paper Company. Since the Se content of *Torula* yeast may vary from batch to batch, it is wise to assay each batch received for total Se content before inclusion of that product in diet formulation. Trace amounts of Se will be present unavoidably in any diet formulation, presumably as contaminants in sulfur compounds. Still, in the formulation listed above, the majority of Se will be supplied by the *Torula* yeast. The typical *Torula* yeast diet will supply Se at less than 0.02 mg/kg. Thus, Se content of the dried *Torula* yeast should be less than 0.05 mg/kg. *Torula* yeast is deficient in methionine, which dictates supplementation of this essential amino acid at the level shown.

TABLE 1.

Composition of the Basal Se-Deficient Diet

Ingredient	Amount (g/kg)
Torula yeast	300.0
Sucrose	590.0
Corn oil	50.0
DL-Methionine	3.0
Calcium carbonate	12.0
Vitamin mixture[1] (Teklad TD 40060)	10.0
Mineral mixture[2] (Teklad TD 80313, with Se omitted)	35.0

[1] Composition of the vitamin mixture (Teklad TD 40060;g/kg vitamin mix): p-Aminobenzoic acid, 11.0132; Ascorbic acid, coated (97.5%), 101.6604; Biotin, 0.0441; Vitamin B_{12} (0.1% trituration in mannitol), 2.9736; Calcium pantothenate, 6.6079; Choline dihydrogen citrate, 349.6916; Folic acid, 0.1982; Inositol, 11.0132; Menadione, 4.9559; Niacin, 9.9119; Pyridoxine HCl, 2.2026; Riboflavin, 2.2026; Thiamin HCL, 2.2026, Dry retinal palmitate (500,000 U/g), 3.9648; Dry cholecalciferol (500,000 U/g), 0.4405; Dry dl-α-tocopheryl acetate (500 U/g), 24.2291; Corn starch, 466.6878.

[2] Composition of the mineral mixture (Teklad TD 80313, with Se omitted; g/kg mineral mix): Calcium phosphate, dibasic, 500.0; Sodium chloride, 74.0; Potassium citrate, monohydrate, 220.0; Potassium sulfate, 52.0; Magnesium oxide, 24.0; Manganous carbonate, 3.5; Ferric citrate, USP, 6.0; Zinc carbonate, 1.6; Cupric carbonate, 0.3; Potassium iodate, 0.01; Chromium potassium sulfate, 0.55; Sucrose, finely powdered, 118.04.

2. *Fat*

The use of corn oil or lard at 5% as the fat source is standard in Se-deficient diets. In certain studies of Se deficiency, it is desirable to induce a concomitant deficiency of vitamin E. This is done by omitting vitamin E from the vitamin mix (see below) and substituting vitamin E-stripped corn oil or lard as the fat source. These products are available from commercial suppliers (e.g., Eastman Chemicals).

3. *Vitamins*

The vitamin mixture listed in the Table above is commercially available (Teklad Premier) and meets established requirements for mice and rats. Burk[7] uses a slightly different formulation, which omits vitamin E. This nutrient, in his formulation, is added separately. Other commercially available vitamin mixes (e.g., AIN Vitamin Mix 76) may also be used. However, some commercially available mixtures do not contain choline. Use of these vitamin mixes dictates addition of choline separately.

4. *Minerals*

The mixture listed in the Table above is made by Teklad Premier. Any mineral mix which omits Se but meets requirements for all other minerals can be used.

We mix together in an industrial size mixer all dry ingredients, then slowly add oil. Mixing is continued only long enough to completely disperse the oil. The mixture is then stored in approximately 1 kg aliquots in double bagged, thick plastic bags, at –80°C. When needed, a bag of diet is thawed and the diet distributed. Diet remaining in the bag is stored at 4°C until used. We usually aliquot diet in such a way that, after initial thawing, diet in a single bag is used within a week. This practice is dictated in our laboratory more by convenience and limited storage capacity than by a careful, methodical study of ingredient stability at various temperatures. Burk[7] reports that the standard Se-deficient formulation may be kept at 4°C for one month. However, Vitamin E-deficient mixtures should not be used after more than 2 weeks storage at this temperature.

As a control diet, Se is added to the level desired in the mineral mix at the expense of carbohydrate. Previous work showed that in the liver of weanling male rats, activity of the Se-dependent GPX and mRNA levels for this enzyme were maximized by feeding diets containing 0.1 mg Se/kg diet.[12] In other work, 0.2 mg Se/kg diet provided greater protection in the retina of male rats against capillary degeneration than did 0.1 mg/kg.[13] The most recent recommendation of the American Institute of Nutrition splits the difference between these two values and establishes 0.15 mg Se/kg diet as the standard adequacy for inclusion in AIN-93 diets.[14]

The AIN recommends the addition of Se as selenate, rather than selenite, since selenate is less likely than selenite to cause oxidation of other dietary components.[15] There is also evidence that selenate is absorbed more efficiently by rats than is selenite.[16]

5. *Alternative Mixtures*

Formulations including casein as the protein source (e.g., AIN-93, with casein at 20%) are reported to provide approximately 0.18 mg Se/kg diet.[14] This level of Se meets minimum dietary requirements and would therefore not induce Se deficiency. Crystalline amino acid mixtures have been used by some investigators to lower the Se content of experimental diets to less than half the lowest level attainable with *Torula* yeast.[17,18] Addition of substances such as silver, which interfere with Se metabolism, has not gained widespread acceptance as a standard method for inducing Se deficiency due to additional biological effects unrelated to Se.

B. Animal Feeding

We use weanling male Sprague-Dawley rats in our studies of Se regulation of gene expression. Since males have a higher requirement for Se than do females, Se deficiency is induced more rapidly in males. Also, since rodents have compensatory mechanisms which conserve body Se during dietary deprivation, we find it useful to start with the youngest possible animals. We weigh each animal directly from the shipping crate, then transfer him to his cage containing the feed dish with the appropriate *Torula* yeast diet. Animals are housed in pairs in suspended stainless steel wire mesh cages in a temperature- and light-controlled room on a 12 hour light-dark cycle. Animals are given free access to food and water.

In most locations, tap water is a negligible source of dietary Se and can be provided to animals being Se-deprived. However, in certain geographic regions, drinking water can contain a significant amount of Se. If there is any question, tap water can simply be assayed directly for total Se. Alternatively, as Burk[7] suggests, a preliminary experiment can be conducted comparing rats given tap water with those given distilled water. Determination of Se deficiency in experimental animals is described below.

Obviously, the farther apart animals are housed, the less likely will be the risk of dietary cross contamination. Where possible, animals being fed different diets should be housed in different rooms. In many cases, this practice is not practical. Housing animals fed different diets in different cage racks on opposite sides of the same animal room is usually sufficient. Occasionally, even this separation is not possible. If animals fed Se-adequate and Se-deficient diets must be housed in the same rack, it is preferable to place the Se-deficient animals on the top rows and house the Se-adequate animals in the bottom rows. Spillage of Se-deficient diet from the top rows into the bottom row feeding cups containing Se-adequate diet will do much less to compromise experimental results than will contamination of Se-deficient diets with spillage from Se-adequate mixtures.

When nutrient deficiencies result in decreased food intake, paired feeding of control animals is necessary. This does not appear to be necessary in the case in Se deficiency, at least in first generation animals. There is one report, now nearly 20 years old, that Se deficiency causes a slight decrease in food intake.[19] However, the vast majority of studies published since that time show no difference in growth rates or feed intake between Se-deficient and Se-adequate first-generation rodents. Impairment of growth is seen in second-generation Se-deficient rats, in which case pair feeding of Se-adequate controls is advisable.

Diet can be provided in powdered form to most rodent species. Guinea pigs are the exception. Our best efforts were insufficient to coax weanling guinea pigs to consume sufficient amounts of the powdered *Torula* yeast mixture to support growth. In many cases, animals lost weight and soon died. This problem was remedied by providing guinea pigs with the *Torula* yeast diet in pelleted form. The composition of the Se-deficient mixture used for guinea pigs in our laboratory is based on the formulation of Burk et al.[20] and has been previously published.[21] This diet was prepared, including pelleting, commercially (TD 91250, Teklad Premier) as was its Se-adequate counterpart (TD 91251). When weanling male guinea pigs were transferred directly from their shipping crates (after weighing) to cages containing the experimental diets, they consumed the pelleted mixtures provided and grew normally.

Prompt introduction of experimental diet to guinea pigs may be crucial in persuading the animals to consume it. Guinea pigs fed standard chow diets for a short period of acclimatization may subsequently reject *Torula* yeast-based mixtures, even if pelleted.[22]

C. Determination of Se Deficiency

As noted above, growth of first generation rats and mice fed Se-deficient diets does not differ significantly from that of Se-adequate control animals. In contrast, we observed in guinea pigs statistically significant differences in average body weights between Se-adequate and Se-deficient animals beginning 2 weeks after feeding and continuing thereafter.[21]

The consequences of pure, uncomplicated Se deficiency in most rodents are mainly biochemical. The most used measure of Se deficiency is activity of liver GPX. Feeding rats the diet described herein results in a reduction of liver GPX activity to a level virtually indistinguishable from background in as short a time as 28 days.[23] In this regard, guinea pigs are also exceptional. Activity of GPX in the liver of this species is less than 10% that in rat liver, even when adequate Se is fed. Accordingly, reductions in liver GPX activity due to dietary Se deficiency in guinea pigs are much less, on a percentage basis, than decreases seen in other rodent species.[20]

III. SELENIUM TOXICITY DIETS

The repeated demonstration of Se's cancer chemopreventive effects has renewed interest in examining the biological effects of Se when fed at high levels. Composition of experimental diets intended to induce Se toxicity in experimental rodents will depend on the form of Se to be tested and the length of the feeding period. Other considerations include (1) the gender of the animals examined (male rats are more resistant than female rats to Se toxicity); (2) the dietary protein level (higher protein levels are more protective against Se toxicity); and (3) the indicator of "toxicity" employed.

A. Measures of Se Toxicity

According to Combs and Combs,[24] "Depressions in the rates of growth of young animals have been shown to be the most sensitive parameters of Se intoxication." Other effects of high Se intake have been observed in young animals even when growth is unaffected. Decreased liver stores of vitamin A,[25] abnormal liver appearance (but normal liver function), and increased portal pressure[26] have all been reported in studies of high Se intake. In our own work, liver glutathione S-transferase activity was significantly higher in rats fed 2.0 ppm Se, compared to rats fed the nutritionally adequate level of 0.1 ppm.[10]

B. Diet Formulation

1. *Protein*

Tinsley et al.[27] fed rats 12 or 22% casein diets and showed that the higher level of protein was protective against Se toxicity when Se was fed as selenate or selenite. More recent work by Salbe and Levander[28] suggests that methionine intake may moderate Se's toxic effects. In their work, rats were fed *Torula* yeast diets supplemented or not supplemented with adequate methionine. All rats fed methionine-inadequate diets had lower weight gains than their methionine-adequate counterparts. In addition, the highest level of Se fed (2.5 ppm) was toxic (as indicated by reduced weight gain) only in methionine-deficient rats.

Clement Ip and his colleagues, in their study of Se's cancer chemopreventive effects, have assayed numerous organic and inorganic Se compounds for their toxicity and inhibitory effects on tumor growth in the rat DMBA-induced mammary tumor model.[6] In their work, the AIN-76A diet was used as the basal regimen. The more recently formulated AIN-93 diets (AIN-93G for growth, pregnancy, and lactation, and AIN-93M for adult maintenance) would serve equally well as the basal diets in studies of Se toxicity. In our studies of gene regulation, we have used the basic *Torula* yeast diet, described in detail above, and simply added Se to the level desired in the mineral mix.

2. *Form of Selenium*

Ip and colleagues defined "maximum tolerable dose" (MTD) as "the dose which produces the first indication of a significant suppression in growth" when 40 day old female rats were fed various organic and inorganic Se compounds for 4 weeks. To illustrate the difference in toxicity between different forms of Se, the MTD for sodium selenite was 4.0 ppm while the MTD for xylene selenocyanate was 20 ppm. There appears to be little difference in toxicity among the forms of Se commonly used in feeding studies (sodium selenite, sodium selenate, selenomethionine) when all other nutrients are supplied in adequate amounts. As described above, selenite may promote oxidation of other dietary components. Obviously, the potential for such oxidation increases with increasing Se content. Hence, sodium selenate is preferable to sodium selenite in studies of high Se intake, unless there is a compelling rationale to study sodium selenite in particular. Metabolism of selenomethionine can be markedly affected by the methionine content of the diet and methionine status of the animal. Therefore, when this form is chosen for supplementation at high levels, the content of methionine in the diet deserves careful attention.

3. *Level of Selenium and Duration of Feeding*

Combs and Combs[24] stated "Animals do not show depressed growth at dietary levels of less than 2 ppm Se when it is provided in the common feeding forms." We have confirmed this observation in our studies of gene regulation, in which we fed *Torula* yeast diets supplemented with 2.0 ppm Se for 13 weeks to male Sprague Dawley weanling rats. Under these conditions, no significant effects on growth were observed. Most work published since the review of Combs and Combs is consistent in showing no effect on growth of male rats fed Se levels up to 4.0 ppm for periods of time ranging form 4 to 10 weeks. However, it was at 4.0 ppm Se that the toxic effects described above (decreased liver vitamin A, increased portal pressure, etc.) were seen after 8 weeks of feeding. This was the level established by Ip and colleagues in their 4 week feeding trials as the maximum tolerable dose in female rats.

Recent studies are also in agreement that growth is inhibited when weanling rats are fed diets containing 5.0 ppm or more Se. Again, these studies employed the more commonly used forms of Se for feeding periods lasting 4 to 10 weeks. As noted above, one of the appealing characteristics of recently synthesized organic Se compounds which are effective in cancer chemoprevention is their markedly reduced toxicity in comparison to sodium selenite or selenate.

IV. SUMMARY

The production of Se deficiency or toxicity in experimental animals by dietary means is a relatively straightforward, uncomplicated proposition. The dietary formulations for inducing Se deficiency or toxicity have been in common use for many years. Measures of Se deficiency are well established, but sensitive indicators of Se toxicity, in addition to growth inhibition, are less well defined. Care must be taken to ensure that all other nutrients are provided in adequate quantities, since deficiency (or excess) of other dietary components may affect metabolism of Se. Different forms of Se will vary in toxicity when fed at high levels.

REFERENCES

1. Schwarz, K. and Folz, C.M., Selenium as an integral part of Factor 3 against dietary necrotic liver degenerating. *J. Am. Chem. Soc.*, 79, 3292, 1957.

2. Rotruck, J.T., Pope, A.L., Ganther, H.E., Swanson, A.B., Hafeman, D., and Hoekstra, W.G., Selenium: biochemical role as a component of glutathione peroxidase. *Science*, 179, 588, 1973.
3. Berry, M.J., Banu, L., and Larsen, P.R., Type I iodothyronine deiodinase is a selenocystein-containing enzyme. *Nature*, 349, 438, 1991.
4. Yang, J.G., Hill, K.E., and Burk, R.F., Dietary selenium intake controls rat plasma selenoprotein P concentration. *J. Nutr.*, 119, 1010, 1989.
5. Vendeland, S.C., Beilstein, M.A., Chen, C.L., Jensen, O.N., Barofsky, E., and Whanger, P.D., Purification and properties of selenoprotein W from rat muscle. *J. Biol. Chem.*, 268, 17103, 1993.
6. Ip, C., El-Bayoumy, K., Upadahyaya, P., Ganther, H., Vadhanavikit, S., and Thompson, H., Comparative effect of inorganic and organic selenocyanate derivatives in mammary cancer chemoprevention. *Carcinogenesis*, 15, 187, 1994.
7. Burk, R.F., Production of selenium deficiency in the rat. *Methods Enzymol.*, 143, 307, 1987.
8. Christensen, M.J. and Burgener, K.W., Dietary selenium stabilizes glutathione peroxidase mRNA in rat liver. *J. Nutr.*, 122, 1620, 1992.
9. Christensen, M.J. and Pusey, N.W., Binding of nuclear proteins to transcription regulatory elements in selenium deficiency. *Biochim. Biophys. Acta*, 1225, 338, 1994.
10. Christensen, M.J., Nelson, B.L., and Wray, C.D., Regulation of glutathione S-transferase gene expression and activity by dietary selenium. *Biochem. Biophys. Res. Commun.*, 202, 271, 1994.
11. Christensen M.J., Cammack, P.M., and Wray, C.D., Tissue specificity of selenoprotein gene expression in rats. *J. Nutr. Biochem.*, 6, 367, 1995.
12. Sunde, R.A., Weiss, S.L., Thompson, K.M., and Evenson, J.K., Dietary selenium regulation of glutathione peroxidase mRNA-implications for the selenium requirement. *FASEB J.*, 6, A1365 (Abs.), 1992.
13. Eckhert, D.D., Lockwood, M.K., and Shen, B., Influence of selenium on the microvasculature of the retina. *Microvasc. Res.*, 45, 74, 1993.
14. Reeves, P.G., Nielsen, F.H., and Fahey, G.C., Jr., AIN-93 purified diets for laboratory rodents: final report of the American Institute of Nutrition Ad Hoc Writing Committee on the Reformulation of the AIN-76A Rodent Diet. *J Nutr.*, 123, 1939, 1993.
15. National Research Council, *Selenium in Nutrition*, rev. ed., p 5, 1985, National Academy Press, Washington, DC.
16. Vanderland, S.C., Butler, J.A., and Whanger, P.D., Intestinal absorption of selenite, selenate, and selenomethionine in the rat. *J. Nutr. Biochem.*, 3, 359, 1992.
17. Thompson, K.M., Haibach, H., and Sunde, R.A., Growth and triiodothyronine levels are modified by selenium deficiency and repleting in second-generation selenium-deficient rats. *J. Nutr.*, 125, 864, 1995.
18. Beckett, G.J., Beddows, S.E., Morrice, P.C., Nicol, F., and Arthur, J.R., Inhibition of hepatic deiodination of thyroxine is caused by selenium deficiency in rats. *Biochem. J.*, 248, 443, 1987.
19. Ewan, R.C., Effect of selenium on rat growth, growth hormone and diet utilization. *J. Nutr.*, 106, 702, 1976.
20. Burk, R.F., Lane, J.M., Lawrence, R.A., and Gregory, P.E., Effect of selenium deficiency on liver and blood glutathione peroxidase activity in guinea pigs. *J. Nutr.*, 111, 690, 1981.
21. Cammack, P.M., Zwhalen, B.A., and Christensen, M.J., Selenium deficiency alters thyroid hormone metabolism in guinea pigs. *J. Nutr.*, 125, 302, 1995.
22. Whanger, P.D., Personal communication, 1995.
23. Lawrence, R.A. and Burk, R.F., Glutathione peroxidase activity in selenium-deficient rat liver. *Biochem. Biophys. Res. Commun.*, 71, 952, 1976.
24. Combs, G.F., Jr. and Combs, S.B., *The Role of Selenium in Nutrition*, 1986, Academic Press, Orlando, FL, 463.
25. Albrecht, R., Pelissier, M.A., and Boisset, M., Excessive dietary selenium decreases the vitamin A storage and the enzymatic antioxidant defense in the liver of rats. *Toxicol. Lett.*, 70, 291, 1994.

26. Bioulac-Sate, P., Dubuisson, L., Bedin, C., Gonzalez, P., Detinguymoreaud, E., Garcin, H., and Balabaud, C., Nodular regenerative hyperplasia in the rat induced by a selenium-enriched diet: study of a model. *Hepatology,* 16, 418, 1992.
27. Tinsley, I.J., Harr, J.R., Bone, J.F., Weswig, P.H., and Yamamoto, R.S., Selenium toxicity in rats. I. Growth and longevity. In: *Selenium in Biomedicine,* 1967, Avi Publishing Co., Westport, CT, 141.
28. Salbe, A.D. and Levander, O.A., Comparative toxicity and tissue retention of selenium in methionine-deficient rats fed sodium selenate of L-selenomethionine. *J. Nutr.,* 120, 207, 1990.

Chapter 9

Selenium in Tissue Culture

Catherine A. Wardle
Alan Shenkin
Department of Clinical Chemistry
Royal Liverpool University Hospital
Liverpool, England

CONTENTS

I. INTRODUCTION

The study of Se in tissue culture has proved difficult in the past for two major reasons. First, serum was required as an essential component of culture media to provide hormones, growth factors, and other nutrients including trace elements. This resulted in the presence of Se as a "contaminant" of culture media (McKeehan et al., 1976; Hamilton and Ham, 1977), and tissues and cells therefore could not be incubated under conditions of Se deficiency. Secondly, methods for the measurement of trace concentrations of Se were not reliable. This introduced an added complication in that the concentration of the element in culture media could not be accurately measured.

0-8493-9611-5/97/$0.00+$.50

Small molecular weight molecules such as Se can be removed from serum by dialysis prior to its addition to culture medium (Ham and McKeehan, 1979). However, a large proportion of the Se present in serum is bound to proteins such as glutathione peroxidase (GPx) and Selenoprotein P (Behne and Wolters, 1983; Combs and Combs, 1986) and is therefore not removed by dialysis. In addition, dialysis of serum removes other small molecular weight nutrients as well as Se — the effects observed using culture media prepared with dialysed serum can therefore not necessarily be attributed simply to the absence of Se.

More recently, the identification of the individual hormones, growth factors, and other nutrients present in serum has allowed the development of chemically defined, serum-free media (Haysahi and Sato, 1976; Barnes and Sato, 1979). These media are useful in the study of trace elements in tissue culture as their composition can be readily altered, and contaminants normally found in serum are absent. The use of such culture media has facilitated the simple design and interpretation of experiments which would otherwise be difficult or impossible to carry out in serum-containing medium. Furthermore, the measurement of low concentrations of Se is now possible. The most commonly used method is graphite furnace atomic absorption with or without Zeeman background correction. Problems previously encountered due to interference by phosphate in biological fluids and culture media can now be avoided either by including palladium as a chemical modifier in the analysis or by using hydride generation. Sensitive fluorometric methods have also been developed for the measurement of Se.

In common with all trace elements, Se can have both beneficial and toxic effects depending on its concentration (Ham and McKeehan, 1979). The concentrations of Se at which signs of deficiency or toxicity occur vary depending on a number of factors including the cell or tissue type used, its selenium status, the composition of the culture medium, the form of selenium added to the culture medium, the duration of the experiment, and the parameter used to study the effects of Se. The variation in response of different *in vitro* systems to the presence of Se is illustrated in Table 1.

In vivo the function of tissues such as cardiac muscle, skeletal muscle, and liver is more rapidly affected by Se depletion than that of other organs (Levander and Burk, 1986). Se is likely to have different functions in different tissues; 13 Se-containing proteins or protein subunits were detected in homogenates of rat tissues (Behne et al., 1988). Most of the proteins were present in all tissues investigated but one was detected only in the testes and spermatozoa and another was present mainly in the thyroid. In weanling rats fed Se-deficient diets, both GPx and type I iodothyronine 5′-deiodinase (5′-DI) activities in liver were decreased, whereas in the same animals thyroid GPx activity was unaffected and thyroid 5′-DI activity was increased (Beech et al., 1995). The implication of these findings is that in Se deficiency the thyroid but not the liver is able to retain sufficient amounts of the trace element to allow continued

TABLE 1.

In Vitro Cell and Tissue Culture Systems Used to Investigate the Role of Selenium

Authors	Cell/Tissue Type Used	[Selenium] (μmol/l)	Measured	Effect
McKeehan et al. 1976	Diploid Human Fibroblasts Chinese Hamster Ovary	0–10	Cell growth	Optimum HDF 0.03 μmol/l CHO 0.10 μmol/l
Kajander et al. 1990	17 Malignant cell lines	Selenomethionine 1–160 μmol/l	Cell growth	Variable Cytotoxic 30–160 μmol/l
Hatfield et al. 1991	Human Myeloid Leukemia (HL-60) Rat Mammary Tumor (RMT)	0.13 63.3	Selenocysteine t-RNA	Increase
Sword et al. 1991	Rat liver and lung slices	Se/Vitamin E deficient diet	Ethane production TBARS Conjugated dienes	Increased in Se deficiency
Mertens et al. 1991b	Adult Rat Hepatocytes	0.1	Reduced and oxidised glutathione	Variable
Leccia et al. 1993	Human Skin Fibroblasts	1266	GPx TBARS	Increase Decrease
Baker et al. 1993	Human Hepatoma (3B and G2)	0.06	GPx GPx mRNA	Increase
Lin et al. 1993	Murine L1210	0.03	GPx Catalase	Increase Decrease
Mertens et al. 1993	Adult Rat Hepatocytes	0.1	GPx Glutathione Reductase	No consistent effect
Oertel et al. 1993	Porcine Kidney Epithelial Cells (LLC-PK1)	$1 \times 10^{-5} - 1$	Cell Growth 5′ Deiodinase expression	Increase
Thomas et al. 1993	Bovine Aortic Endothelial Cells	0.13	GPx Glutathione	Increase Decrease
Ueki et al. 1993	Rat Adipose Tissue	1000	Lipoprotein Lipase	Increase
Lu et al. 1994	Rat Aortic Rings	2000 4000	ACh-dependent relaxation	None
Beech et al. 1995	Human Thyrocytes	<1–1	Iodothyronine 5′-Deiodinase GPx	Increased by TSH addition None

expression of 5′-DI and GPx, with 5′-DI receiving a preferential supply of selenium.

Most studies have used Se supplements in the form of sodium selenite although neutralized selenious acid and selenomethionine have also been used (McKeehan et al., 1976; Kajander et al., 1990). The uptake of ^{75}Se-labeled selenite and selenomethionine has been compared in transformed mouse lung fibroblasts (White and Hoekstra, 1979). Selenite was rapidly absorbed into the cell wall but slowly incorporated into the soluble protein fraction. In contrast, selenomethionine was more slowly absorbed but once absorbed was rapidly incorporated into soluble cytoplasmic protein. Selenite was incorporated into GPx whereas selenomethionine was incorporated into a wide spectrum of proteins and only later into GPx. In cultured phytohemagglutinin-transformed lymphocytes incubated in the presence of ^{75}Se-labeled selenite, selenate, selenocysteine, and selenomethionine, Karle et al. (1983) demonstrated preferential uptake of the organic forms of Se and increased GPx activity in the presence of such compounds.

Selenomethionine is known to be an excellent analog for methionine in biochemical reactions (Sunde, 1984) and can substitute for methionine during protein synthesis (McConnell and Hoffman, 1972). It has been suggested that when methionine is limiting, Se sequestered as selenomethionine would not be used in a functional role for Se until these proteins were degraded, but could potentially serve as a ready source of Se at a later time (Waschulewski and Sunde, 1988).

Oxygenation of the medium is another important consideration in experiments which involve the study of Se. A minimal amount of oxygen is essential for the multiplication of most cell types in culture. However, excess oxygen may be inhibitory due to increased free radical production. GPx is known to protect against free radical damage and Se is essential for GPx activity. The inhibitory effects of excess oxygen have been shown to be much more severe when cells are marginally deficient in Se than when adequate amounts are supplied (Ham and McKeehan, 1979) and GPx activity was lower when oxygen-saturated Krebs Henseleit buffer was used to isolate adult rat hepatocytes than when nongassed Hepes buffer was used (Mertens et al., 1993).

II. SELENIUM DEFICIENCY

The effects of Se deficiency on cells or tissue in culture can be studied by comparing tissue from animals fed a Se-deficient diet with that from animals fed a control or Se-supplemented diet (Sword et al., 1990; Lu et al., 1994). Alternatively, cells or tissue from animals receiving standard diets can be used and the effects of the addition of Se as a supplement to the culture medium studied (Hatfield et al., 1991; Baker et al., 1993; Leccia et al., 1993). The effects observed vary depending on the duration of Se deficiency, the cell or tissue type, and the culture conditions used. The effect of Se supplementation to the culture medium of two cell types is illustrated in Figure 1.

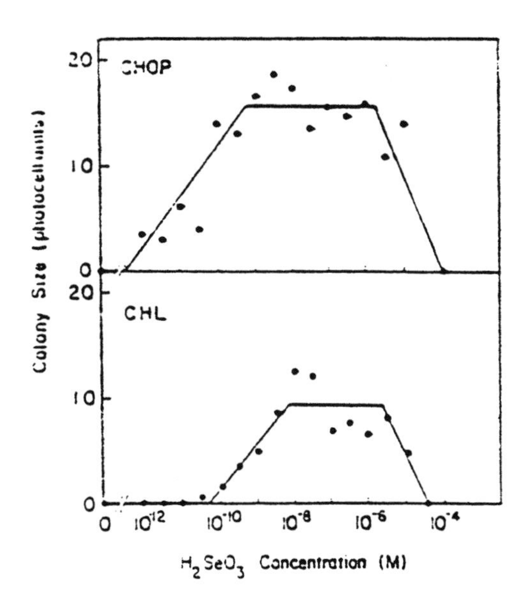

Figure 1
The effect of Se supplementation on clonal growth of chinese hamster ovary (CHOP) and chinese hamster lung (CHL) cells cultured using F12 medium. (From Hamilton and Ham, 1977. With permission.)

The presence of Se in culture medium has been shown to be essential for clonal growth of cell lines such as WI-38 diploid human fibroblasts, chinese hamster ovary, and chinese hamster lung cells (McKeehan et al., 1976; Hamilton and Ham, 1977). Another study demonstrated that the addition of Se to chemically-defined culture medium of rat mammary tumor cells increased steady-state concentrations and relative distributions of selenocystyl transfer-RNA (t-RNA) species which are responsible for the donation of selenocysteine to protein in the synthesis of Se-containing proteins such as GPx (Hatfield et al., 1991).

Se is known to be essential for the activity of GPx in the protection of cells and tissues from damage caused by free radical generation (Forstrom et al., 1978). Many studies have therefore used either GPx activity (Baker et al., 1993; Thomas et al., 1993; Leccia et al., 1993) or markers of oxidative damage as measures of the effect of Se supplementation or deficiency in cultured cells or tissues (Sword et al., 1991; Leccia et al., 1993). Figure 2 illustrates the effect of Se supplementation on GPx activity in Hep 3B cells.

GPx activity in pure cultured adult rat hepatocytes was found to increase during the first 2 to 3 days of culture and then to decrease slowly

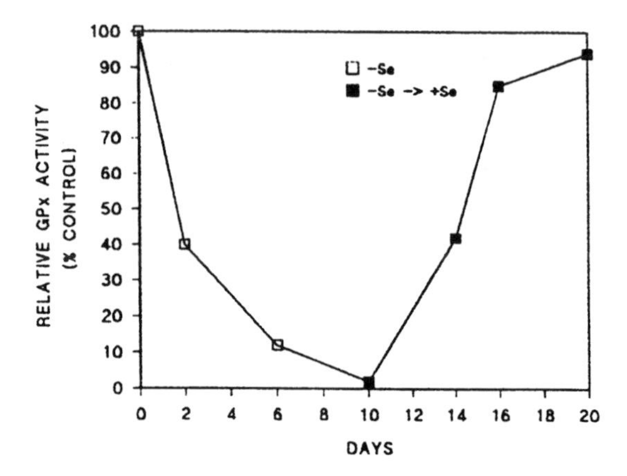

Figure 2
Glutathione peroxidase activity in Hep 3B cells in the absence (□) and presence (■) of Se (60 nmol/l). (Adapted from Baker et al., 1993. With permission.)

over the next 3 to 4 days (Mertens et al., 1993), suggesting that the timing of the measurement of GPx activity is an important consideration.

A method has been developed for the measurement of reduced and oxidized glutathione in cultures of adult rat hepatocytes by reverse phase high performance liquid chromatography (hplc) with spectrophotometric detection at 365 nm (Mertens et al., 1991a). However, supplementation of culture medium with Se (0.1 μmol/l) had no effect on reduced and oxidized glutathione concentrations in adult rat hepatocytes (Mertens et al., 1991b).

Hepatocytes from rats fed Se-deficient diets have been used as a model for the study of the effects of Se deficiency on the metabolism and detoxification of xenobiotics (Hill and Burk, 1982). Se-deficient hepatocytes were more susceptible to damage induced by cumene hydroperoxide and diaquat than were Se-replete hepatocytes. However, Se deficiency had no effect on diisopropylene toxicity and was protective against acetaminophen toxicity (Hill and Burk, 1984). The effect of Se deficiency on acetaminophen toxicity is thought to be due to the fact that glutathione synthesis is increased during conditions of Se depletion. Liver and lung tissue slices from rats fed Se- and vitamin E-deficient or supplemented diets have also been used to study oxidative damage caused by *Salmonella typhimurium* endotoxin (Sword et al., 1991).

Se is also known to be involved in thyroid hormone metabolism, being required for activity of iodothyronine 5′-deiodinase (5′-DI). The p27 5′-DI subunit involved in the catalytic step of type I 5′-deiodination has been identified as a selenoprotein in rat liver microsomes (Behne et al., 1990). It has subsequently been shown that selenite concentration regulates activity and expression of the p27 substrate-binding subunit of type I 5′-DI in a porcine-kidney epithelial cell line (LLC-PK1) cultured in a serum-free medium (Oertel et al., 1993). Compared with GPx, expression of the p27 5′-DI subunit was observed at 10-fold lower concentrations of Se in the growth medium, suggesting an intracellular hierarchy of selenite utilization.

In agreement with previous findings, and illustrated in Figure 3, 5′-DI activity of human thyrocytes in culture could be stimulated by the addition of thyroid-stimulating hormone (TSH) under conditions of Se deficiency (<1 nmol/l), and further increased in the presence of sodium selenite (50 to 1000 nmol/l). GPx activity in the same cells was unaffected by Se deficiency, suggesting that the thyroid is able to retain sufficient amounts of Se to allow continued expression of 5′-DI and GPx in the face of Se deficiency (Beech et al., 1995).

The effects of Se on rat adipose tissue metabolism *in vitro* have also been studied. Sodium selenate has been shown to mimic actions of insulin such as the stimulation of glucose uptake and cyclic AMP phosphodiesterase activity (Ezaki, 1990) and to increase the activity of lipoprotein lipase in isolated rat adipose tissue. Lipoprotein lipase activity was shown to be increased in response to selenate via amiloride- and monensin-sensitive processes, involving Ca^{++} mobilization linked to a rapid increase in inositol 1,4,5-triphosphate (IP_3) content in the tissue (Ueki et al., 1993).

III. SELENIUM TOXICITY

Selenium toxicity has been demonstrated in several types of cells and tissues. Vernie et al. (1983) demonstrated inhibition of protein synthesis and subsequent cell death in L1210 (rat liver) cells by selenite at concentrations greater than 2 μmol/l and several studies have demonstrated chromosomal changes in *in vitro* systems (Lu et al., 1978; Whiting et al., 1980; Siranni and Huang, 1983). Bell et al. (1991) showed that 25 μmol/l selenite reduced cell viability of rat hepatocytes by 43%, whereas Park and Whanger (1995) reported an LD_{50} value of 500 μmol/l for selenite in isolated rat hepatocytes and suggested that the toxic effects of the element may depend on the concentrations of other nutrients in the culture medium and the duration of the incubation. In cultured HeLa 53 cells, 25 μmol/l Se inhibited DNA and RNA synthesis but the presence of Se at concentrations greater than 200 μmol/l were needed to induce cell lysis (Greunwedel and Cruikshank, 1979).

Inorganic Se is detoxified by its conversion to less toxic methylated compounds such as dimethylselenide and trimethylselenonium (McConnell, 1952; Palmer, 1969). In a study investigating the role of vitamin B_{12} as a cofactor of methionine synthetase, the ability of primary hepatocytes to methylate and detoxify radiolabeled ^{75}Se was significantly decreased when the animals were fed Vitamin B_{12}-deficient diets (Chen and Whanger, 1992).

Se compounds are known to act as anticancer agents both in intact animals and cellular systems (Medina and Morrison, 1988; Vernie, 1984)

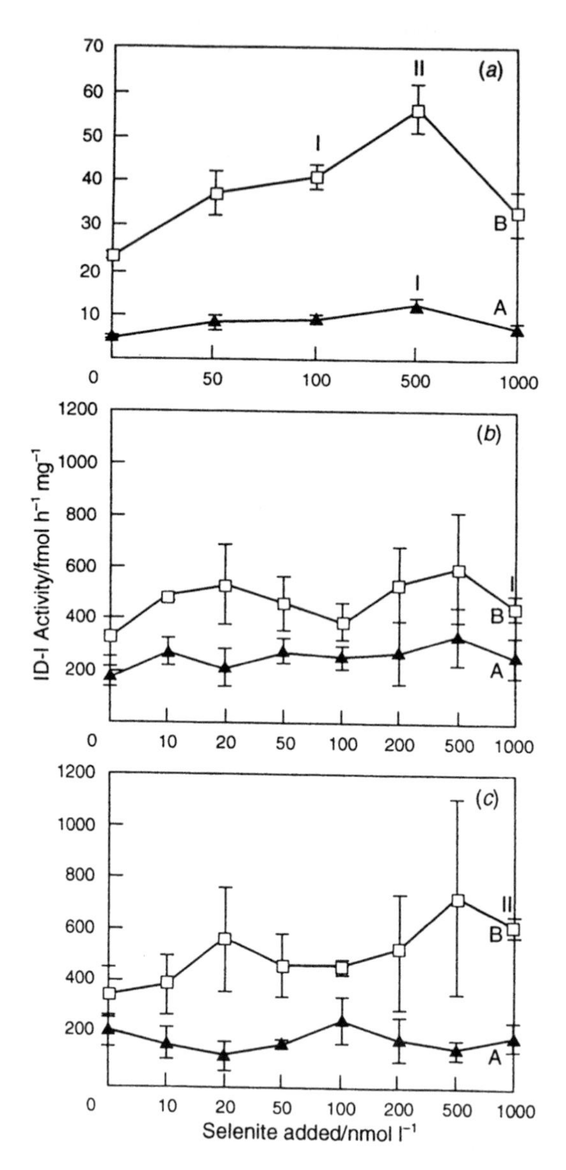

Figure 3
The effects of TSH and selenite addition on the activity of 5′-DI in human thyrocytes cultured from tissue obtained from a patient with Graves' disease. A. No. TSH added: B. TSH added at 1 U/l. Significant increase in 5′-DI activity resulting from addition of selenite. I. $p < 0.05$: II. $p < 0.01$. (From Beech et al., 1995. With permission.)

and compounds such as selenite induce a number of cytotoxic effects in tumor cells *in vitro* including inhibition of cell growth and proliferation (LeBoeuf et al., 1985; Medina and Oborn, 1981; Milner and Hsu, 1981). In rat mammary tumor cells, selenite inhibited cell proliferation, protein synthesis, and DNA replication in a dose-dependent manner at concentrations greater than 1 μg/l (Lewko and McConnell, 1985).

Kajander et al. (1990) studied the effects of selenomethionine on the growth of 17 cultured normal and malignant cell lines. Toxicity was observed in all 17 cell lines after 3 to 10 days of culture, 50% growth inhibition being observed at selenomethionine concentrations between 30 and 160 μmol/l. Cytotoxicity was common to normal and malignant cell lines, suggesting that selenomethionine does not have specific antineoplastic activity. In another study, drug-resistant human ovarian tumor (NIH:OVCAR3) cells derived from a patient refractory to treatment with doxorubicin, cisplatin, and cyclophosphamide were shown to be more sensitive to the toxic effects of selenite (5 to 20 μmol/l) than drug-sensitive cells (Caffrey and Frenkel, 1992), suggesting that selenite may have potential as a treatment for drug-resistant tumors.

IV. CONCLUSION

The study of the effects of Se in tissue culture readily allows investigation of the effects of the element in the absence of the complicating factors which are present *in vivo*. The development of chemically defined culture media and more reliable methods for the measurement of trace concentrations of Se has enabled the systematic investigation of the effects of Se deficiency or supplementation in cell or tissue culture. The evidence available so far suggests that the effects of Se supplementation are dependent on the cell or tissue type studied, the concentration of Se used and the culture conditions. Future work with Se in cell or tissue culture should provide more information about the role of Se at the cellular and molecular level and about its interaction with other nutrients.

REFERENCES

Baker, R.D., Baker. S.S., LaRosa, K., Whitney, C., and Newburger, P.E., Selenium regulation of glutathione peroxidase in human hepatoma cell line Hep3B. *Arch. Biochem. Biophys.*, 304, 53, 1993.

Barnes, D. and Sato, G., Methods for growth of cultured cells in serum-free medium. *Anal. Biochem.*, 102, 255, 1980.

Beech, S.G., Walker, S.W., Beckett, G.J., Arthur, J.R., Nicol, F., and Lee, D., Effect of selenium depletion on thyroidal type-I iodothyronine deiodinase activity in isolated human thyrocytes and rat thyroid and liver. *Analyst*, 120, 827, 1995.

Behne, D., Hilmert, H., Scheid, S., Gessner, H., and Elger, W., Evidence for specific selenium target tissues and new biologically important selenoproteins. *Biochim. Biophys. Acta*, 966, 12, 1988.

Behne, D., Kyriakopoulos, A., Meinhold, H., and Kohrle, J., Identification of Type I iodothyronine 5′-deiodinase as a selenoenzyme. *Biochem. Biophys. Res. Comm.*, 173, 1143, 1990.

Behne, D. and Wolters, W., Distribution of selenium and glutathione peroxidase in the rat. *J. Nutr.*, 113, 456, 1983.

Bell, R.R., Nonavinakere, V.K., Soliman, M.R.I., and Early, II, J.L., Effect of *in vitro* treatment of rat hepatocytes with selenium and/or cadmium on cell viability, glucose output and cellular glutathione. *J. Toxicol.*, 69, 111, 1991.

Caffrey, P.B. and Frenkel, G.D., Selenite cytotoxicity in drug resistant and non-resistant human ovarian tumour cells. *Cancer Res.*, 52, 4812, 1992.

Chen, C.L. and Whanger, P.D., Effect of vitamin B_{12} status on selenium methylation and toxicity in rats: *in vivo* and *in vitro* studies. *Toxicol. Appl. Pharmacol.*, 118, 65, 1993.

Combs, G.F. and Combs, S.B., Absorption, excretion and metabolism of selenium. In: *The Role of Selenium in Nutrition*, 179, 1986.

Ezaki, O., The insulin-like effects of selenate in rat adipocytes. *J. Biol. Chem.*, 265, 1124, 1990.

Forstrom, J.W., Zakowski, J.J., and Tappel, A.L., Identification of the catalytic site of rat liver glutathione peroxidase as selenocysteine. *Biochem.*, 17, 2639, 1978.

Greunwedel, D.W. and Cruishank, M.K., The influence of sodium selenite on the viability and intracellular synthetic activity (DNA, RNA and protein synthesis) of HeLa 53 cells. *Toxicol. Appl. Pharmacol.*, 50, 1, 1979.

Ham, R.G. and McKeehan, W.L., Media and growth requirements. In: *Methods in Enzymology;* Eds. Jacoby, W.B. and Pastan, I.H., 43, 44, 1979.

Hatfield, D., Lee, B.J., Hampton, L., and Diamond, A.M., Selenium induces changes in the selenocysteine tRNA$^{[Ser]Sec}$ population in mammalian cells. *Nucleic Acids Research*, 19, 939, 1991.

Hayashi, I. and Sato, G.H., Replacement of serum by hormones permits growth of cells in a defined medium. *Nature*, 259, 132, 1976.

Hill, K.E. and Burk, R.F., Effect of selenium deficiency and vitamin E deficiency on glutathione metabolism in isolated rat hepatocytes. *J. Biol. Chem.*, 10668, 1982.

Hill, K.E. and Burk, R.F., Toxicity studies in isolated hepatocytes from selenium-deficient rats and vitamin E deficient rats. *Toxicol. Appl. Pharmacol.*, 72, 32, 1984.

Kajander, E.O., Harvima, R.J., Kauppinen, L., Akerman, K.K., Martikainen, H., Pajula, R.L., and Karenlampi, K.K., Effects of selenomethionine on cell growth and on S-adenosylmethionine metabolism in cultured malignant cells. *Biochem. J.*, 267, 767, 1990.

Karle, J.A., Kull, F.J., and Shrift, A., Uptake of selenium-75 by PHA-stimulated lymphocytes. Effect on glutathione preoxidase. *Biol. Trace Elem. Res.*, 5, 17, 1983.

LeBoeuf, R., Laishes, B., and Heokstra, W., Effects of selenium on cell proliferation in rat liver and mammalian cells as indicated by cytokinetic and biochemical analysis. *Cancer Res.*, 45, 5496, 1985.

Leccia, M.T., Richard, M.J., Beani, J.C., Faure, H., Monjo, A.M., Cadet, J., Amblard, P., and Favier, A., Protective effect of selenium and zinc on UV-A damage in human skin fibroblasts. *Photochem. Photobiol.*, 58, 548, 1993.

Levander, O.A. and Burk, R.F., Report on the 1986 A.S.P.E.N. research workshop on selenium in clinical nutrition. *J. Parent Ent. Nutr.*, 10, 545, 1986.

Lin, F., Thomas, J.P., and Girotti, A.W., Hyperexpression of catalase in selenium-deprived murine L1210 cells. *Arch. Biochem. Biophys.*, 305, 176, 1993.

Lo, L.W., Koropatnik, J., and Stich, H.F., The mutagenicity and cytotoxicity of selenite, "activated" selenite and selenate for normal and DNA repair-deficient human fibroblasts. *Mutat. Res.*, 49, 305, 1978.

Lu, X., Liu, S.Y., and Man, R.Y.K., Enhancement of endothelium dependent relaxation in the rat aortic ring by selenium supplement. *Cardiovasc. Res.*, 28, 345, 1994.

McConnell, K.P., Selenium toxicity in rats as influenced by choline, betaine and methionine. *Fed Proc.*, 11, 255, 1952.

McConnell, K.P. and Hoffman, J.L., Methionine-selenomethionine parallels in rat liver polypeptide chain synthesis. *FEBS Lett.*, 24, 60, 1972.

McKeehan, W.L., Hamilton, W.G., and Ham, R.G., Selenium is an essential trace element for growth of WI-38 diploid human fibroblasts. *Proc. Natl. Acad. Sci.*, 73, 2023, 1976.

Medina, D. and Morrison, D., Current ideas on selenium as a chemopreventative agent. *Pathol. Immunopathol. Res.*, 7, 187, 1988.
Medina D. and Oborn, C., Differential effects of selenium on the growth of mouse mammary cells *in vitro*. *Cancer Lett.*, 13, 333, 1981.
Mertens, K., Rogiers, V., Sonck, W., and Vercruysse, A., Measurement of reduced and oxidised glutathione in cultures of adult rat hepatocytes. *J. Chromatog.*, 565, 149, 1991a.
Mertens, K., Rogiers, V., Sonck, W., and Vercruysse, A., Reduced and oxidised glutathione contents in adult rat hepatocytes under various culture conditions. *Cell Biol. Toxicol.*, 7, 101, 1991b.
Mertens, K., Rogiers, V., and Vercruysse, A., Glutathione dependent detoxification in adult rat hepatocytes under various culture conditions. *Arch. Toxicol.*, 67, 680, 1993.
Milner, J. and Hsu, C., Inhibitory effects of selenium on the growth of L1210 leukemic cells. *Cancer Res.*, 41, 1652, 1981.
Oertel, M., Gross, M., Rokos, H., and Kohrle, J., Selenium-dependent regulation of type I deiodinase expression. *Am. J. Clin. Nutr. Suppl.*, 57, 313s, 1993.
Palmer, I.S., Fischer, D.D., Halverson, A.W., and Olson, O.E., Identification of a major selenium excretory product in rat urine. *Biochim. Biophys. Acta*, 177, 336, 1969.
Park, Y.C. and Whanger, P.D., Toxicity, metabolism and absorption of selenite by isolated rat hepatocytes. *J. Toxicol.*, 100, 151, 1995.
Siranni, S.R. and Huang, C.C., Induction of sister chromatid exchange by various selenium compounds in Chinese hamster cells in the presence and absence of S9 mixture. *Cancer Lett.*, 18, 109, 1983.
Sunde, R.A., The biochemistry of the selenoproteins. *Am. J. Oil Chem.*, 61, 1891, 1984.
Sword, J.T., Pope, A.L., and Hoekstra, W.G., Endotoxin and lipid peroxidation *in vitro* in selenium- and vitamin E-deficient and -adequate rat tissues. *J. Nutr.*, 121, 258, 1991.
Thomas, J.P., Geiger, P.G., and Girotti, A.W., Lethal damage to endothelial cells by oxidised low density lipoprotein: role of selenoperoxidases in cytoprotection against lipid hydroperoxide- and iron-mediated reactions. *J. Lipid Res.*, 34, 479, 1993.
Ueki, H., Ohkura, Y., Motoyashiki, T., Tominaga, N., and Morita, T., Increase in lipoprotein lipase activity in isolated rat adipose tissue by selenate. *Biol. Pharm. Bull.*, 16, 6, 1993.
Vernie, L., Selenium in carcinogenesis. *Biochim. Biophys. Acta*, 738, 203, 1984.
Waschulewski, I.H. and Sunde, R.A., Effect of dietary methionine on utilization of tissue selenium from dietary selenomethionine for glutathione peroxidase in the rat. *J. Nutr.*, 118, 367, 1988.
Whiting, R.F., Wei, L., and Stich, H.F., Unscheduled DNA synthesis and chromosome aberrations by inorganic and organic selenium compounds in the presence of glutathione. *Mutat. Res.*, 78, 159, 1980.

COPPER

Chapter 10

COPPER-DEFICIENT AND EXCESS DIETS: THEORETICAL CONSIDERATIONS AND PREPARATIONS

Moshe J. Werman
Department of Food Engineering and Biotechnology
Technion — Israel Institute of Technology
Haifa, Israel

Sam J. Bhathena
Metabolism and Nutrient Interaction Laboratory/ARS
Beltsville Human Nutrition Research Center
U.S. Department of Agriculture
Beltsville, Maryland

CONTENTS

0-8493-9611-5/97/$0.00+$.50

I. INTRODUCTION

Dietary copper is an essential trace element required by all living organisms, from bacteria to humans, for continued growth and development. Copper forms complexes with certain proteins and serves as a cofactor for enzyme activation. Since copper is biologically active at very low concentrations, very rigidly controlled experimental conditions are essential. In excess of cellular needs, copper promotes free radical production and oxidation and can become toxic to the organism. To understand and explore the nutritional roles of copper, experiments need to be performed in laboratory animals with diets supplemented or deprived of the element. Currently, a wide variety of experimental diets differing in amounts and types of ingredients are being used which results in inconsistency and conflicting data. There is increasing awareness among experimental nutritionists of a need for a nutritionally adequate purified diet that could be used to standardize studies among laboratories. At the same time, since copper interacts with different ingredients, the diet needs to be modified to suit the specific area of research. The goal of this chapter is to provide the reader with guidelines toward the standardization of procedures in experiments that utilize rodents as an animal model for evaluating the nutritional importance of copper. We will focus on diet ingredients and preparation, animal care, and equipment used to perform copper-deficient studies and the modes of dietary copper intake. A brief consideration will also be given to the methods used to alter nutritional copper status, and the role of age and gender on copper nutriture in rodents. This, we hope, will help experimental nutritionists in the area of copper nutriture to feel confident with the dietary aspects of their studies and to ease the interpretation of results between experiments and laboratories.

II. THEORETICAL CONSIDERATIONS

The most difficult problem in formulating animal diets low in copper undoubtedly lies in the provision of ingredients low in the element. The diets used three decades ago to produce copper deficiency faced problems of copper impurities and poor performance.[1] Another disadvantage was that the type of component used did not permit variations of ingredients to study the influence of such changes on copper uptake and utilization by the animal. These problems were solved by the introduction of semi-synthetic diets of low copper content which when adequately supplemented with copper, gave excellent growth and breeding performance. The composition of the diet given in Table 1 for the study of copper deficiency and toxicity is consistent and reproducible and is based on the general guidelines for nutrient requirements of laboratory animals issued by the National Research Council.[2]

TABLE 1.

Composition of a Typical Diet for Producing Copper Deficiency or Overload

Ingredients	%
Carbohydrate (starch, sucrose, fructose, glucose)[a]	60–75
Protein (egg white, lactalbumin, casein, amino acid mixture)[b]	12–20
Fat (soybean oil, corn oil)[c]	4–10
Non-nutritive fiber (cellulose)	3–5
Mineral mix (AIN-93M-MX)[d]	3.5
Vitamin mix (AIN-93M-VX)	1.0

[a] To produce severe copper deficiency, high amounts of sucrose or fructose are recommended.

[b] With egg white, supplementation with biotin is required. With casein, supplementation with DL-methionine or L-cysteine is required.

[c] Higher amount of corn oil is required than soybean oil.

[d] Concentration of copper in copper-deficient diet is usually 0.2–0.6 mg/kg diet. For copper-adequate (control) diet, the recommended copper concentration is 6.0 mg/kg diet. To produce copper overload, 10–100 mg copper needs to be added/kg diet.

A. Rodents as Models of Copper Deficiency and Excess

Rodents have been extensively used as models for the study of genetic disorders of copper metabolism in humans. Two well recognized genetic disorders of copper metabolism in humans are Wilson's disease and Menke's disease. Wilson's disease is an inherited disorder of abnormal copper metabolism leading to progressive copper toxicity. It is characterized by potentially lethal cerebral degeneration and severe liver pathology. Excess copper accumulates in several tissues in addition to the liver and brain, notably in the kidney and cornea. The primary defect appears to be in ceruloplasmin synthesis and in biliary excretion of copper. Copper absorption is normal. Though no true rodent model of Wilson's disease is available showing central nervous system and liver pathology, Long-Evans cinnamon (LEC) rats have been extensively used as a model for Wilson's disease to study liver toxicity.

Menke's disease is an X-linked inherited condition present in male infants characterized by neurological deterioration early in life and abnormal appearance of hair. The disorder is due to malabsorption of copper and defective intracellular copper utilization. These defects lead to decreases in tissue copper levels especially in the liver and brain, while in the kidney, intestine, spleen, lung, muscle, pancreas, and skin they are increased. There is a reduction in serum copper and ceruloplasmin and a postulated defect in metallothionein. The brindled mutant mouse and macular mutant mouse most closely resemble Menke's disease and have been extensively used as models.

It is difficult to produce copper toxicity in normal rats and mice by dietary means. Normal rats can tolerate high levels of copper rather well. Parenteral administration of copper is more effective in producing copper toxicity than with dietary means. Rats are more sensitive to acute copper toxicity than mice. Among macular mutant mice, females are more sensitive to copper induced toxicity while among normal mice, males are more sensitive than females.

Most of the toxic effects of copper are due to its oxidative property by increasing lipid peroxidation which is reversed by antioxidants such as vitamin E. Thus, to produce copper toxicity, the level of dietary vitamin E needs to be lowered in high copper diets. Copper deficiency can be easily produced if the diet is properly selected with regards to carbohydrate and protein content, and if the right age and sex of the animal is used as described below.

B. Altering Copper Status by Dietary Means

Though it is not pertinent to the aims of this chapter, it is important to point out that copper deficiency can be achieved by means other than depriving animals of dietary copper. Since copper can be easily stored in many organs, notably in the liver, it is difficult to produce copper deficiency in adult animals by feeding them low copper diets. Copper status of an animal can be affected by altering absorption, transport, or cellular uptake.[3] The substances that alter copper status include metal ions, amino acids and proteins, carbohydrates, ascorbic acid, and fiber. Metals such as zinc, iron, cobalt, and molybdenum, which have similar chemical and physical properties to copper, generally act as antagonist to each other. Metallothioneins play an important role in metal absorption and transport. Large doses of dietary zinc interfere with copper absorption by inducing intestinal metallothionein, which binds copper and makes it unavailable for transfer.[4] Iron also decreases the relative availability of copper and plasma copper levels, but has little effect on copper absorption. The detailed mechanism of the interaction is not clear.[5] Molybdenum also interferes with copper metabolism. Molybdenum salts decrease liver copper stores but increase plasma copper levels. Thiomolybdate is in fact used as a treatment for reducing excess liver copper and copper toxicity. Subcutaneous administration of cobalt in rats causes redistribution of copper in rat liver and kidney and increases urinary copper excretion.[6] Other metals that interact with copper include manganese, mercury, lead, nickel, tin, magnesium, selenium, cadmium, and boron.

Excess amounts of ascorbic acid reduce copper absorption by decreasing the concentration of soluble copper in the lumen of the small intestine in rats.[7] Iron utilization is also affected by ascorbic acid, more so in copper-deficient male animals. Amino acids and proteins also play a role in copper metabolism. Copper-histidine complex increases copper uptake in cells. Albumin-copper complex also increases copper uptake. However, in excess amounts, sulfur containing amino acids, histidine, methionine, cysteine, and homocysteine,[8–10] inhibits copper uptake from intestinal lumen. Diets high in fiber and phytate also lower intestinal copper uptake.[3]

Our laboratory and others have demonstrated that copper status of rats[11,12] and mice[13] depend on gender and the nature and amount of dietary carbohydrate. Thus, fructose or sucrose as compared to starch, when fed with low copper, produces more severe copper deficiency, including death due to cardiac rupture in male but not female rats.[11] Fructose and sucrose appear to limit copper absorption and increase copper requirement in rats.[14] Alcohol consumption has a different effect on copper status in rodents. In rats fed copper-deficient diet, ethanol lowers hepatic copper levels and aggravates copper deficiency,[15] but in mice it causes hepatic copper overload.[16]

Among the nondietary factors involved in causing copper deficiency is the use of penicillamine, a sulfhydryl drug, which chelates copper and increases the rate of copper excretion in the urine.[17] Penicillamine is used to reduce copper toxicity in patients with Wilson's disease. The copper chelator triethylene-tetramine is often used in studies that require a facilitated depletion of plasma and tissue copper stores.[18]

C. Role of Age, Gender, and Physiological Conditions

Age and sex of rodents are important in producing copper deficiency by dietary means. The severity of the deficiency depends on the type of carbohydrate, gender, physiological conditions, and growth rate. Adequate dietary copper is necessary for normal development of the embryo, neonate, and weaned animal. Regardless of the dietary carbohydrate, induction of copper deficiency several weeks prior to conception and during mating and pregnancy results in fetal abnormalities and fetal death. However, if dietary copper is depleted during pregnancy, the severity of the deficiency is ameliorated only if starch, but not sucrose or fructose, is used as a carbohydrate source. Interestingly, this copper-carbohydrate interaction spares the pregnant animal.[19] Our laboratory and others have shown that copper deficiency is difficult to produce in adult rats because of large stores of copper, especially in the liver. Although plasma copper levels are lowered in adult rats by feeding low copper diets, they do not show typical signs of copper deficiency. The time required for the appearance of copper deficiency symptoms in rats fed copper-deficient diets increases with the age of the animal, unless a copper chelating agent is used.[18] Weanling rats, on the other hand, are vulnerable to dietary copper deprivation and can be made copper-deficient by feeding diets low in copper as their hepatic copper stores are low. Although both male and female weanling rats show similar low plasma copper levels when fed a low copper diet, the signs of copper deficiency are more severe in male than in female rats. If the weaned rats are fed low copper diets high in fructose, the males but not females show increased mortality after 5 to 7 weeks due to cardiac rupture at the apex. Apparently, gonadal hormones are not involved in this sexual dimorphism, but pituitary or hypothalamic sex hormones may be responsible.

D. Forms and Modes of Copper Administration

Regardless of the dietary copper regimen, e.g., deficient, adequate, or excess, copper is delivered to the animal through the feed or in drinking water. A less documented copper delivery system is the subcutaneous injection. While rodents on a copper-deficient regimen receive no copper supplementation, their counterpart controls are provided with a diet containing 6 mg Cu/kg diet. Copper carbonate is the recommended source for copper since it is better absorbed than copper sulfate.[20] The salt is premixed with part of the carbohydrate source before being added to the total diet. Copper-adequate controls may be maintained on a copper-deprived diet if given copper in drinking water (4 μg Cu/ml water) as copper sulfate or acetate. It is important to note that absorption of copper is greater from solid foods than from liquid foods or drinking water. When copper is given by injection to a control group, maintained on a copper-deprived diet, the amount injected is calculated based on two parameters: (1) the amount of diet consumed per day and (2) the amount of copper absorbed by the rodent from a copper-deficient diet. The parenterally administered copper is more effective in producing copper overload since it bypasses intestinal absorption which is affected by factors other than copper, such as metals, type and levels of carbohydrates, proteins, etc., as described earlier.

Excess copper is provided to laboratory rodents by supplementing either a purified laboratory diet or a commercial chow diet with various copper levels. Supplementation of copper depends on the desired copper overload. Copper overload can also be achieved by supplying copper in the drinking water.

E. Copper Absorption

The absorption of copper from the gut depends primarily on dietary copper levels. Less copper is absorbed when dietary copper levels are high and vice versa. As described above, copper absorption also depends on the presence of metal ions, and the type and amount of carbohydrates, proteins, and amino acids. The majority of absorbed copper binds to albumin and amino acids in plasma and is transported to the liver where it is incorporated in enzymes and metalloproteins primarily in ceruloplasmin and metallothionein. Ceruloplasmin then enters plasma where it is the main copper complex. Copper is excreted primarily through bile.

III. COMPOSITION OF A TYPICAL DIET

A typical purified basal diet fed to induce copper deficiency in laboratory rodents includes the following nutrients: carbohydrate, 60 to 75%; protein, 12 to 20%; fat, 4 to 10%; nonnutritive fiber, 3 to 5%; mineral mixture (free of copper), 3.5%; vitamin mixture, 1%; and choline bitartrate, 0.25%, to supply 1 g of choline. At this point it is important to note that with all the precautions taken (e.g., purchasing high quality dietary ingredients and a careful diet preparation), an absolute copper-free diet is impossible to achieve. Traces of copper are always present in the diet and copper content may vary from lot-to-lot. Thus, it is not surprising to find a wide range of copper levels in copper-deficient diets among laboratories. Dietary copper requirements vary among laboratory animals. Among rodents, the rat requires 5 mg Cu/kg diet, while the mouse and rabbit require less copper, 4.5 and 3 mg Cu/kg diet, respectively. The pig, sheep, and cat require 5 mg Cu/kg diet, and the dog 7.3 mg Cu/kg diet. The proposed copper-deprived diet provides less than 1 mg Cu/kg diet, while the recommended copper-adequate diet is formulated to provide 6 mg Cu/kg diet. Diets with copper levels between these concentrations are considered as copper-marginal diets. To produce copper toxicity, levels of 10 to 100 mg Cu/kg diet are generally used. Various components of the recommended diet are described below.

A. Individual Ingredients

1. *Carbohydrate*

The major role of dietary carbohydrates in rodent feed is to provide energy. For the last two decades, it has been repeatedly shown by various laboratories that dietary carbohydrates play a significant role in expression of the severity of copper deficiency in rats.[21–24] Two types of carbohydrates are commonly used, simple and complex carbohydrates. Simple dietary carbohydrates include sugars such as sucrose, glucose, fructose, and lactose, while the major complex carbohydrate is cornstarch. Carbohydrates are added to the diet either as a single source or as a combination of two or more sources in various proportions. Copper deficiency syndrome in rats is more severe when simple sugars such as sucrose,[21] fructose,[22,23] or lactose[24] are used rather than starch. Male weaned rats consuming a copper-deficient diet containing sucrose or fructose begin to die from the fifth week after feeding. In contrast, none of the rats fed starch die during this time period.[21–23] We therefore suggest that for long-term copper-deficient studies, the animals should be provided with starch as the major or sole carbohydrate source. Although 60 to 65% fructose produces severe and fatal copper deficiency, in order to extrapolate the results to humans, it is necessary to formulate a diet with up to 20% fructose or sucrose and the balance of carbohydrate as starch which resembles more closely the human consumption.

2. *Protein*

The amount of protein in the proposed diet is 12 to 20% depending on the nature of protein used. The dietary protein component is the main source for introducing copper impurities into a purified copper-deficient diet. Assuming that high quality proteins are used, high-protein cereals[25] are a rich source for copper contamination, as compared with animal-derived proteins.[26] Therefore, we suggest the use of either egg-white solids, vitamin-free casein, or lactalbumin as the preferred dietary protein sources for copper deficiency studies. Egg-white solids provides less copper impurities than lactalbumin or vitamin-free casein.[26] However, uncooked egg-white solids contain avidin which binds and prevents the absorption of both dietary biotin and biotin synthesized by intestinal bacteria.[27] Hence, copper-deficient diets that contain egg-white solids as their protein source should be supplemented with a 2 mg biotin/kg diet.[28] In addition, egg-white solids contain large amounts of sodium, potassium, and chloride and low amounts of phosphorus. Therefore, the protein source needs to be analyzed for these minerals and the mineral mixture may be reformulated to meet the recommended levels of these nutrients. In choosing vitamin-fee casein as the protein source, it should be noted that casein is low in sulfur amino acids. Dietary supplementation of either DL-methionine or L-cysteine, at a level of 3 g/kg diet, is recommended.

Substituting the protein source with a mixture of pure amino acids reduces the impurities of copper. Using amino acid mixtures as a protein source, copper levels of less than 0.2 mg/kg diet were reported by Farquharson and Robins.[29] Replacing dietary proteins with amino acid mixtures had no effect on plasma copper concentration, but lowers liver copper levels.[30] High levels of dietary amino acids may also affect kidney performance.

3. *Fat*

The fat in the diet is needed to supply essential fatty acids of the n-6 and n-3 series. Corn oil has been used by our group, but other oils, such as soybean oil which has a good balance between n-6 and n-3 fatty acids and polyunsaturated to saturated fatty acid ratio,[31,32] can also be used. The amount of dietary oil can be varied depending on the amount of carbohydrate present in the diet. A minimum of 5% soybean oil is sufficient to supply the necessary fatty acids for rodents. However, a higher amount of corn oil is needed.

4. *Non-Nutritive Fiber*

The amount of non-nutritive fiber in the proposed copper-deficient diet is 30 to 50 g/kg diet. Since the composition of commercially available fiber sources can vary in their components, especially that of essential minerals, it is recommended that fiber source be analyzed for its copper and other mineral content.

5. *Mineral Mix*

Mineral mix is a critical element in the copper-deficient and copper-excess diets. Copper interacts with several metals (see above), most notably zinc and iron. It is therefore essential to control the concentration of these metals. Since many commercially available salts contain traces of other metals as impurities, the ingredients should be carefully selected for the preparation of mineral mix. The most commonly used salt mix for rodents is AIN-93M-MX, which is modified from AIN-76.[33] It includes recently recognized beneficial minerals such as selenium, molybdenum, silicon, chromium, boron, nickel, lithium, and vanadium. For preparation of a copper-deficient diet, copper carbonate should be eliminated. Starch or a non-nutritive fiber should replace sucrose as the mixture filler for excess copper diets. At a mineral mixture level of 35 g/kg diet, the final concentration of mineral elements (mg/kg diet) in the recommended copper-deficient diet are: calcium, 5000.0; phosphorus, 3000.0; potassium, 3600.0; sulfur, 300.0; sodium, 1030.0; chloride, 1610.0; magnesium, 511.0; iron, 45.0; zinc, 35.0; manganese, 10.0; iodine, 0.2; molybdenum, 0.15; selenium, 0. 15; silicon, 5.0; chromium, 1.0; fluoride, 1.0; nickel, 0.5; boron, 0. 5; lithium, 0. 1; and vanadium, 0.1.

6. *Vitamin Mix*

The vitamin mixture for copper-deficient studies is prepared according to the recommended formula of the AIN-93-VX purified diet for laboratory rodents,[33] but with one important modification. Starch or a non-nutritive fiber should replace sucrose as the mixture filler. At a level of 10 g/kg diet, the final vitamin concentration in the recommended copper-deficient diet are: nicotinic acid, 30 mg; pantothenate as calcium salt, 15 mg; pyridoxine, 6 mg; thiamin, 5 mg; riboflavin, 6 mg; folic acid, 2 mg; vitamin K, 800 to 900 μg; biotin, 200 μg; vitamin B-12, 25 μg; vitamin A, 4000 IU; vitamin D, 1000 IU; and vitamin E, 75 IU. Since most of the choline salts are hygroscopic, choline is not added to the vitamin mixture, but incorporated into the total diet to provide the required level (1 g/kg diet).

7. *Ingredient Sources*

The dietary ingredients should be of the best quality available and purchased from reliable and reputable suppliers. Nutritional analysis data of each dietary ingredient lot should be requested from the suppliers or the analysis be performed by the investigator. Special care should be taken to avoid the use of dietary ingredients with high copper impurities. To avoid batch variations, purchasing large amounts of each ingredient needed for the entire study is highly recommended. To minimize the introduction of copper impurities by dust accumulation, dietary ingredients should be stored in tightly closed plastic containers under the conditions recommended by the supplier. Since several minerals and vitamins are hygroscopic, they should be stored desiccated and dried before use. Dietary ingredients susceptible to light deterioration should be stored in the dark.

B. Preparation of the Diets

1. *Preparation of Mineral and Vitamin Mixes*

Because very small quantities of individual salts and vitamins are required in the diet, it is advisable to prepare mineral mix and vitamin mix separately and add them as a complex mixture to macro ingredients. In order to avoid contamination from copper, the mineral mixture for copper-deficient diets needs to be prepared using porcelain or stainless-steel utensils. The required salts are added into a large porcelain jar mill equipped with porcelain pellets for grinding. The jar is placed on a rotor and rotated overnight to prepare a homogeneous powder-like mixture. The mixture should be stored at room temperature in a desiccator to avoid clumping. The jar and pellets should be thoroughly cleaned with dilute HCl between different batch preparations to avoid cross contamination.

Vitamin mix can be prepared in the same way as mineral mix or can be bought from several commercial sources. If prepared in the laboratory, it should be done in a mixer equipped only with a stainless-steel bowl and rotor, rotating at a low speed to avoid excess heat generation. The vitamin mixture needs to be stored in the dark and refrigerated.

2. *Diet Mixing, Pelleting, and Storage*

All dietary components, except the oil, are placed in a covered stainless-steel bowl and mixed with a stainless-steel rotor at low speed to reduce dust formation and heat generation. Ingredients required in small quantities, such as choline, biotin, and amino acids (methionine or cysteine), are first mixed with a small quantity of carbohydrate, preferably starch. This is then added to the total diet and mixing is continued. After several minutes, the oil is slowly added and mixing continues for about 15 to 20 minutes at the same low speed. For copper-deficient studies, diet pelleting is not recommended, unless pelleting equipment is made of stainless-steel to prevent introducing impurities.

Diets should be stored either refrigerated (4°C) or frozen (–20 to –80°C) in tightly closed plastic containers for no more than 3 to 6 months, depending on storage conditions.

C. Analysis of the Diets

Accurate analysis of the exact amount of copper in dietary ingredients and in the diets is critical for the success of the study. Analytical techniques applied to the measurement of copper concentrations in biological materials include flame and flameless atomic absorption spectrometry, spectrophotometry, emission spectrography, neutron activation analysis, and anodic stripping voltammetry. A reference method for copper has not been published, and the currently preferred and most used method is flame atomic absorption. As for any trace metal, analysis of copper will be inaccurate unless strict precautions against contamination are exercised from diet preparation and storage to the final step of the analysis. Samples are pretreated by dry ashing and acid digestion. A certified copper standard and a blank, digested and analyzed along with samples, should be used to verify accuracy and to assess the contribution of background contamination.

IV. ANIMAL CARE AND EQUIPMENT

Besides the formulation and preparation of the diet, care in animal handling is equally important for the success of studies involving copper deficiency. Poor animal handling can affect the outcome of the experiment. These include housing conditions (e.g., lighting, temperature, and humidity), bedding, caging, drinking water, and diet supply. During copper-deficient studies, special care should be taken to prevent the introduction of external copper contaminations. Stainless-steel or plastic cages fitted with stainless-steel wire mesh bottoms are preferred and the use of equipment made of other alloys that may rust or contain copper should be avoided. Animal rooms should be temperature and humidity controlled and equipped with dust collecting air filters.[34] The use of cage bedding should be avoided whenever possible. Feed should be provided in stainless-steel dishes and drinking water in glass or plastic bottles. Drinking water and its delivery system are potential sources of mineral contamination, including copper. In recognition of this possibility, animals on a dietary copper-deficient regimen should be provided only with distilled or deionized water, either through an automatic watering system made of stainless-steel pipes or from glass or plastic bottles fitted with plastic caps and stainless-steel sipper tubes. Use of rubber stoppers should be avoided as commercially available rubber stoppers generally contain significant amounts of copper, which may leach into drinking water. If rubber stoppers must be used in copper-deficient studies, Kennedy and Beal[35] suggest the use of black, pure gum rubber stoppers (Daigger Scientific, Wheeling, IL), soaked in either dilute hydrochloric acid, nitric acid, and/or EDTA prior to use. In animal facilities where distilled or deionized water is supplied from a central system or prepared in the laboratory, water should be analyzed periodically for copper content.

V. PARAMETERS TO ASSESS COPPER STATUS

Before studying the effect of copper-deficient and copper excess diets, it is necessary to confirm that the animals are copper-deficient or copper-overloaded. There are several common and some novel parameters to assess the copper status. Liver copper in animals is at present the most reliable marker. Plasma copper levels are also commonly used. Since most of the copper in plasma is bound to ceruloplasmin, a copper containing enzyme, plasma ceruloplasmin activity also indicates copper status. Another copper containing enzyme used as a marker of copper status is superoxide dismutase. It is measured in red cells and liver. Since copper deficiency increases iron in liver, liver iron can be a good measure. Since copper is either an integral part of, or a cofactor in enzymes involved in processing neuropeptides, the levels of the latter in plasma and in the brain have also been suggested as markers of copper status. These include β-endorphin, leu-enkephalin, and met-enkephalin.[36] However, the use of neuropeptides to assess copper status needs to be validated in other studies. Peptidyl-glycin-α-amidating monooxygenase,[37] atrial natriuretic peptide,[12] and lysyl oxidase[38] have also been suggested as possible markers of copper status. Since copper deficiency produces anemia, hemoglobin level is a good marker of copper status especially to determine the extent of the deficiency.

VI. CONCLUSIONS

Rodents have been used extensively for the study of dietary copper deficiency and toxicity. Moderate to severe copper deficiency in rats can be achieved by selecting weanling rats and feeding them diets high in simple sugars (sucrose or fructose) and metals such as iron and zinc and low in antioxidant vitamins such as vitamin E. Copper deficiency in adult rats can be achieved by feeding low copper diets and treating the animals with copper chelators such as penicillamine. On the other hand, it is difficult to produce copper overload by dietary means. Feeding very high copper diets decreases copper absorption. Fortunately, several genetic models are available to study copper overload and copper toxicity. Most commonly used rodent models are LEC rats and brindled mice and macular mice. The copper-deficient or copper-overloaded diets should be carefully formulated, prepared, and stored. The amount of macronutrients (e.g., carbohydrate, protein, and fat) can be varied within wide range, and different types of micronutrients can also be selected depending on the specific aims of the study.

REFERENCES

1. Mills, C.F. and Murray, G., The preparation of a semi-synthetic diet low in copper for copper deficiency studies with the rat, *J. Sci. Food Agric.*, 11, 547, 1960.
2. National Research Council, Nutrient requirements of laboratory animals, 3rd review edition, National Academy Press, 1978, Washington DC.
3. Cousins, R.J., Absorption, transport, and hepatic metabolism of copper and zinc: special reference to metallothionein and ceruloplasmin. *Physiol. Rev.*, 65, 238, 1985.
4. Fischer, P.W.F., Giroux, A., and L'abbe, M.R., Effects of zinc on mucosal copper binding and on the kinetics of copper absorption, *J. Nutr.*, 113, 462, 1983.
5. Yu, S., West, C.E., and Beynen, A.C., Increasing intakes of iron reduce status, absorption and biliary excretion of copper in rats, *Br. J. Nutr.*, 71, 887, 1994.
6. Rosenberg, D.W. and Kappas, A., Trace metal interaction *in vivo*: Inorganic cobalt enhances urinary copper excretion without producing an associated zincuresis in rats, *J. Nutr.*, 119, 1259, 1989.
7. Van Den Berg, G.J., Yu. S., Lemmens, A.G., and Beynen, A.C., Dietary ascorbic acid lowers the concentration of soluble copper in small intestinal lumen of rats, *Br. J. Nutr.*, 71, 701, 1994.
8. Aoyama, Y., Mori, M., Hitomi-Ohmura, E., and Yoshida, A., Effects of dietary excess histidine and varying levels of copper on the metabolism of lipids and minerals in rats, *Biosci. Biotech. Biochem.*, 56, 335, 1992.
9. Strain, J.J. and Lynch, S.M., Excess dietary methionine decreases indices of copper status in the rat, *Ann. Nutr. Metab.*, 34, 93, 1990.
10. Brown, J.C.W. and Strain, J.J., Effect of dietary homocysteine on copper status in rats, *J. Nutr.*, 120, 1068, 1990.
11. Fields, M., Lewis, C., Scholfield, D.J., Powell, A.S., Rose, A.J., Reiser, S., and Smith, J.C., Female rats are protected against the fructose induced mortality of copper deficiency, *Proc. Soc. Exp. Biol. Med.*, 183, 145, 1986.

12. Bhathena, S.J., Kennedy, B.W., Smith, P.M., Fields, M., and Zamir, N., Role of atrial natriuretic peptides in cardiac hypertrophy of copper-deficient male and female rats, *J. Trace Elem. Exp. Med.*, 1, 199, 1988.
13. Fullerton, F.R., Greenman, D.L., and Kushmaul, R.J., Sex specificity of myocardial damage in mice fed a purified diet, *Lab. Anim. Sci.*, 36, 650, 1986.
14. Fields, M., Holbrook, J., Scholfield, D., Smith, J.C., and Reiser, S., Effect of fructose or starch on copper-67 absorption and excretion by the rat, *J. Nutr.*, 116, 625, 1986.
15. Fields, M., Lewis, C.G., and Lure, M., Alcohol consumption mimics the effects of a high-fructose, low-copper diet in rats, *Alcohol*, 11, 17, 1994.
16. Reimers, E.G., Moreno, F. R., Aleman, V.C., Fernandez, F.S., Martin, L.G., Torres, R.F., and Martell, A.C., Effect of ethanol on hepatic iron, copper, zinc and manganese contents in the male albino mouse, *Drug Alcohol Dependence*, 26, 195, 1990.
17. McQuaid, A., Lamand, M., and Mason, J., The interaction of penicillamine with copper in vivo and the effect on hepatic metallothionein levels and copper/zinc distribution: the implications for Wilson's disease and arthritis therapy, *J. Lab Clin. Med.*, 119, 744, 1992.
18. Jankowski, M.A., Uriu-Hare, J.Y., Rucker, R.B., and Keen, C.L., Effect of maternal and dietary copper on fetal development in rats, *Reproductive Toxicol.*, 7, 589, 1993.
19. Fields, M., Lewis, C.G., and Beal, T., Copper-carbohydrate interaction in maternal, fetal and neonate rat, *Neurotoxicol. Teratol.*, 10, 555, 1989.
20. DiSilvestro, R.A. and Cousins, R.J., Physiological ligands for copper and zinc, *Ann. Rev. Nutr.*, 3, 261, 1983.
21. Fields, M., Ferretti, R.J., Smith, J.C., and Reiser, S., Effect of copper deficiency on metabolism and mortality in rats fed sucrose or starch diets. *J. Nutr.*, 113, 1335, 1983.
22. Fields, M., Ferretti, R J., Reiser, S., and Smith, J.C., The severity of copper deficiency is determined by the type of dietary carbohydrate, *Proc. Soc. Exp. Biol. Med.*, 175, 530, 1984.
23. Redman, R.S., Fields, M., Reiser, S., and Smith, J.C., Dietary fructose exacerbates the cardiac abnormalities of copper deficiency in rats, *Atherosclerosis*, 74, 203, 1988.
24. Strain, J.J., Milk consumption, lactose and copper in the aetiology of ischaemic heart disease, *Med. Hypotheses*, 25, 99, 1988.
25. Lo, G.S., Setle, S.L., and Steinke, F.H., Bioavailability of copper in isolated soybean protein using the rat as an experimental model, *J. Nutr.*, 114, 332, 1984.
26. Fields, M., Lewis, C.G., and Lure, M.D., Cooper deficiency in rats: The effect of type of dietary protein, *J. Am. College Nutr.*, 12, 303, 1993.
27. Dakshinamurti, K. and Chauhan, J., Biotin, *Vitam. Horm.*, 45, 337, 1989.
28. Klevay, L.M., The biotin requirement of rats fed 20% egg white, *J. Nutr.*, 12, 303, 1973.
29. Farquharson, G. and Robins, S.P., Female rats are susceptible to cardiac hypertrophy induced by copper deficiency: The lack of influence of estrogen and testosterone, *Proc. Soc. Exp. Biol. Med.*, 188, 272, 1988.
30. Veenendaal, M., Zhang, X., Lemmens, A.G., and Beynen, A.C., Liver and plasma copper concentrations in rats fed diets containing various proteins, *Biol. Trace Elem. Res.*, 34, 213, 1992.
31. Bourre, J.M., Francois, M., Youyou, A., Dumont, O., Piciotti, M., Pascal, G., and Durand, G., The effect of dietary α-linolenic acid on the composition of nerve membrane, enzymatic activity, amplitude of electrophysiological parameters, resistance to poisons and performance of learning tasks in rats, *J. Nutr.*, 119, 1880, 1989.
32. Lee, J.H., Fukumoto, M., Nishida, H., Ikeda, I., and Sugano M., The interrelated effects of n-6/n-3 and polyunsaturates/saturated ratio of dietary fats on the regulation of lipids metabolism in rats, *J. Nutr.*, 119, 1893, 1989.
33. Reeves, P.G., Nielsen, F.H., and Faye, G.C., AIN-93 purified diets for laboratory rodents: final report of the American Institute of Nutrition ad hot writing committee on the reformulation of the AIN-76A rodent diet, *J. Nutr.*, 123, 1939, 1993.
34. Klevay, L.M., Petering, H.G., and Steemer, K.L., A controlled environment for trace metal experiments on animals, *Environ. Sci. Technol.*, 5, 1196, 1971.

35. Kennedy, B.K. and Beal, T.S., Mineral leached into drinking water from rubber stoppers, *Lab. Animal Sci.*, 41, 233, 1991.
36. Bhathena, S.J., Recant, L., Voyles, N.R., Timmers, K.I., Reiser, S., Smith, J.C., Jr., and Powell, A.S., Decreased plasma enkephalins in copper deficiency in men, *Am. J. Clin. Nutr.*, 43, 42, 1986.
37. Prohaska, J.R., Bailey, W.R., and Lear, P.M., Copper deficiency alters rat peptidylglycine-α-amidating monooxygenase activity, *J. Nutr.*, 125, 1447, 1995.
38. Werman, M.J., Barat, E., and Bhathena, S.J., Gender, dietary copper and carbohydrate source influence cardiac collagen and lysyl oxidase in weanling rats, *J. Nutr.*, 125, 857, 1995.

Chapter 11

Copper in Tissue Culture

Edward D. Harris
Evelyn Tiffany-Castiglioni
Yongchang Qian
Department of Veterinary Anatomy and Public Health
Texas A & M University
College Station, Texas

CONTENTS

0-8493-9611-5/97/$0.00+$.50

I. BACKGROUND

The application of tissue and cell culture to the study of trace metals is gaining in popularity. When it was still in its developmental stages, tissue culture systems came to rely on cells from rodents. Not only were rodent cells more accessible, but the establishment of permanent cell lines was more readily achievable with rodent cells than with cells from other species. The development of some useful human cells lines, e.g., W138 human embryonic lung fibroblasts[1] came about the same time as the very popular hamster kidney fibroblasts,[2] and was preceded some 20 years earlier by the commonly used L-cell, a mouse subcutaneous fibroblast.[3] In time, refinements in media composition, the identification of the essential growth factors, and improvements in culturing conditions brought a wide variety of different cell types into the culturing sphere. A culture system approach, therefore, affords Cu workers access to a variety of isolated cell types in an environment relatively free of interaction with other cells and hormonal factors. The movement of Cu across the cell membrane, its transport in the cytosol, storage, and movement to organelles is more easily charted. Culture systems have also permitted insights into the toxic effects of heavy metals on cells. For all of their advantages, however, a Cu worker must always bear in mind that homogeneous cells in culture do not faithfully duplicate conditions *in vivo* and could misrepresent or over simplify the reactions taking place. In this chapter, we will explore some of the ways tissue cultures have been used to study Cu transport and metabolism as well as learn some of the precautions that must be taken to assure the processes under study mimics conditions *in vivo.*

II. GENERAL METHODOLOGY

Successful employment of cell cultures to the study of copper metabolism requires an ultra clean environment with ultra pure solutions. Quantitative studies especially mandate that utensils and solutions be free from Cu contamination. An investigator must have a culture facility for growing and maintaining cells. Below are some considerations when dealing with Cu and cells. The reader is referred to a number of excellent monographs and manuals on cell culture procedures. The book by Freshney is especially recommended.[4]

A. The Laboratory Environment

The laboratory environment is especially critical. An immediate danger of all culture operations is contamination from air-borne and surface-borne bacteria, yeast, and other microorganisms. A closed room dedicated to culture operations is a necessity. The room should contain a laminar flow positive-pressure hood fitted with a special filter (HEPA) to sterilize circulating air and a uv sterilization lamp to decontaminate when personnel are not in the room. Away from the hood should be incubators for growing cells (6 ft^3 /person is recommended). The incubator should provide a humidified atmosphere at an appropriate temperature (e.g., 37°C) and should be equipped with temperature and CO_2 sensors and an alarm system. A backup CO_2 system is recommended but not required. Most culture media are bicarbonate-buffered and require CO_2 gas for pH stabilization. Pulse-injected CO_2 incubators are recommended for economic reasons. A refrigerator or cold box should be available for storing media and perishables. A sealed container of liquid nitrogen for storing stocks of seed cultures is desirable. The room should have vacuum lines to allow filter-sterilization of liquids. Cabinets for storing pipettes, forceps, flasks, dishes, filters, and gloves should be readily available. Utility items include medium speed table-top centrifuges for collecting cells and microbalances for weighing medium components. Finally, an autoclave capable of producing superheated steam at 120 to 150°C is necessary.

B. Culture Apparatus and Media

Basic culture apparatus such as plates, flasks, filters, pipettes, and petri dishes are disposable and may be purchased presterilized. Nutrient-rich growth medium may be purchased in powdered form or in ready-made sterile solutions. Powdered media should be dissolved in deionionized distilled water (Milli-Q system, Millipore, or equivalent). With the exception of hemopoietic cells and some transformed or tumor-derived cell lines, cells require a surface to adhere to if they are to grow. Anchorage-independent cells can be maintained as suspension cultures. Growth substrates, therefore, become an essential component of successful culturing systems. These may be plastic or glass that is usually treated or coated to enhance cell attachment. Most media must be supplemented with 10 to 20% (v/v) fetal or newborn calf serum for cells to survive. A more chemically defined "serum-free" medium will sustain cells for long periods if the proper supplements are added to optimize growth. Antibiotics are sometimes included if sterile conditions are suboptimal. A typical defined medium is shown in Table 1.

Defined media vary from Eagle's and Dulbecco's minimum essential medium (MEM) which are saline-based buffered solutions containing essential amino acids and vitamins to the more complex media such as 199, CMRL 1066, and F12, which contain a larger variety of amino acids, vitamins, and extra metabolites. One of the richest complex media combines Dulbecco's MEM with F12.[5]

1. Culture Plates

Plastic dishes and utensils generally do not present problems with Cu contamination. Glass surfaces, however, are more prone to absorb Cu ions nonspecifically. Soaking glass in 5% (v/v) HNO_3 or HCl followed by a thorough rinsing in deionized distilled water is recommended to remove surface Cu contamination. Contamination by extraneous Cu ions is more likely when an organic coating such as collagen is used. Ways to render solutions free of Cu are described below.

2. Copper Chelators and Culture Medium

A variety of different media have been employed in Cu studies. Media contain organic agents that bind Cu. Amino acids, the most likely offenders, differ in their potential to bind Cu. As shown in Table 2, chelators of cupric Cu [Cu(II)] can be ranked in order of chelating potential at their respective concentrations in plasma.[6] Histidine, glutamine, threonine, and cystine are excellent Cu chelators. Experiments with individual amino acids alone or in combination with histidine, however, have shown that histidine with either threonine, asparagine, or glutamine provides a stronger chelating group for Cu than histidine alone.[6] One cannot predict if chelation by amino acids will hinder or facilitate Cu uptake. For example, histidine both stimulates and blocks Cu^{2+} uptake by cells; the exact effect will depend on the cell type.[7]

The investigator is wise to determine through experiment the effects of all potential chelators on the particular cell type being studied. Besides amino acids, Cu^{2+} is also subject to chelation by proteins and chemical agents in the medium. Chemical agents such as EDTA (ethylenediamine-tetratactetate), and NTA (nitrilotriacetate) are common divalent cation chelators that could interfere with uptake. For example, we have shown that EDTA blocks the uptake of Cu^{2+} into aortic tissue.[8] A moderately alkaline medium (pH > 8) favors complexation of Cu ions with the peptide bonds of proteins. Serum albumin, a component of most serum-supplemented growth media, is know to have specific binding sites for Cu^{2+} on the N-terminal of the protein.[9]

3. *Solubility and Valance State*

The most common salts of Cu ($CuCl_2$, $CuSO_4 \cdot 5\,H_2O$, and Cu acetate) are cupric salts [Cu(II)]. An aqueous complex of Cu(II) has a slightly bluish tint that is enhanced strongly by amine compounds. Cu^{2+} ions precipitate as the insoluble hydroxide $Cu(OH)_2$ from mildly alkaline solutions. Chelators such as EDTA and NTA prevent formation of $Cu(OH)_2$. Stock solutions of $CuCl_2$ ions are quite stable when kept in dilute (0.1 *M*) HCl. The cuprous form [Cu(I)] is considerably more unstable than cupric. In solutions typical of those used in culture studies, Cu(I) is immediately converted to Cu(II) by the dissolved O_2 in the solution. Cu(I) also tends to be very insoluble and will precipitate as CuO (recall Benedict's test for reducing sugars where CuO as a brick red precipate indicates a positive test). Compounds with sulfhydryl groups (–SH) tend to retain Cu(I) in the cuprous state. Similarly, –SH compounds tend to reduce Cu(II) to Cu(I) in solution. Thus, common medium components such as cysteine, glutathione, and dithiothreitol, will generally reduce Cu(II) to Cu(I).

4. *Sources of Copper Contamination*

Organic chemicals such as amine-based organic buffers like TRIS (trihydroxyaminoethylmethane) and HEPES (*N*[12-hydroxyethyl]piperazine-*N'*[2- ethanesulfonic acid]), that are commonly used in growth media, add little if any Cu to the media. On the other hand, phosphate salts used as buffers contain high amounts of inorganic Cu (and zinc). Serum proteins obtained from fetal or newborn calves also are rich in organic Cu ions. In addition, serum contains ceruloplasmin and albumin which are the major protein Cu carriers in serum. In our experience, growth medium containing 10 or 15% serum will have between 0.1 and 0.5 ppm (μg/ml) Cu. Cu in undiluted animal serum generally ranges between 0.8 and 1.2 ppm (μg/ml).

TABLE 1.

Components of Typical Growth Medium: Minimal vs. Enriched (in mg/L)

	MEM*	F12		MEM*	F12
Amino Acids					
L-Alanine	—	8.90	L-Leucine	105	3.94
L-Arginine (HCl)	84.0	211	L-Isoleucine	105	13.1
L-Asparagine H_2O	—	15.0	L-Lysine HCl	146	36.5
L-Aspartic acid	—	13.3	L-Methionine	30.0	4.48
L-Cystine	48.0	—	L-Phenylalanine	66.0	4.96
L-Cysteine HCl H_2O		35.1	L-Proline	—	34.5
L-Glutamic acid		14.7	L-Serine	42.0	10.5
L-Glutamine	584	146	L-Threonine	95.0	11.9
Glycine	30.0	7.50	L-Tryptophan	16.0	2.04
L-Histidine HCl H_2O	42.0	21.0	L-Tyrosine	72.0	5.40
			L-Valine	94.0	11.7
Vitamins					
Biotin	—	0.0073	Pyridoxal HCl	4.00	0.062
D-Ca pantothenate	4.00	0.480	Riboflavin	0.40	0.038
Choline chloride	4.00	14.0	Thiamin HCl	4.00	0.34
Folic acid	4.00	1.30	Vitamin B_{12}	—	1.36
i-inositol	7.20	18.0	Pyridoxine HCl	—	0.062
Nicotinamide	4.00	0.04			
Inorganic Salts					
$CaCl_2$ (anhyd.)	200	—	$NaHCO_3$	3,700	1,176
$CaCl_2$ $2H_2O$	—	44.0	NaH_2PO_4 H_2O	125	—

TABLE 1. (Continued)

Components of Typical Growth Medium: Minimal vs. Enriched (in mg/L)

	MEM*	F12		MEM*	F12
$Fe(NO_3)_3$ 9 H_2O	0.10	—	NaH_2PO_4 $7H_2O$	268	
KCl	400	224	$CuSO_4$ $5H_2O$	—	0.00249
$MgCl_2$ $6H_2O$	—	122	$FeSO_4$ $7H_2O$	—	0.834
$MgSO_4$ $7H_2O$	200	—	$ZnSO_4$ $7H_2O$	—	0.863
NaCl	6,800	6,400			
Other Components					
D-Glucose	4,500	1,802	Linoleic acid	0.084	—
Lipoic acid	—	0.21	Putrescine 2HCl	—	0.161
Phenol red	15.0	12.0	Thymidine	—	0.73
Sodium pyruvate		110	CO_2 (gas phase)	10%	5% (pH 7.0)
Hypothanine	—	4.10			2% (pH 7.4)

Note: * Values are represented in mg/L

TABLE 2.

Amino Acid Chelators of Copper in Serum

Amino Acid	Approximate Concentration in Serum		
	Mg/100 ml	μM	Relative Binding Affinity[a]
Histidine	2.1	135.5	1.00
Glutamine	9.7	664.4	0.53
Theonine	3.1	260.5	0.26
Cystine	2.0	83.3	0.17
Glycine	2.8	373.3	0.15
Glutamate	4.4	299.3	0.15
Asparagine	1.4	106.1	0.15
Alanine	4.2	471.9	0.13
Valine	3.4	290.6	0.13
Leucine	2.4	183.2	0.11
Isoleucine	1.8	137.4	0.08
Methylhistidine	0.2	11.7	0.08
Methionine	0.9	60.4	0.08
Serine	1.3	123.8	0.08
Ornithine	0.8	60.6	0.06
Tyrosine	1.5	82.9	0.06
Lysine	3.0	205.4	0.04
α-Aminobutyrate	0.3	25.6	0.02
Proline	2.9	252.2	0.02
Phenylalanine	1.9	115.2	-0-
Arginine	2.5	143.7	-0-
Tryptophan	1.7	83.3	-0-
Citrulline	0.5	28.7	-0-

[a] Relative affinity is based on percentage of ^{64}Cu remaining in the supernatant after ultracentrifugation of predialyzed serum (Cu/albumin = 2.0) containing the amino acid at the concentration in serum. Histidine (135 m*M*) is rated 1.0.

Data from Neumann and Sass-Kortsak.[6] With permission.

5. *Preparation of Copper-Free Solutions*

Extraneous Cu ions can be removed from buffers by chelators or polystyrene resins containing carboxyl groups. Growth medium should not be treated with chelators lest the chelators remove growth factors and essential mineral elements indiscriminantly. Efforts to obtain chelators selective for a specific metal have met with some success. Messer et al. coupled the chelator iminodiacetate to an insoluble agarose and succeeded in producing a Zn-free medium that produced a Zn deficiency in lymphocytes.[10] The authors suggest this approach can be applied to other metals including Cu. Others have added sterile Chelex ion-exchange resin to media supplemented with calf serum and succeeded in removing Zn, Sr, Al, Cu, Mn, Ni, and Cr.[11]

The extent of removal for each metal was not the same, however. Another promising method uses the chelator tetraethylenepentamine (TEPA). According to Percvial and Layden-Patrice, TEPA added to cultures of HL-60 cells reduced cellular Cu levels and the activities of two Cu enzymes, Cu/Zn superoxide dismutase and cytochrome c oxidase, without affecting cell viability or differentiation potential.[12] These results show that approaches to learning the effects of Cu deficiency in a purely cell cultural environment may some day be possible if it is not possible already.

When it is essential to have buffer solutions as Cu-free as possible, the chelator *dithizone* (diphenylthiocarbazone) is very effective in ridding all traces of Cu from solutions. The reagent, a bluish-black crystalline powder, is highly insoluble in water. In the procedure, 1 ml of 0.05% dithizone in $CHCl_3$ is added to 1 l of a buffer solution (generally phosphate buffer) that had been placed in a 2 l separatory funnel with a stop cock. The funnel is sealed with a glass stopper and shaken vigorously for 30 seconds, occasionally inverting the funnel and releasing air by opening the stop cock. After two 30-second shakes, the funnel is set in a ring stand to allow the chloroform to settle towards the bottom. The presence of Cu is denoted by a reddish-brown color. Zinc, calcium, and magnesium form pink complexes with dithizone. After all the chloroform has settled to the bottom, the stop cock is opened to remove the dithizone-metal complex. One ml of fresh chloroform (without dithizone) is added and the operation is repeated. A green solution settling to the bottom indicates that all Cu ions have been removed. Two ml aliquots of fresh chloroform are added until the excess (green color) dithizone is removed from the flask. Often, this takes more than 20 extractions. This procedure, although laborious, is very effective in preparing buffer solutions that are virtually free of Cu.

III. PREPARATION OF DEFINITION OF CELL LINES

Hepatocytes, fibroblasts, and cells cultured from specialized tissues have been used in Cu metabolic studies. Procedures for preparing and using each of these cell types are available in the literature (Table 3). When faced with cells that do not survive in culture medium for more than a few days, an investigator must prepare the cells fresh for each experiment. An alternative is to use continuous cell lines. The most convenient commercial source of rodent cell lines is the American Type Culture Collection (ATCC) in Bethesda, Maryland. ATCC has a vast variety of different cell types from both normal and mutant animals. Cells in the collection can be perused on the internet address: http://www.atcc.org/. For an investigation that requires the establishment of a cell line or cells from a particular tissue, the options are not as straightforward.

A. Primary Cultures and Cell Lines

Cell cultures are classified as primary cultures and cell lines, the latter including finite, continuous, and clonal cell lines. Primary cultures are obtained from explants of excised tissue that have been disturbed

TABLE 3.

Examples of Rodent Cell Types Used in Copper Research

Cell Type	Objective and Reference
Ascites tumor (Ehrlich)	Cu interactions[65]
Astroglia	Pb-Cu interactions[66]
Astroglia	Cu toxicity[67]
Astroglia	Metallothione levels[68]
B-lymphocytes	Immunogenicity[69]
Fetal lung cells	Cu enzymes[70]
Fibroblasts	Cu kinetics and transport[71–73]
Fibroblasts	Genetic regulation[52]
Fibroblasts	Cu toxicity[74]
Glioma	Cu transport[34]
Hepatocytes	Cu transport and uptake[75–79]
Hepatocytes	Cu kinetics and transport[51,80–82]
Hepatocytes	Cu-Zn regulation[83]
Hepatocytes	Ceruloplasmin synthesis[84]
Hepatocytes	Cu metabolism, toxicity[85]
Hepatoma cells	Cu metabolism, toxicity[85,86]
Hepatoma cells	Cu deficiency[87]
Hepatoma cells	Metallothionine synthesis[88]
Kidney epithelial cells	Cu/Cd interactions[89]
Leukemia cells	Cu toxicity[90]
Leukemia cells	Growth and proliferation[91]
Lymphocytes	Cell proliferation[92,93]
Macroglia	Ceruloplasmin secretion[94]
Macrophages	Cu deficiency[95]
Myeoblasts (HL60)	Cu deficiency[12]
Neutrophils	Cu deficiency[96]
Neurons	Metallothionein levels[68]
Neurons	GABA receptor[97]
Neurons	Glutathione levels[67,98]
Respiratory epithelium	Cu toxicity[99]
Splenic T-cells	Immunogenicity[100]
Splenic T-cells	Immune functions[101]

gently so as to allow the cells to grow out from the tissue or from cells mechanically or enzymatically dissociated from tissue. After several days or weeks, the cells that survive and proliferate may be collected and passaged (i.e., reseeded) into fresh culture vessels; usually trypsinization is used to remove them from the culture vessel. Reseeding allows faster growing cells to overtake the slower growing and eventually become the dominant cells in the culture. Primary cultures are initially a heterogenous mixture of cells from a tissue, although a number of procedures is available to enrich the culture for specific cell types. After the first passage of a culture, the daughter cultures become known as a cell line. Most cell lines derived from normal, differentiated cells are amenable to only a limited number of passages *in vitro* before dying out or becoming senescent and are referred to as finite cells lines. Finite cell lines when propagated tend to maintain a constant chromosome number (euploidy) and hence show a typical life cycle. The life time of finite cells consequently is very short. In contrast, continuous cell lines have extended life times, and can be cultivated indefinitely. Continuous cell lines, which are derived either from tumors or from cell lines that have undergone a process of transformation, are often aneuploid, possessing a chromosome complement between diploid and tetraploid. Individual cells may be selected and propagated from a continuous cell line by cloning, yielding a clonal cell line or cell strain that exhibits a high degree of population uniformity.

B. Continuous Cell Lines

Normal cells in culture have two disadvantages: slow growth rates and limited survival times. With each passage cells lose some of their growth potential as senescence sets in. This is true of most terminally differentiated cells which literally are passing through a life cycle of divide, thrive, age, and die. Investigators have long sought to establish cell lines that have practically unlimited survival characteristics, i.e., immortalized cell lines. To be immortalized, however, cells must lack growth control mechanisms and therefore usually resemble cells in early stages of differentiation. Continuous cells lines are thus frequently established from tumors. Alternatively, continuous cell lines derived from normal tissue may undergo transformation, either as a spontaneous, multistep process or by chemical or viral induction, resulting in the immortalization of the cell line. Transformation often leads to increased tumorigenicity. Perhaps the most common method for the induction of transformation in a continuous cell line is via infection with a tumor virus specific for that cell type. For example, rouse sarcoma virus is relatively specific for infecting fibroblasts and muscle cells. Mouse mammary tumor virus or Epstein-Bar virus can infect a host of mammalian tissues. A second common method is to use tools of genetic engineering and infect the cells with a retrovirus containing a DNA construct that specifically disrupts the genetic factors associated with cell cycling. Arresting cells in one of the dividing stages prohibits them from entering the resting stage and keeps the cells dividing continuously.

IV. MONITORING COPPER

Nutritional studies will be limited by the ability of the investigator to detect Cu which is often present only in trace amounts. Detection can be accomplished in a number of ways. The way chosen will be dependent on the desired results. For example, stable copper pool size can be estimated by atomic absorption spectrometry (AAS), neutron activation, or colorimetic analysis. Metabolic studies where the intent is to identify the metabolic fate of newly absorbed copper will require radioactive copper. The procedures and application of these techniques are described below.

A. Quantitative Analysis of Copper

1. Colorimetric tests for copper ions exploit chelators that form colorful complexes that can be measured spectrophotometrically. Some chelators, e.g., bathocuproine sulfonate and 2,2′-biquinoline react only with Cu(I).[13,14] In contrast, cuprizone is specific for Cu(II).[15] Many chelators of Cu tend to be nonspecific and interference by other divalent metal ions (Zn^{2+}, Fe^{2+} in particular) is likely and should be determined. Colorimetic analysis has lower sensitivity than other methods and interference by reducing agents, detergents, and acids limits its application. A recent study by our laboratory found bichinchoninic acid (BCA) to be a superior reagent for quantitating small amounts of copper in biological samples with minimal interference by detergents and other metals.[16]
2. Atomic absorption spectrophotometric analysis (AAS) is perhaps the most common method for quantitating Cu in extracts from cells and tissues. AAS is highly sensitive, specific, and the instrumentation is of modest cost and can be adapted to most laboratories. Before the analysis, organic matter is "wet-ashed" with boiling $HClO_4$-HNO_3. The clear hydrolysate is injected directly into an air-acetylene flame or graphite furnace atomizer connected to an atomic absorption spectrophotometer (AAS) that is equipped with a hollow-cathode lamp radiating at 324.8 nm (maximum for the copper). Limits of detection are 0.01-0.1 μg/ml for flame AAS and 0.01-1.0 ng/ml for graphite furnace AAS. A complete description of the methodology has recently been published.[17]
3. Other: As a variation to the flame or furnace, Inductively Coupled Plasma (ICP) applies radio frequency energy to argon gas in a sealed chamber to atomize the sample. ICP has the advantage of heating the sample uniformly down to the core. Used with Atomic Emission Spectrophotometry (AES), ICP extends linearity over a concentration range and at least four orders of magnitude. Neutron Activation Analysis (NAA) uses neutron bombardment of copper to generate ^{64}Cu/^{66}Cu, two radioactive nuclides which are identified and quantitated by their gamma emissions. The process uses thermoneutron capture and is very sensitive because of the large capture capability of the Cu atom. NAA, however, requires special equipment and shielded environment and thus is not a common tool of most laboratories.
4. Radioactive Monitoring: Copper has two radioactive nuclides: ^{64}Cu and ^{67}Cu. The two are distinguishable by their half lives and type of emissions. ^{64}Cu can be detected by its beta emissions. The half-life of ^{64}Cu, 12.8 hours, has precluded long term studies with the isotope. ^{67}Cu has a half-life of 62 hours and is a strong gamma emitter. This isotope is favored for isotopic studies. ^{64}Cu and ^{67}Cu can be purchased through the U.S. Atomic Energy Commission which currently has production facilities in Los Alamos, NM and Brookhaven, NY.

V. COPPER TOXICITY

Toxicity is always a concern when exposing cells to heavy metals. As a redox metal, Cu readily catalyzes the oxidation of thiol groups in membrane and soluble proteins, and as a prooxidant, Cu will dismutate H_2O_2 causing formation of destructive active oxygen species. Cells vary in tolerance to Cu and for this reason an investigator should conduct tolerance curves to determine safe levels. Our experience has been that concentrations between 1 to 50 μM Cu do little harm in short term cultures. On the other hand, fibroblasts and lymphoblasts in culture for two or three days will show growth impairments with Cu above 10 μM. Cells do not respond to Cu uniformly and different cell types will have systems for rendering copper ions less toxic (see below). The different salts of copper (chloride, sulfate, gluconate) are not as major a factor in toxicity as are levels of protein in the medium, the presence of other divalent metals, chelators, peptides, and reducing agents. Below, we discuss factors that promote toxicity and ways to lessen Cu toxicity in culture media. It is especially important to note that toxicity of Cu can be both direct and indirect, the latter caused by interactions with components in the growth medium. While not specifying Cu in particular, Christie and Costa have recently reviewed the toxic effects of metal compounds on cells.[18]

A. Copper-Sensitive Systems in Cells

A cell in a rapidly growing state is quite sensitive to chemicals in its immediate environment. Cu ions can readily penetrate the plasma membrane of most cells and enter the cytosol where they are rapidly transported to the nucleus and other organelles. In both the internal and external environment, Cu(II) has the potential to oxidize exposed thiol (–SH) groups. Structures of membrane bound enzymes, such as acetylcholine esterase[19] or cytochrome P450, are vulnerable to the toxicity of Cu, the latter by interchain crosslinking.[20] Disrupting the function of the membrane-bound enzymes can lead to further disruptions in cell function. For example, 10 ppm $CuCl_2$ inhibits the activity of Na,K-ATPase.[21] An IC50 of 51 μM was reported for the enzyme in cultured muscle cells.[22] The toxicity to the muscle cells was mediated by an increase in intracellular Ca^{2+}, presumably because the unexpelled Na^+ ions forced the Ca/Na exchanger to run in the reverse direction. Internally, free Cu ions will uncouple oxidative phosphorylation[23] and interact with ascorbate, causing irreversible inhibition of catalase.[24] The damaging effects of Cu ions result in lysis of the membranes,[25] a clear indication that if left unsequestered, Cu ions have the potential to cause major structural modifications to membrane ultrastructure.

Valence state of Cu is another concern. A puzzling observation was that reducing agents such as ascorbic acid appeared to exacerbate the toxic effects of Cu on cells.[19,24] This observation points to Cu(I) as potentially a more serious threat to cells than Cu(II). Epitomizing the danger, erythrocytes will hemolyze when Cu(II) is added to the medium. Hemolysis, however, is delayed for at least 1 h after Cu addition, conceivably to allow the Cu(II) to be reduced to Cu(I). Cells kept under N_2 or in a medium presaturated with carbon monoxide did not hemolyze when Cu(II) was added.[25] Only Cu(I), the reduced species, can catalyze peroxidation reactions through generation of free radicals.[19] Hydroxyl radical, the most destructive oxygen radical in a biological system, is formed when hydrogen peroxide reacts with Cu(I). The widespread damage to cell constituents explains the toxicity of Cu(I). Its participation in numerous toxic responses has been well established.[26] Bathocuproine sulfonate, a chelator that binds to Cu(I) and not Cu(II), has been shown to moderate toxic effects. The findings imply that even small amounts of Cu(II) in a medium rich in reducing agents risks cell death.

B. Indirect Effect of Copper Ions on Cells

An indirect effect of Cu implies an interaction with media components that are themselves nontoxic but capable of potentiating a toxicity when exposed to Cu ions. As noted above, the toxicity of ascorbate to a culture of malignant melanoma cells was associated with the generation of Cu(I) in the medium.[27] The vitamin caused a 50% decrease in colony formation, but only when added with 5 μM Cu(II) in solution, a combination that lowered the ID_{50} more than 20 times.[27] A rather interesting and perhaps disturbing observation is that chelators designed to ameliorate toxic effects actually assist the toxicity. As an example, Mohindru et al. reported that micromolar amounts of diethyldithiocarbamate (DDC), KTS, and 2,9-dimethyl-1,10-phenolthroline (DMP) potentiated the toxic effects of copper in a culture of L1210 murine leukemic cells; Cu alone (100 μM) showed no growth arrest.[28] These observations show that even as a complex with chelators, Cu can still be lethal to cells.

C. Cellular Targets of Copper Toxicity

Beyond the plasma membranes, nucleic acids and cytosolic proteins are also susceptible to toxic effects of copper. Damage to DNA is a special concern. Studies *in vitro* have clearly established the potential for Cu(II) (as a complex with 1,10-phenanthroline) in the presence of a reducing agent (ascorbate, mercaptoethanol) to cleave DNA.[29] The scission reaction appeared to be mediated by hydroxyl radical and formed through a disproportionation of hydrogen peroxide.

D. Defense Against Copper Toxicity

Slight modifications to culture media can sometimes prove effective in lessening toxic effects of Cu on cells. The addition of acetate, proline, or cysteine a few minutes before or after 80 μM Cu reduced death and Cu binding to *Pseudomonas syringae*. These agents were as effective as EDTA and nitrilotriacetate (NTA) in protecting the bacteria.[30] Cells themselves have mechanisms for protecting against copper toxicity. Synthesis of internal sequestering protein such as the metallothioneins and an ATP-dependent efflux system come into play in protecting the cell. Indeed, it has been possible to develop high resistance to Cu by culturing Morris rat hepatoma cells in a medium with gradually increasing Cu.[31] Such cells will tolerate Cu and Cu complexes greater than 500 μM. Superior tolerance to Cu in hepatoma cells has been attributed to increased levels of intracellular metallothionein, a metal binding protein, which sequesters Cu ions via thiol groups along the protein surface.[32,33] An important observation we made was that rat glioma cells in culture possess very efficient (Km = 0.21 μM) efflux systems for expelling Cu. The efflux system was perhaps four times more sensitive than the influx system.[34] This finding suggests that tolerance to Cu in glioma and perhaps other cells is built around a mechanism where input is countered by a very effective system for effluxing the Cu taken in.

VI. APPLICATION OF TISSUE CULTURE TO COPPER TRANSPORT STUDIES

In our studies, we use both human and rodent cells to study the Cu transport mechanism. The studies are predicated on the supposition that cells contain Cu-specific membrane transport systems that recognize plasma copper carriers or free Cu ions and are capable of relocating Cu from the medium or plasma membrane to the interior of the cell. We also assume that Cu is expelled from cells to maintain homeostasis and reduce toxicity. Inside the cells, Cu is free to interact with transport and storage factors or to be transferred to specific organelles. The latter appears to involve ill-defined transport factors that convey Cu ions to the nucleus, mitochondria, or other intracellular locales. Transport studies are aided by the use of ^{67}Cu. Below is a brief outline of methodology used to characterize the Cu transport system in rat glioma cells.

A. Cells and Culture Conditions

The C6 rat glioma cell line we use was purchased from the American Type Culture Collection. We culture the cells in 75-cm^2 uncoated flasks

(Corning) under a 1:1 solution of Dubecco's modified Eagle's medium/nutrient F12 mixture (DMEM/F12, Sigma) which is supplemented with 10% (v/v) fetal bovine serum (GIBCO). A test of the medium showed 2.5 n*M* Cu. Cysteine is present at 0.05 m*M*. The cells are grown at 37°C under 5% CO_2 in humidified incubators with fresh medium supplied every 2 to 3 days. We normally passaged the cells every 3 to 5 days after detaching the cells with 1 m*M* ethylenediaminetetraacetate (EDTA) in Pucks saline solution. When these cells are used in transport studies, we seed them at high density in 35-mm tissue culture dishes (Corning) and grow them for at least for 2 to 3 days until they reached a final density of about 3×10^6 cells per dish.

B. Incorporation of ^{67}Cu Into Cells

^{67}Cu can be obtained from the Atomic Energy Commission laboratories at either Brookhaven or Los Alamos. The two sources alternate nonoverlapping production schedules during the year. The Los Alamos linear accelerator facility produces a higher quality grade of ^{67}Cu with specific activity >8000 Ci/mmole and minimal contamination with ^{64}Cu. The ^{67}Cu could be added directly to the cells or first incorporated into ceruloplasmin which then serves as the copper donor.[35] In the ceruloplasmin procedure, ascorbate is used to catalyze the exchange of the radiolabeled Cu with the protein.

C. Transport Studies

Cells are incubated with either free ^{67}Cu or ^{67}Cu bound to ceruloplasmin for an appropriate length of time. At predetermined times, the medium is decanted, the cells washed twice with isotonic PBS and then once with an acidic isotonic saline buffer (0.05 *M* Na acetate, 0.15 *M* NaCl, pH 5.0) to remove loosely bound ^{67}Cu ions. The amount of ^{67}Cu absorbed into the cytosol is measured with the aid of a gamma counter. Controls are run at colder temperatures (4 to 10°C) to assure that ^{67}Cu retained by the cells penetrated the cell membrane. Total cell ^{67}Cu can be determined by removing the cells from the dish with the aid of a cell scraper and taking appropriate aliquots for radioactivity and protein determinations.

D. Subcellular Localization of Copper

The potential for Cu to enter the cell via vesicles has been established.[36] Cells thus can absorb Cu into both cytosolic and membrane compartments. We characterize the membrane location by density-gradient centrifugation in Percoll. Percoll is a polysaccharide matrix that separates organelles on the basis of density. In the experiments, a low-speed supernatant from the cells is taken, mixed with a pre-made Percoll solution in sucrose, and placed in the centrifuge tubes. A short spin of about 30 min at 32,000 × *g* will cause the Percoll to establish a gradient. Membranes and other particulate components that contain the radioactive Cu will be resolved by virtue of their buoyant density in Percoll. The lighter fraction near the top of the tube represents the lightest membrane fraction or the one with the highest lipid to protein ratio. The separated membranes can be collected by displacing the Percoll from the tube bottom with a 60% sucrose solution. Alternatively, the tube can be punctured with a 20-gauge needle and the Percoll drawn into a 5 ml syringe. After diluting 1:1 with buffer, the membranes can be collected by centrifugation and analyzed further. In the procedure, colored marker beads signifying the various densities are included with the Percoll-extract mixture. This step will assist in identifying the precise density of the membrane fraction. A typical Percoll fractionation is shown in Figure 1.

VII. APPLICATION OF CELL CULTURE TO DISEASES OF COPPER METABOLISM

Two prominent human diseases associated with copper metabolism owe their current understanding in part to the use of rodent models and cell cultures. Wilson's disease and Menkes disease have been reviewed extensively.[37–40] The Long Evans Cinnamon (LEC) rat, so named because of its cinnamon-like coat, has been a model of Wilson's disease. A second model, the so-called "toxic milk" mouse, is less utilized.[41,42] A Menkes-like condition is expressed in the mottled mutant mouse and the various allelic phenotypes (Table 4). Cultured cells from rodents have contributed to the identity of the defect.

A. Menkes Disease

This X-linked disease of Cu metabolism has its counterpart in the mottled mutant heterozygous male mouse (*Mo*/y). Typically, heterozygous females have light and dark variegated coat colors. The mutation in these mice, although lethal, was not understood until Hunt showed that mutant males had impaired copper transport.[43] The brindled mouse, one of the allelic variants, has been studied extensively. Both neonatal male and adult female brindled mice have impaired copper homeostasis[44] and require Cu intraperitoneally 7 days postnatally to survive. Hemizygous mutant males will otherwise die around 14 days postpartum. Injected mice grow rapidly and apparently normally for 60 days, despite slowly declining Cu levels in liver and brain to a near deficient status at 25 days post injection.[45] Post injection Cu (10 μg/g) will also raise cerebral Cu levels and maximize activity of brain cytochrome *c* oxidase activity.[21,46]

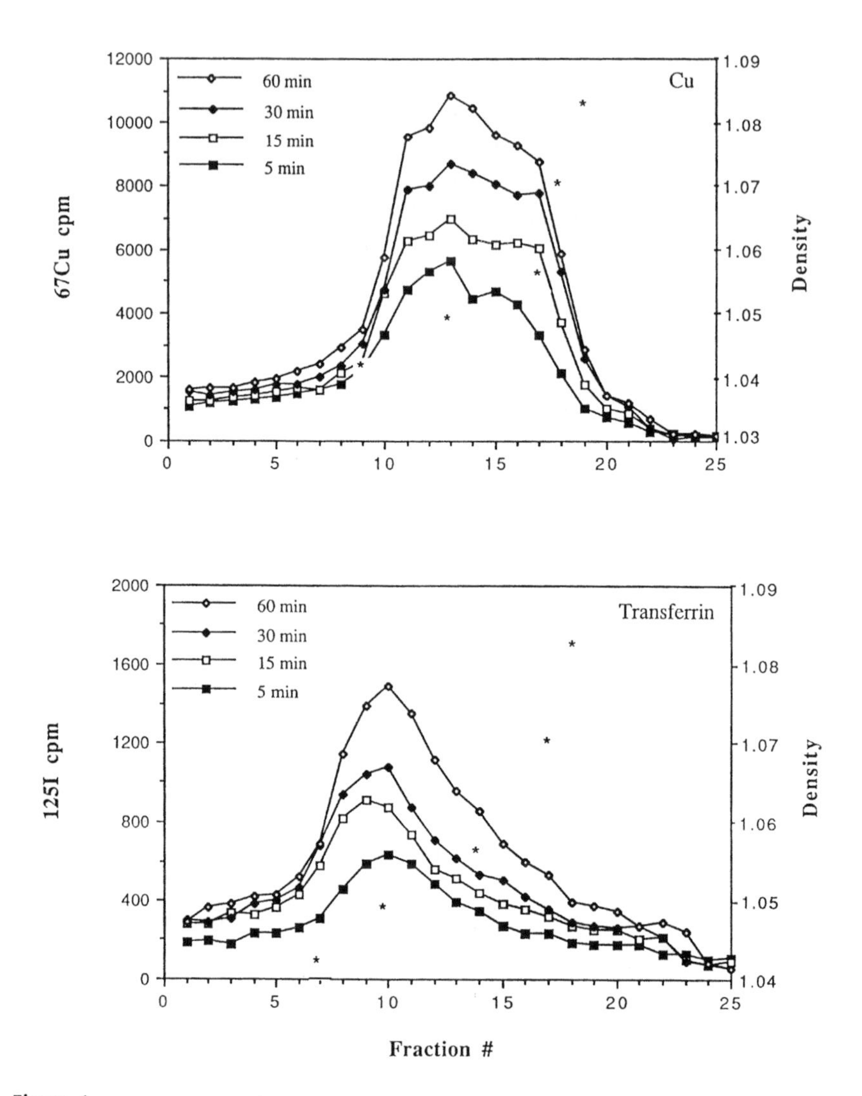

Figure 1
Percoll Density-Gradient Centrifugation of Particulate Fraction from Cells. Upper panel shows a time-course study of ^{67}Cu distribution in membranes of erythroleukemic cells at 5, 15, 30, and 60 min after adding ^{67}Cu-labeled ceruloplasmin to the cells. Lower panel shows another group of cells treated with ^{125}I-transferrin. The Percoll gradient is indicated by the asterisks. The data show two membrane components of different density retaining ^{67}Cu whereas ^{125}I- transferrin is found in only one. For further details, see Reference 36.

Again, effects are not permanent and Cu levels decline steadily.[21] Typically, a brindled mouse will display the same severe Cu malabsorption of a Menkes child and retain excess Cu in the kidney.[47] Mutant mouse cells have established that a low hepatic Cu level is not a consequence of low Cu intake, but instead a rapid loss of Cu after loading. Cells from mottled

TABLE 4.

Phenotypic Variability of the Mottled Locus in Heterozygous Male Mice

Allele	Name	Phenotypic Expression
Mo^{pew}	Pewter	Changes in coat color
Mo^{blo}	Blotchy	Defects in connective tissue
Mo^{vbr}	Viable brindled	Defects in connective tissue
Mo^{br}	Brindled	Neurological defects, perinatal lethality
Mo^{ml}	Macular	Neurological defects, perinatal lethality
Mo^{dp}	Dapled	Prenatal lethality
Mo^{to}	Tortoise shell	Prenatal lethality

Data taken from Levinson et al.[64] With permission.

mice expressing the various phenotypic variations have been used to characterize the site of the lesion (see Table 2).

1. Cell Culture Studies

Early studies by Goka et al.,[48] Horn,[49] and Chan et al.[50] established that fibroblasts from Menkes patients display the copper-retention phenotype of the disease when grown in culture medium and thus could be a diagnostic tool. Cultured mouse hepatocytes from heterozygous males and hemizygous females have been very beneficial in elucidating the defect. Hepatocytes from mutant mice show no difference in the rate of ^{64}Cu accumulation or efflux of normal hepatocytes. Mutant cells, however, accumulate 25% less total copper during the culture period, a difference not seen in normal cells from copper-deficient animals.[51] Mutant hepatocytes, therefore, are apparently expressing the primary defect. Likewise, fibroblasts from mottled mice show impaired copper transport leading to an accumulation of cytosolic copper. Cellular sequestration is associated with the absence of a Cu-transporting ATPase. Fibroblasts from mottled mice fail to incorporate copper into lysyl oxidase and show suppressed levels of mRNA for the enzyme and for connective tissue proteins in general.[52]

More recently, attention has been given to the macular mouse as a model of Menkes disease.[53] Macular mice accumulate large amounts of Cu in organelle-free cytoplasm and low amounts in mitochondria.[54] Brain deterioration is common in this mutant and single injections of 50 μg Cu subcutaneously on day 7 postnatal raises brain Cu to near normal levels and elevates brain cytochrome c oxidase activity to near-normal levels in 180 days.[55] Cultures of astrocytes from macular mice show excessive accumulation of Cu which has been interpreted as showing a block in the passage of Cu from astrocytes to neurons.[56] Although information from cultures of macular mouse cells is still premature, it is clear that tissue from this mutant can also be applied to learning more about the basic mechanism of Menkes disease.

B. Wilson's Disease

For many years, the molecular cause of Wilson's disease was unknown despite a thorough documentation of the disease symptoms. The LEC rat accumulates Cu in liver spontaneously up to 40 times more than the Long Evans Agouti (LEA), the control, and develops jaundice and hepatitis. Hepatocytes from LEC rats have elevated Cu levels, which is apparently the cause of the hepatitis, but show no defect in the ceruloplasmin gene despite the fact that LEC rats have low levels of ceruloplasmin in the plasma.[57] The Cu is distributed randomly, but tends to be more focused in the central lobe.[58] The build-up of Cu is continuous with a more intensified staining being evident in 8 vs. 3 months; hepatocytes from LEA rats barely show a positive stain for Cu.[58] Levels of metallothionein mRNA are higher than controls and apparently correlate with the elevated copper levels. Nuclei from hepatocytes have elevated levels of a protein that recognizes the metal responsive element (MRE) for metallothionein gene.[59] Hepatocytes from LEC rats show morphological changes such as steatosis in the cytosol and pleomorphism of mitochondria similar to Wilson's disease.[60] The mutant cells are also impaired in the transfer of Cu to various organelles intracellularly and thus have been invaluable in giving insight into the intracellular transport mechanism.[61] One of the more interesting applications that exploited hepatocyte cell cultures was a study by Nakamura in which cDNA from human ceruloplasmin was used to transfect the cells. The results showed that LEC hepatocytes processed ceruloplasmin normally and were unimpaired in incorporating Cu into the protein.[62] These data counter the studies that suggest holoceruloplasmin cannot be produced in LEC hepatocytes. Cultured hepatocytes from LEC rats have been shown to incorporate Cu into ceruloplasmin in the Golgi apparatus, an event that is preceded by the synthesis of a polypeptide with a 19 amino acid leader sequence that is apparently removed during processing.[63] These data point to the utility of the LEC hepatocytes to decipher the defect in Wilson's disease.

VIII. SUMMARY AND CONCLUSIONS

Tissue and cell culture systems afford investigators the opportunity to study Cu at its most elementary biological level, the cell. Cellular transport proteins, membranes, and cytosolic factors that normally take part in the movement of copper can be identified and their precise role in a delivery mechanism determined. Cell cultures also allow investigators to study copper interactions with other metals and potentially toxic agents. There are numerous highly sensitive colorimetric and physical tests available to quantitate Cu in cell compartments. Investigations into the metabolism of Cu have necessitated the use radioisotopes of copper, ^{67}Cu and ^{64}Cu, as indispensable markers of Cu metabolism. ^{67}Cu with a longer half-life tends to be preferred. Because of tissue culture, it has been possible to learn how individual cells metabolize this essential metal. From such studies has come the realization that there are distinct differences in the manner in which each cell metabolizes Cu. Tissue culture approaches to diseases of Cu metabolism, Menkes and Wilson's disease, have provided crucial insight into genes that regulate Cu movement. The diseases have shown that there are important and specific factors under genetic regulation that play decisive roles in Cu transport and maintaining Cu homeostasis in cells.

IX. ACKNOWLEDGMENTS

Funding for this review was provided in part by grants HD29952 and ES05781 from the National Institutes of Health.

REFERENCES

1. Hayflick, L. and Moorhead, P.S., The serial cultivation of human diploid cell strains. *Exp. Cell Res.*, 25, 585, 1961.
2. MacPherson, I. and Stoker, M., Polyoma transformation of hamster cell clones-an investigation of genetic factors affecting cell competence. *Virology*, 16, 147, 1962.
3. Earle, W.R., Schilling, E.L., Stark, T.H., Straus, N.P., Brown, M.F., and Shelton, E., Production of malignancy *in vitro*. IV. The mouse fibroblast cultures and changes seen in living cells. *J. Nat. Cancer Inst.*, 4, 165, 1943.
4. Freshney, R.I., *Culture of Animal Cells. A Manual of Basic Technique*. Third ed. New York: Alan R. Liss, Inc., 1994.
5. Barnes, W.D. and Sato, G., Method for growth of cultured cells in serum-free medium. *Anal. Biochem.*, 102, 255, 1980.
6. Neumann, P.Z. and Sass-Kortsak, A., The state of copper in human serum: evidence for an amino acid-bound fraction. *J. Clin. Invest.*, 46, 646, 1967.
7. Harris, E.D., Copper transport: an overview. *Proc. Soc. Exp. Biol. Med.*, 196, 130, 1991.
8. Dameron, C.T. and Harris, E.D., Regulation of aortic Cu,Zn-superoxide dismutase with copper. Ceruloplasmin and albumin transfer copper and reactivate the enzyme in culture. *Biochem. J.*, 248, 669, 1987.
9. Bradshaw, R.A. and Peters, T.J., The amino acid sequence of peptide (1-24) of rat and human serum albumin. *J. Biol. Chem.*, 244, 5582, 1969.
10. Messer, H.H., Murray, E.J., and Goebel, N.K., Removal of trace metals from culture media and sera for in vitro deficiency studies. *J. Nutr.*, 112, 652, 1982.
11. Rayner, M.H. and Suzuki, K.T., A simple and effective method for the removal of trace metal cations from a mammalian culture medium supplemented with 10% fetal calf serum. *Blood*, 86, 2807, 1995.
12. Percival, S.S. and Layenden-Patrice, M., HL-60 cells can be made copper deficient by incubating with tetraethylenepentamine. *J. Nutr.*, 122, 2424, 1992.
13. Blair, D. and Diehl, H., Bathophenanthrolinedisulphonic acid and bathocuproinedisulphonic acid, water soluble reagents for iron and copper. *Talanta*, 7, 163, 1961.
14. Hanna, P.M., Tamilarasan, R., and McMillin, D.R., Cu(I) analysis of blue copper proteins. *Biochem. J.*, 256, 1001, 1988.

15. Peterson, R.E. and Bollier, M.E., Spectrophotometric determination of serum copper with biscyclohexanoneoxalyldihydrazone. *Anal. Chem.*, 27, 1195, 1955.
16. Brenner, A.J. and Harris, E.D., A quantitative test for copper using bicinchoninic acid. *Anal. Biochem.*, 226, 80, 1995.
17. Tiffany-Castiglioni, E., Legare, M.E., Schneider, L.A., Harris, E.D., Barhoumi, R., Zmudzki, J., Qian, Y.C., and Burghardt, R.C., Heavy metal effects on glia. In: Perez-Polo, J.R., Ed. *Methods in Neurosciences*, 1996 (in press).
18. Christie, N.T. and Costa, M., *In vitro* assessment of the toxicity of metal compounds. IV. Disposition of metals in cells: interactions with membranes, glutathione, metallothionein, and DNA. *Biol. Trace Element Res.*, 6, 139, 1984.
19. Shinar, E., Navok, T., and Chevion, M., The analogous mechanisms of enzymatic inactivation induced by ascorbate and superoxide in the presence of copper. *J. Biol. Chem.*, 258, 14778, 1983.
20. McIntosh, P.R. and Freedman, R.B., Characteristics of a copper-dependent cross-linking reaction between two forms of cytochrome P-450 in rabbit-liver microsomal membranes. *Biochem. J.*, 187, 227, 1980.
21. Hansen, J.I., Mustafa, T., and Depledge, M., Mechanisms of copper toxicity in the shore crab, Carcinus maenas. I. Effects on Na,k-ATPase activity, haemolymph electrolyte concentrations and tissue water contents. *Mar. Biol.*, 114, 253, 1992.
22. Benders, A.A.G.M., Li, J., Lock, R.A.C., Bindels, R.J.M., Wendelaar Bonga, S.E., and Veerkamp, J.H., Copper toxicity in cultured human skeletal muscle cells: The involvement of Na^+/K^+-ATPase and the Na^+/Ca^{2+}- exchanger. *Pflugers Arch.*, 428, 461, 1994.
23. Leblondel, G. and Allain, P., A thiol oxidation interpretation of the Cu^{2+} effects on rat liver mitochondria. *J. Inorgan. Biochem.*, 21, 241, 1984.
24. Davison, A.J., Kettle, A.J., and Fatur, D.J., Mechanism of the inhibition of catalase by ascorbate. Roles of active oxygen species, copper and semidehydroascorbate. *J. Biol. Chem.*, 261, 1193, 1986.
25. Barnes, G. and Frieden, E., Oxygen requirement for cupric ion induced hemolysis. *Biochem. Biophys. Res. Commun.*, 115, 680, 1983.
26. Mohindru, A., Fisher, J.M., and Rabinovitz, M., Endogenous copper is cytotoxic to a lymphoma in primary culture which requires thiols for growth. *Experientia*, 41, 1064, 1985.
27. Bram, S., Froussard, P., Guichard, M., et al., Vitamin C preferential toxicity for malignant melanoma cells. *Nature*, 284, 629, 1980.
28. Mohindru, A., Fisher, J.M., and Rabinovitz, M., Bathocuproine sulphonate: a tissue culture-compatible indicator of copper-mediated toxicity. *Nature*, 303, 64, 1983.
29. Dizdaroglu, M., Aruoma, O.I., and Halliwell, B., Modification of bases in DNA by copper ion-1,10-phenanthroline complexes. *Biochemistry*, 29, 8447, 1990.
30. Cabral, J.P.S., Influence of organic ligands on the toxicity of copper to *Pseudomonas syringae*. *FEMS Microbiol. Lett.*, 117, 341, 1994.
31. Freedman, J.H., Weiner, R.J., and Peisach, J., Resistance to copper toxicity of cultured hepatoma cells. Characterization of resistant cell lines. *J. Biol. Chem.*, 261, 11840, 1986.
32. Freedman, J.H., Ciriolo, M.R., and Peisach, J., The role of glutathione in copper metabolism and toxicity. *J. Biol. Chem.*, 264, 5598, 1989.
33. Czaja, M.J., Weiner, F.R., and Freedman, J.H., Amplification of the metallothionein-1 and metallothionein-2 genes in copper-resistant hepatoma cells. *J. Cell. Physiol.*, 147, 434, 1991.
34. Qian, Y.C., Tiffany-Castiglioni, E., and Harris, E.D., Copper transport and kinetics in cultured C6 rat glioma cells. *Am. J. Physiol. Cell Physiol.*, 269, C892, 1995.
35. Percival, S.S. and Harris, E.D., Copper transport from ceruloplasmin: Characterization of the cellular uptake mechanism. *Am. J. Physiol.*, 258, C140, 1990.
36. Davidson, L.A., McOrmond, S.L., and Harris, E.D., Characterization of a particulate pathway for copper in K562 cells. *Biochim. Biophys. Acta*, 1221, 1, 1994.
37. Sternlieb, I., Perspectives on Wilson's disease. *Hepatology*, 12, 1234, 1990.

38. Yarze, J.C., Martin, P., Muñoz, S.J., and Friedman, L.S., Wilson's disease: Current status. *Am. J. Med.*, 92, 643, 1992.
39. Harris, E.D., Menkes' disease: perspective and update on a fatal copper disorder. *Nutr. Rev.*, 51, 235, 1993.
40. Vulpe, C.D. and Packman, S., Cellular copper transport. *Ann. Rev. Nutr.*, 15, 293, 1995.
41. Porter, H., Neonatal hepatic mitochondrocuprein. IV. Sulfitolysis of the cystine rich crude copper protein and isolation of a peptide containing more than 35% half-cystine. *Biochim. Biophys. Acta*, 229, 143, 1971.
42. Mercer, J.F.B., Grimes, A., Danks, D.M., and Rauch, H., Hepatic ceruloplasmin gene expression is unaltered in the toxic milk mouse. *J. Nutr.*, 121, 894, 1991.
43. Hunt, D.M. and Johnson, D.R., An inherited defect in noradrenaline biosynthesis in the brindled mouse. *J. Neurochem.*, 19, 2811, 1972.
44. Evans, G.W. and Reiss, B.L., Impaired copper homeostasis in neonatal male and adult female brindled (Mobr) mice. *J. Nutr.*, 108, 554, 1978.
45. Mann, J.R., Camakaris, J., Danks, D.M., and Walliczek, E.G., Copper metabolism in mottled mouse mutants. Copper therapy of brindled (Mo^br) mice. *Biochem. J.*, 180, 605, 1979.
46. Fujii, T., Ito, M., Tsuda, H., and Mikawa, H., Biochemical study on the critical period for treatment of the mottled brindled mouse. *J. Neurochem.*, 55, 885, 1990.
47. Prins, H.W. and van den Hamer, C.J.A., Primary biochemical defect in copper metabolism in mice with a recessive X-linked mutation analogous to Menkes' disease in man. *J. Inorgan. Biochem.*, 10, 19, 1979.
48. Goka, T.J., Stevenson, R.E., Hefferan, P.M., and Howell, R.R., Menkes disease: a biochemical abnormality in cultured human fibroblasts. *Proc. Nat. Acad. Sci. (USA).*, 73, 604, 1976.
49. Horn, N., Copper incorporation studies on cultured cells for prenatal diagnosis of Menkes' disease. *Lancet*, 1, 1156, 1976.
50. Chan, W., Cushing, W., Coffman, M.A., and Rennert, O.M., Genetic expression of Wilson's disease in cell culture: A diagnostic marker. *Science*, 208, 299, 1980.
51. Darwish, H.M., Hoke, J.E., and Ettinger, M.J., Kinetics of Cu(II) transport and accumulation by hepatocytes from copper-deficient mice and the brindled mouse model of Menkes disease. *J. Biol. Chem.*, 258, 13621, 1983.
52. Gacheru, S., McGee, C., Uriu-Hare, J.Y., et al., Expression and accumulation of lysyl oxidase, elastin, and type I procollagen in human Menkes and Mottled Mouse Fibroblasts. *Arch. Biochem. Biophys.*, 301, 325, 1993.
53. Kodama, H., Recent developments in Menkes disease. *J. Inherit. Metab. Dis.*, 16, 791, 1993.
54. Kodama, H., Abe, T., Takama, M., Takahashi, I., Kodama, M., and Nishimura, M., Histochemical localization of copper in the intestine and kidney of macular mice: Light and electron microscopic study. *J. Histochem. Cytochem.*, 41, 1529, 1993.
55. Meguro, Y., Kodama, H., Abe, T., Kobayashi, S., Kodama, Y., and Nishimura, M., Changes of copper level and cytochrome c oxidase activity in the macular mouse with age. *Brain Dev. (Tokyo)*, 13, 184, 1991.
56. Kodama, H., Meguro, Y., Abe, T., et al., Genetic expression of Menkes disease in cultured astrocytes of the macular mouse. *J. Inherited Metab. Dis.*, 14, 896, 1991.
57. Sone, H., Maeda, M., Gotoh, M., et al., Genetic linkage between copper accumulation and hepatitis/hepatoma development in LEC rats. *Mol. Carcinog*, 5, 199, 1992.
58. Fujii, Y., Shimizu, K., Satoh, M., et al., Histochemical demonstration of copper in LEC rat liver. *Histochemistry*, 100, 249, 1993.
59. Yamada, T., Suzuki, Y., Agui, T., and Matsumoto, K., Elevation of metallothionein gene expression associated with hepatic copper accumulation in Long-Evans Cinnamon mutant rat. *Biochim. Biophys. Acta Gene Struct. Expression*, 1131, 188, 1992.
60. Li, Y., Togashi, Y., Sato, S., et al., Spontaneous hepatic copper accumulation in Long-Evans Cinnamon rats with hereditary hepatitis. A model of Wilson's disease. *J. Clin. Invest.*, 87, 1858, 1991.

61. Yamada, T., Agui, T., Suzuki, Y., Sato, M., and Matsumoto, K., Inhibition of the copper incorporation into ceruloplasmin leads to the deficiency in serum ceruloplasmin activity in Long-evans Cinnamon mutant rat. *J. Biol. Chem.*, 268, 8965, 1993.
62. Nakamura, K., Endo, F., Ueno, T., Awata, H., Tanoue, A., and Matsuda, I., Excess copper and ceruloplasmin biosynthesis in long-term cultured hepatocytes from Long-Evans Cinnamon (LEC) rats, a model of Wilson disease. *J. Biol. Chem.*, 270, 7656, 1995.
63. Terada, K., Kawarada, Y., Miura, N., Yasui, O., Koyama, K., and Sugiyama, T., Copper incorporation into ceruloplasmin in rat livers. *Biochim. Biophys. Acta Mol. Basis Dis.*, 1270, 58, 1995.
64. Levinson, B., Vulpe, C., Elder, B., et al., The mottled gene is the mouse homologue of the Menkes disease gene. *Nature Genet.*, 6, 369, 1994.
65. Kraker, A., Krezoski, S., Schneider, J., Minkel, D., and Petering, D.H., Reactions of 3-ethoxy-2-oxoburyraldehyde bis(thiosemicarbazonato) Cu(II) with ehrlich cells. *J. Biol. Chem.*, 260, 13710, 1985.
66. Tiffany-Castiglioni, E., Zmudzki, J., Wu, J.-N., and Bratton, G.R., Effects of lead treatment on intracellular iron and copper concentration in cultured astroglia. *Metabol. Brain. Dis.*, 2, 61, 1987.
67. Pileblad, E., Eriksson, P.S., and Hansson, E., The presence of glutathione in primary neuronal and astroglial cultures from rat cerebral cortex and brain stem. *J. Neural Transm.*, 86, 43, 1991.
68. Hidalgo, J., Garcia, A., Oliva, A.M., et al., Effect of zinc, copper and glucocorticoids on metallothihonein levels of cultured neurons and astrocytes from rat brain. *Chemico-Biol. Interact.*, 93, 197, 1994.
69. Failla, M.L. and Bala, S., Cellular and biochemical functions of copper in immunity. In: Chandra, R.K., Ed. *Nutrition and Immunity.* St. John's, Canada: ARTS Publishers, 1992, 129.
70. Randhawa, P., Hass, M., Frank, L., and Massaro, D., Dexamethasone increases superoxide dismutase activity in serum-free rat fetal lung organ cultures. *Pediatr. Res.*, 20, 895, 1986.
71. Waldrop, G.L. and Ettinger, M.J., The relationship of excess copper accumulation by fibroblasts from the brindled mouse model of Menkes disease to the primary defect. *Biochem. J.*, 267, 417, 1990.
72. Aasa, R., Malmström, B.G., Saltman, P., and Vänngard, T., The specific binding of iron (III) and copper (II) to transferrin and conalbumin. *Biochim. Biophys. Acta*, 75, 203, 1963.
73. McArdle, H.J., Guthrie, J., Ackland, M.L., and Danks, D.M., Albumin has no role in copper uptake by fibroblasts. *J. Inorgan. Biochem.*, 31, 123, 1987.
74. Camakaris, J., Danks, D.M., Ackland, L., Cartwright, E., Borger, P., and Cotton, R.G.H., Altered copper metabolism in cultured cells from human Menkes' syndrome and mottled mouse mutant. *Biochem. Genet.*, 18, 117, 1980.
75. Weiner, A.L. and Cousins, R.J., Copper accumulation and metabolism in primary monolayer cultures of rat liver parenchymal cells. *Biochim. Biophys. Acta*, 629, 113, 1980.
76. van den Berg, G.J., De Goeij, J.J.M., Bock, I., et al., Copper uptake and retention in liver parenchymal cells isolated from nutritionally copper-deficient rats. *J. Nutr.*, 121, 1228, 1991.
77. McArdle, H.J., Gross, S.M., and Danks, D.M., Uptake of copper by mouse hepatocytes. *J. Cell. Physiol.*, 136, 373, 1988.
78. McArdle, H.J., Gross, S.M., Creaser, I., Sargeson, A.M., and Danks, D.M., Effect of chelators on copper metabolism and copper pools in mouse hepatocytes. *Am. J. Pathol.*, 256, G667, 1989.
79. McArdle, H.J., Gross, S.M., Danks, D.M., and Wedd, A.G., Role of albumin's copper binding site in copper uptake by mouse hepatocytes. *Am. J. Physiol. Gastrointest. Liver Physiol.*, 258, G988, 1990.

80. Schmitt, R.C., Darwish, H.M., Cheney, J.C., and Ettinger, M.J., Copper transport kinetics by isolated rat hepatocytes. *Am. J. Physiol.*, 244, G183, 1983.
81. Darwish, H.M., Cheney, J.C., Schmitt, R.C., and Ettinger, M.J., Mobilization of copper(II) from plasma components and mechanism of hepatic copper transport. *Am. J. Physiol.*, 246, G72, 1984.
82. Darwish, H.M., Schmitt, R.C., Cheney, J.C., and Ettinger, M.J., Copper efflux kinetics from rat hepatocytes. *Am. J. Physiol.*, 246, G48, 1984.
83. Weiner, A.L. and Cousins, R.J., Differential regulation of copper and zinc metabolism in rat liver parenchymal cells in primary cultures. *Proc. Soc. Exp. Biol. Med.*, 173, 486, 1983.
84. Weiner, A.L. and Cousins, R.J., Hormonally produced changes in ceruloplasmin synthesis and secretion in primary cultured rat hepatocytes. *Biochem. J.*, 212, 297, 1983.
85. Steinbach, O.M. and Wolterbeek, H.T., Effects of copper on rat hepatoma HTC cells and primary cultured rat hepatocytes. *J. Inorg. Biochem.*, 53, 27, 1994.
86. Steinbach, O.M. and Wolterbeek, H.T., Role of cytosolic copper, metallothionein and glutathione in copper toxicity in rat hepatoma tissue culture cells. *Toxicology*, 92, 75, 1994.
87. Renault, E. and Deschatrette, J., Inductive effect of copper deficiency on the reversion of dedifferentiated rat hepatoma cells and on gene amplification. *J. Cell Sci.*, 107, 3251, 1994.
88. Steinbach, O.M. and Wolterbeek, H.T., Determination of zinc-65, copper-64 and sulphur-35 labelled rat hepatoma tissue culture metallothioneins by high-performance liquid chromatography with on-line radioactivity detection. *J. Chromatog.*, 619, 199, 1993.
89. Endo, T. and Shaikh, Z.A., Cadmium uptake by primary cultures of rat renal cortical epithelial cells: influence of cell density and other metal ions. *Toxicol. Appl. Pharmacol.*, 121, 203, 1993.
90. Rabinovitz, M. and Fisher, J.M., Evidence for a copper: S-(methylthio)-L-homocysteine complex in L1210 murine leukemia cells. *Mol. Pharmacol.*, 34, 401, 1988.
91. Oblender, M. and Carpentieri, U., Control of the growth of leukemic cells (L1210) through manipulation of trace metals. *Anticancer Res.*, 11, 1561, 1991.
92. Anderson, W.L. and Tomasi, T.B., Suppression of lymphocyte proliferation by copper-albumin chelates. *J. Biol. Chem.*, 259, 7602, 1984.
93. Hopkins, R.G. and Failla, M.L., Chronic intake of a marginally low copper diet impairs *in vitro* activities of lymphocytes and neutrophils from male rats despite minimal impact on conventional indicators of copper status. *J. Nutr.*, 125, 2658, 1995.
94. Zahs, K.R., Bigornia, V., and Deschepper, C.F., Characterization of "plasma proteins" secreted by cultured rat macroglial cells. *Glia*, 7, 121, 1993.
95. Babu, U. and Failla, M.L., Respiratory burst and candidacidal activity of peritoneal macrophages are impaired in copper-deficient rats. *J. Nutr.*, 120, 1692, 1990.
96. Babu, U. and Failla, M.L., Copper status and function of neutrophils are reversibly depressed in marginally and severely copper-deficient rats. *J. Nutr.*, 120, 1700, 1990.
97. Ma, J.Y. and Narahashi, T., Differential modulations of GABAa receptor-channel complex by polyvalent cations in rat dorsal root ganglion neurons. *Brain Res.*, 607, 222, 1993.
98. Raps, S.P., Lai, J.C.K., Hertz, L., and Cooper, A.J.L., Glutathione is present in high concentrations in cultured astrocytes but not cultured neurons. *Brain Res.*, 493, 398, 1989.
99. Gabridge, M.G., Dougherty, E.P., Gladd, M.F., and Meccoli, R.A., Effects of heavy metals on structure, function, and metabolism of ciliated respiratory epithelium *in vitro*. *In Vitro*, 18, 1023, 1982.
100. Bala, S. and Failla, M.L., Copper repletion restores the number and function of CD4 cells in copper-deficient rats. *J. Nutr.*, 123, 991, 1993.
101. Flynn, A. and Yen, B.R., Mineral deficiency effects on the generation of cytotoxic T-cells and T-helper cell factors *in vitro*. *J. Biol. Chem.*, 907, 1980.

ZINC

Chapter 12

ZINC DIETS: DEFICIENCY AND EXCESS

Myna Panemangalore
Frederick N. Bebe
Nutrition Research Program
Kentucky State University
Frankfort, Kentucky

CONTENTS

0-8493-9611-5/97/$0.00+$.50

I. INTRODUCTION

The nutritional essentiality of zinc is based on its biochemical functions which in turn are related to the chemical properties of the element. The more than 200 zinc metalloenzymes have a unique catalytic role in metabolism; their high stability ensures their catalytic function.[1,2] Among the zinc binding proteins, metallothionein is a primary intracellular cytosolic zinc binding protein which can bind seven moles of zinc per mole of protein because of its unique amino acid sequence; the 20 cysteine molecules form two zinc clusters. Metallothionein has important functions in the localization of zinc in tissues, as a major intracellular zinc binding protein, in the intestinal absorption of zinc, in the binding of copper, and in the detoxification of nonessential and toxic element cadmium.[3] The other critical and important functions of zinc and zinc finger proteins involved in gene expression have been summarized in several review papers and books.[4–7]

Normal absorption of zinc is essential for the maintenance of zinc status and is an important consideration when formulating zinc diets and designing zinc experiments. Zinc uptake by high molecular weight proteins in the intestinal mucosa is an active process requiring ATP, and the amount of available zinc for mucosal uptake depends on the relative zinc binding affinity of the zinc binding ligands in the intestinal lumen; zinc is transferred by a membrane associated carrier.[8] Several factors are known to influence net zinc absorption in the intestines; these include zinc status, zinc binding ligands which reduce zinc absorption (plant proteins [soy], phytates, fiber, EDTA used while processing vegetables, and alcohol ingestion), and high levels of calcium or iron intake. The inhibitory effects of the ligands may be due to the formation of zinc-ligand complexes that reduce the binding of zinc to the mucosal receptor.[8] Zinc absorption is enhanced by histidine and other amino acids of endogenous origin, citric acid, immunoglobulins, and lactoferrins.[5]

Another important aspect of zinc metabolism is the interaction of other essential elements with zinc at the site of absorption.[9] Low dietary zinc can alter the metabolism of other elements and modify responses to toxic elements. Trace elements in the diet with physicochemical properties similar to zinc may compete for common pathways of absorption. The inhibitory effect of iron on zinc absorption depends on the amounts of zinc and iron in the diet, although it is clear that high dietary or supplemental iron can reduce zinc absorption. High levels of copper may not interfere with zinc absorption, but can disturb zinc metabolism. However, interactions between zinc and other essential elements are significant and can reduce zinc absorption when zinc in the diet is deficient or marginal and those of other elements are high.[10,11] The systemic metabolism of zinc is largely influenced by the interaction between zinc and a variety of biological molecules such as proteins, hormones, and zenobiotics.[12] Zinc absorption has been found to increase in zinc deficient and pregnant rats and decline in rats fed excess zinc when other essential elements in the diet are at normal levels.[13]

The assessment of zinc status is difficult because of the lack of specific and sensitive biochemical changes that respond to zinc deficiency or lowered zinc intake. Moreover, animals adapt to dietary zinc changes efficiently by maintaining homeostasis using physiological and biochemical means. The feeding of zinc deficient diet decreases growth in young animals which enables them to maintain whole body zinc concentrations or zinc homeostasis.[1] The progression of steps in the development of both mild and severe zinc deficiency are (1) reduced growth or excretion of zinc; (2) avid tissue zinc conservation; in mild zinc deficiency zinc homeostasis is reestablished; (3) mobilization of zinc from the exchangeable zinc pool (in severe zinc deficiency); (4) general tissue dysfunction.[14] The singular feature of moderate or severe zinc deficiency is that there is no significant reduction in tissue concentrations of zinc except in the bone, though during prolonged zinc depletion there is a small decline in liver and pancreas zinc.[15] In adult animals or humans, zinc excretion declines and adjusts to low or deficient zinc intake.[14] These effects of zinc deficiency are an incongruity that must be considered by researchers who study zinc deficiency or marginal zinc status since overt signs of deficiency are not comparable with tissue zinc levels or zinc related biochemical changes, though plasma zinc concentrations decline. The manipulation of zinc or other trace element levels in the diet can produce different types of deficiencies which have been elaborated and classified by Golden:[14] Type 1 is characterized by "a reduction in tissue concentrations of the nutrient and a defect in one of the metabolic pathways," while Type 2 is characterized by "a primary diminution or cessation of growth without reduction in tissue concentration of the limiting nutrient and is not associated with diagnostically specific signs and symptoms of deficiency." The deficiency of zinc falls into the latter category along with other elements such as magnesium, phosphorus, and sulfur. The anorexia or reduced food intake and cyclical intake induced by zinc deficient diets in rats could be due to the impaired function of neuropeptide Y, a known stimulator of appetite and food consumption.[16,17] Thus, either pair-feeding of a normal zinc diet or some other method is essential to differentiate the effects of zinc deficiency from partial inanition (starvation).

In contrast, high levels of zinc in the diet do not modify food intake or produce weight gain or loss. Though zinc is considered relatively nontoxic, overt symptoms of toxicity such as nausea and fatigue have been found in humans consuming very high levels of zinc. The severity of high zinc intakes in animals is related to the excess level of dietary zinc, the duration of exposure, animal age, sex, species, and diet composition. The most significant effects of excess dietary zinc are changes in tissue trace element levels: zinc concentrations increase in the liver, intestines, pancreas, kidney, bone, testes, and heart; copper (even when zinc intake is slightly higher than RDA) and iron concentrations decline in the liver and kidney, while manganese declines in the liver; there is increased excretion of calcium.[18] Also, high zinc intake can impair trace element homeostasis, immune function, and lipoprotein profiles.[19,20] High levels of zinc in the diet induce intestinal metallothionein synthesis to bind the excess zinc for a short period, but on long term feeding adaptive regulatory responses reduce metallothionein synthesis and body zinc levels to normal.[21] Although the adaptive responses to a chronic high zinc intake have not been identified, they merit consideration when studying trace element interaction in long term experiments. The toxicity of very high, pharmacological levels of zinc can cause a decline in food consumption followed by reduced growth or weight loss.[18]

Thus, it is apparent that both amounts and proportions of essential trace and macro elements as well as other dietary constituents are critical when designing diets to study functions, metabolism, interaction, or effects of zinc in experimental animals. Herein, we discuss the major points to consider when designing and conducting zinc experiments in rodents.

II. ANIMAL EXPERIMENTATION

In general, rats are the most commonly used experimental rodent element for studying trace metal nutrition. The advantages of using rats or mice over other species of rodents are (1) the availability of a large number of animals; (2) the ease of handling; (3) their versatility; (4) adaptability to different caging systems; (5) the smaller amount of space needed to house more animals; (6) the relatively short life cycle is suitable for life span, aging, and generational studies.

A. Animals

Albino rats, and to a lesser extent mice, rabbits, hamsters, and guinea pigs, have been used in zinc nutritional studies. Information on different species of experimental animals is detailed in the CRC Handbook of Toxicology.[22] Zinc studies have been conducted primarily in rats and to some extent in mice and rabbits. Mice are suitable for aging and immunological studies. Some characteristics of gastrointestinal zinc absorption during aging, whole body zinc retention in genetically obese mice at various ages,[23] whole body half life of zinc in male and female C57BL/6J mice, and plasma zinc levels during aging in CBA/J mice[24,25] are available for basing newer research. Since rats are the most frequently used animals in zinc research, extensive and detailed data are available on most aspects of zinc nutrition and metabolism. Sprague Dawley rats are not suitable for aging studies as they continue to gain weight almost throughout their life span. Wistar is another strain of rats that is used in nutritional research, although they are not as common as the Sprague Dawley rats. Another species of rats, Fischer 344, are used in genetic and aging studies because their body weight gain slows down considerably as they age in contrast to Sprague Dawley rats. Fischer 344 rats are also useful for dietary restriction studies or for experiments which seek to eliminate alterations in adipocyte cell size.[26] There are very few studies in rabbits and hamsters for practical reasons such as high cost (large number of animals are needed for nutrition studies), housing, diet, and maintenance expenses. Some studies have shown that guinea pigs or rabbits can also exhibit classical signs and symptoms of zinc deficiency.[27,28] In general, male rodents are used for zinc nutritional studies because of their faster growth rate and a relatively uniform response to nutrient or dietary changes; the response of female rodents is erratic and less uniform and thus more variable than male rodents.

B. Acclimation

Purchase a few extra experimental animals than required within a 10 g weight range. During the adjustment period rats do not respond similarly to the new environment and diet, so body weights can vary toward the end of the acclimation period. Normally, rats that arrive from animal supply companies are stressed because of limited access to water, crowding in the shipping boxes, exposure to hot or cold weather, and a disruption of the light cycles of the nocturnal animals. All experimental animals should be acclimated to the environment of the animal facility. In commercial facilities, rats are housed in groups and for most nutritional experiments, especially zinc studies, animals should be housed individually after they arrive at the animal facility. The recommended duration for acclimation is 2 to 3 days. If a longer duration of 7 days is needed, then purchase rats that are about a week younger than needed for a particular study. The major criticism against a 7–day period is that the animals will be a week older and body weights much higher when the experiment begins; using rats a week younger will overcome this criticism.

Feed rats the experimental control diet and distilled drinking water. The feeding of regular "chow" diet (Wayne blocks or similar formulations) can introduce a variation in the experiment as these diets usually contain high levels of zinc. Moreover, tap water can be contaminated with zinc, so give rats deionized or distilled water. Drinking water bottles with fresh water should be replaced at weekly intervals. If the water is contaminated excessively with diet and/or saliva, replace water bottles every 3 to 4 days. This type of problem can occur when feeding powdered diets. In most nutritional studies, rats are weighed weekly; under special experimental conditions where body weight is a critical variable, they should be weighed more frequently.

C. Housing

In general, correct and proper housing of the experimental animals is important when conducting zinc studies. Housing should facilitate animal well-being, prevent infection, and minimize variation to meet research requirements. Animal facilities should meet the standards set down by the U.S. Department of Health and Human Services (National Institute of Health, Publication # 86-23, 1985). The research requirements are summarized below:[29]

1. Adequate space, comfortable, and regulated temperature and humidity (20°C and 50 to 55% relative humidity).
2. Adequate ventilation; avoid crowding the room with too many cages of rats. House rats in individual cages.
3. Automated light switching for 12 hour light and dark cycles and adequate illumination.
4. Easily accessible feeders and watering devices.
5. Stainless steel racks; suspended polypropylene cages with stainless steel wire floors are recommended to prevent coprophagy that is common in rats given any type of diet, but more so in rats fed imbalanced diets. Raised (37.5 mm high) stainless steel wire racks can be used in polypropylene cages if wire screen floor cages are not available.
6. Fiber boards with low trace element content (Shepherd Speciality Papers, Kalamazoo, MI) is the recommended contact bedding for cages that have raised stainless steel screen floors. Rats have a tendency to chew bedding, even fiber board. Replace fiber boards every 2 to 3 days. Any bedding material can be used if polypropylene cages have stainless wire mesh floors since the animals are not in contact with the bedding.
7. Minimize noise in the animal facility.

D. Grouping

The assignment of rats/mice to various groups (different experimental diets) is important. Randomized distribution of experimental animals minimizes variation introduced by differences in body weight, age, sex, and adaptation of animals to the new environment. Random allotment of animals to experimental groups can be done on a computer using an appropriate software program. A manual method for randomized assignment of rats/mice is given below:

1. Weigh rats the morning of the day an experiment is to begin.
2. Eliminate unsuitable animals — those which have not grown normally, have not adapted to the environment or weigh much lower or higher than the rest of the group.
3. From the experimental design, ascertain the number of groups and number of rats in each group.
4. Organize the body weights in either ascending or descending order.
5. The number of groups equals the number of blocks. Divide the body weights into blocks. For example, there are 32 rat weights arranged in descending order, and there are four diet groups. So there will be eight blocks of four rats each.
6. Then, using random number tables, allot one rat in each block to one group. This should be done for each block in sequence — for the above example this will be repeated eight times.
7. Group the rats by old ID number, new sequential ID number, and weight for each rat. The average body weight in all the diet groups should be similar. Note, however, that if a large number of rats with highly varied body weights are grouped by this method, some of the groups can have average body weights that are either higher or lower. Redistribute the heavier or lighter rats between groups to achieve similar average body weights. This blocked randomization can be accounted for in the statistical design of the experiment.
8. When pair-fed groups are included in an experiment, body weight matched pairs of rats should be distributed to the zinc deficient and the corresponding pair-fed control group.
9. Distribute rats in the animal facility to the diet groups according to the randomization.
10. Renumber rats and attach labels indicating diet groups.
11. Initiate experiment — feed experimental diets.

E. Feeding

Purified or semi-purified diets are generally in powder form and are put in either glass or polypropylene food cups with a fitted stainless steel disk with 1.5 cm diameter holes and a lid with a 5 cm diameter hole in the center that allows easy access to the diet and reduces spillage. If the diet is pelleted, food pellets can be put in stainless steel slatted food holders. Since rodents have a tendency to hold small pieces of pellets in their forepaws and eat, there could be more spillage of food as the smaller pellets can fall through the screen floor. When an accurate food intake record is necessary, give a weighed amount of diet in excess of what the animal is expected to eat, and determine intake by weighing the left over diet. If there is excess spillage, collect the spilled food and add the weight to the weight of the left over diet. When food intake record is not needed, give the animals more food to ensure that they have excess food in the cup or holder.

Different types of feeding are used for rodents. ***Ad Libitum*** feeding allows the animals to regulate their own dietary intake to meet their energy requirement. This is the most common method of feeding rodents and ensures the optimum intake of all nutrients when the diet is correctly balanced and they are given free access to the food. **Restricted** feeding is used when limited food intake is necessary for regulating the growth rate or energy and nutrient intake of animals for experimental purposes. This type of feeding is not used in zinc nutritional studies, but it can be used if body weight regulation is a necessary part of the experimental design. Generally, moderate restriction of about 10% of control intake does not often reduce body weight gain, but 20% restriction can decrease body weight in young adult rats.[30,31]

Pair-feeding is essential when zinc deficient groups are included in experiments because feeding of zinc deficient diet decreases food intake, and as mentioned earlier pair-feeding of the control diet is required to differentiate the effects of zinc deficiency and lowered food intake. Sometimes certain low zinc diets may also require pair-feeding because of reduced food intake. Monitoring of food intake of the low zinc group will help ascertain the need for such feeding. Pair-feeding is done in two ways. (1) **Group pair-feeding**: determine the average daily food intake of the zinc deficient group over the previous 24 hours and feed all rats in the pair-fed group the same amount of control diet that is eaten by the zinc deficient group each day. On the very first day of the experiment the pair-fed group will eat *ad libitum*; this is unavoidable. (2) **Matched pair-feeding**: determine individual food intake of rats in the zinc deficient group and feed an equal amount of control diet to the corresponding weight-matched pair in the pair-fed group. All paired rats are fed individually in this manner daily. Group pair-feeding is less time consuming and is used more often than matched pair-feeding.

Fasting of experimental animals for different periods of time prior to termination is customarily used in nutritional or biochemical studies. Rats can tolerate 18 to 24 hours of fasting with no severe physiological stress. But the optimum duration for mice is 4 to 6 hours; longer durations from 16 to 24 hours can cause severe debility, dehydration, and even death.[22] For zinc experiments, fasting prior to termination of animals is not recommended as there is evidence that it can alter zinc metabolism. Fasting can cause the transfer of zinc from the peripheral tissues into the liver. This in turn induces the synthesis of metallothionein which binds the excess zinc. This redistribution of zinc between tissues can distort or influence the results of any experiment.[12]

F. Experimental Design

Biological variation necessitates that a large or appropriate number of animals be used to minimize variation and to obtain data that are valid.

Appropriate statistical experimental designs must be used to interpret the data and draw valid conclusions. The power of an experiment can be increased by minimizing variation, for example, by using genetically similar experimental animals and a uniform environment. It is possible to improve the precision of an experiment by making the treatment differences distinctly different. In zinc studies, 6 to 8 animals are needed per group.

Variation and confounding effects in any animal study are minimized by using (1) proper randomized distribution of rats to experimental groups; (2) a good statistical experimental design; the design for statistical analysis of data obtained from the animal experiment should be decided when planning the experiment so that all variables are included and accounted for in the design; (3) correct housing and handling of rats; and (4) ensuring sanitation and avoiding contamination of the environment with zinc. Zinc deficiency and bioavailability studies are particularly sensitive to these variables.

Standard statistical experimental designs for the analysis of variance include one way or two way multifactorial designs. When an experiment requires a large number of animals, and space is a limiting factor or a large number of rats cannot be handled at one time, an experiment can be conducted as two or four replicates or blocks. However, utmost care should be taken to minimize variation when an experiment is split into replicates by ensuring similar conditions for each replicate. Blocking and/or replication should maintain the homogeneity of the experimental groups and each block/replicate should contain all treatment groups. Factorial designs permit the measurement of more information and the inclusion of more than one independent variable at a time. The statistical design should compensate for blocking or replication of the experiment.[32]

Example: an experiment to determine zinc and copper interaction may have the following dietary groups: (1) basal zinc deficient group (<1 mg Zn/kg); (2) pair-fed group; (3) high zinc group; or (4) control zinc group. The level of copper in the diet may be: (1) low; (2) high; or (3) normal. The factorial design for the analysis of variance is four levels of factor one (zinc) and three levels of factor two (copper) or a 4 × 3 factorial design. This experimental design is given in Table 1:

TABLE 1.

Experimental Design

	Zinc in the Diet							
Cu in Diet	**Zn DEF**		**Zn H**		**Zn CON**		**Zn PFC**	
	Rp1	Rp2	Rp1	Rp2	Rp1	Rp2	Rp1	Rp2
Low								
High	8 rats/group; total of 96 rats for this experimental design							
Normal	(Rp = replication; 4 rats/group/replicate)							

If the 96 rats that will be used in this experiment cannot be handled at one time, but 48 can be managed, the experiment can be replicated twice with 48 rats, or four rats/group with all groups included in each of the two replications. If 24 is the largest number that can be accommodated, then the experiment will have to be blocked or replicated four times with two rats per group. The experiment cannot be blocked in any other way as confounding effects will be introduced in the data collected from the experiment. The data from the replicates are merged after statistical analysis and indicates no significant differences between the replicates.

G. Special Considerations/Limitations

1. The commonly used pair feeding technique may not adequately differentiate the effects of zinc deficiency from lowered food intake. Moreover, pair-feeding can lead to meal feeding in rats in long-term studies, and there are suggestions that pair-feeding is an inadequate control for experiments or research involving zinc status.[33] The other method, force feeding of low zinc or zinc deficient diets, can be used mainly to control food intake in short-term (1 week) studies. This is not a normally acceptable physiological procedure, and it does not work when diets contain very low levels of zinc; force feeding has been used in some studies.[34,35] An apparatus that can largely simulate normal feeding patterns by uniformly and continuously supplying small amounts of food to rats has been developed by Quarterman et al.[36] However, this method has not been used widely. Meal feeding by splitting the amount of food given and feeding equal amounts two/or more times a day to rats in all groups in an experiment has been suggested as a means of overcoming pair-feeding problems in long term experiments.[37] However, all the feeding techniques have weaknesses. The type of feeding used in a particular experiment will depend on the objectives, the experimental design, and the duration of the study. The researcher will have to choose an appropriate method. For the present, pair-feeding is commonly used when deficient or low zinc diets are included in an experiment.
2. Distilled drinking water in individual glass or polypropylene bottles is preferred over the more commonly used watering devices which will introduce zinc contamination. Distilled or deionized water should be given to all rats in an experiment, not just the zinc deficient groups.
3. Standard body weight and food consumption data tables are available for the life spans of experimental rodents,[22] and they are a good guide for researchers. It is interesting that body weights peak in male Sprague Dawley rats at the age of about 71 weeks and food consumption stabilizes at 51 weeks, while in female rats, body weight increases throughout their life span and food consumption declines slowly. In contrast, male Fischer 344 rats stabilize body weight at 59 weeks and food consumption at 39 weeks.

4. When zinc metabolism or balance studies are conducted for determining zinc bioavailability in experimental animals, many precautions are important. First of all, the animals should be adapted to the restrictive metabolism cages by placing them in the cages for increasing periods of time at least a week prior to the beginning of the balance study. In general, balance studies are tedious and imprecise because of contamination and the inability to make total collection of urine and feces. Very careful collection of urine and feces without contamination from spilt diet is critical. Nonetheless, balance studies are fairly good indicators of bioavailability, since a reliable and good biochemical index of zinc utilization is not available.[38]
5. There are suggestions that for long term studies it is not advisable to allow rats free access to food and that dietary restriction should be used to decrease mortality and incidence of disease such as tumor development and nephropathy.[39] However, researchers should be aware that dietary restriction can induce meal eating in rodents which habitually eat small amounts of food frequently. If a meal feeding pattern is induced then the control, as well as all the experimental groups, should also be adapted to a similar pattern. One way to overcome the problems with restricted feeding is to provide a nutritionally adequate diet with low nutrient density,[40] but extreme caution should be used in zinc studies if such nutrient restriction techniques are employed to attain food restriction as they could influence the outcome of the experiment. This method of restricted feeding is not recommended for zinc studies.

III. DIETS

A. Formulation/Preparation

The evaluation of the biochemical or physiological effects of zinc involves the development of deficiency or marginal zinc states by feeding a diet containing the lowest or graded amounts of zinc in a semi-purified or purified diet. In zinc bioavailability studies, different types of diets or sources of dietary zinc may be used, but the concentration of zinc in the diet in relation to other trace elements should be similar to that in the control diet. This is critical because of trace element interaction and the presence of chelating agents in diets.[41] The major assumption in developing zinc deficiency or marginal zinc status or in formulating diets for bioavailability studies is that the diet contains specific or the desired amount of zinc and that the environment in which the experimental animals are housed has minimal levels of zinc. It is preferable to formulate the basal zinc deficient diet and to then modify zinc concentrations by adding zinc salts to the diet at the desired level to obtain the control and experimental diets. The salt mixture that is added to purified or semi-purified diets should contain all the essential macro and trace elements, but it should be free of zinc or any other trace element of interest that is to be modified in the diet.

The nutrient requirements of laboratory animals (rats and mice) were published in 1972.[42] These have been the basis for the formulation and composition of the first AIN 1976[43] and the latest AIN 1993 diets.[44] The requirements of trace elements are enumerated in Table 2. The AIN 93 G and M formulations have in general overcome problems with the AIN 76 formula; after major alterations in carbohydrate source-starch and dextrin instead of sucrose, other changes include casein with cystine supplementation as the protein source, soybean oil as fat source at 7% (more polyunsaturated fatty acids), Solka Floc as the standardized fiber source, a modified mineral mixture as a source of elements, and a modified vitamin mixture providing higher levels of vitamin K (phylloquinone), vitamin B_{12} (cyanocobalamin), vitamin E (α-tocopherol) and choline (see Reference 44 for details).

TABLE 2.

Macro and Trace Element Levels in Rat and Mice Diet: Comparison of AIN 76 and 93 M Element Levels When Mineral Mixtures are Used at 3.5% of the Diet

Element	NAS-NRC Recommendation	AIN 76 A	AIN 93 M[a]
	mg/kg diet		
Calcium	5000.00	5200.00	5000.00
Phosphorus	4000.00	4000.00	1992.00
Sodium	500.00	1020.00	1019.00
Potassium	1800.00	3600.00	3600.00
Magnesium	400.00	500.00	507.00
Manganese	50.00	54.00	10.00
Iron	35.00	35.00	35.00
Copper	5.00	6.00	6.00
Zinc	12.00	30.00	30.00
Iodine	0.15	0.20	0.20
Selenium	0.04	0.10	0.15
Chromium	—	2.00	1.00
Chloride	500.00	1560.00	1571.00
Sulfate	—	1000.00	300.00[b]

[a] Contains ultra trace elements.
[b] Inorganic sulfur.

Data taken from references 42, 43, 44.

The differences between the NAS/NRC recommended allowances and the AIN 76 and AIN 93 element mixes are apparent upon scrutiny of Table 2. The levels of certain nutrients are higher in the AIN 93 G diets in consideration of additional demands for pregnancy, lactation, and growth in rodents[44] (not presented in Table 2). The AIN 93 M mineral mixture has lower levels of phosphorus and manganese, altered level and form of selenium, and ultra trace elements in the following amounts (mg/kg diet): molybdenum = 0.15; silicon = 5.0; fluoride = 1.0; nickel = 0.5; boron = 0.5; lithium = 0.1; and vanadium = 0.1 as compared to the AIN 76 mineral mixture. The problems with the AIN 76 mineral mixture and modifications of the AIN 1993 mineral mixture have been discussed in detail by Reeves et al.[44] We have found that in purified diets using egg white as a protein source, 30 to 35 mg/ kg diet zinc is not necessary and young rats (125 g and above) grow normally with 12-15 mg zinc/kg diet.[20] If zinc and copper or other trace element interactions are being investigated, the level of zinc in the diet will be critical, so researchers should be cautious about the concentration of zinc in the experimental diet. The higher amount of zinc, 30 to 35 mg/kg diet (AIN 93 diets), would be essential in diets containing other sources of protein such as soy protein isolate which contains phosphorus in the form of phytate, or dietary fiber at higher levels. The mineral mixtures should be prepared as a premix of the mineral salts ground to a powder and mixed with a carbohydrate source such as sucrose, glucose, or starch to get a uniform distribution of the elements in the premix and the diet. Though carbohydrate sources such as corn starch, dextrinized starch, and cellulose do not contain much zinc, they may contain other elements, as compared to simple sugars (sucrose, glucose). To avoid hardening of the mineral premix, the least hygroscopic form of an element should be used and it is preferable to store the premix in a desiccator at room temperature packed in plastic zip lock bags from which all air has been expelled to prevent the lumping of the premix.[44]

The design of certain zinc experiments may preclude the need for ultra trace elements that are added to the AIN 93 mineral mixture. If the ultra trace elements are excluded from the mineral mixture, an equal amount of powdered sucrose should be added to the mineral mix to maintain the concentration of the other elements. The essentiality of ultra trace elements has not been adequately established, but they may be needed in purified diets under certain experimental conditions when animals are fed imbalanced diets or exposed to toxic substances.[44]

B. Formulation of Zinc Deficient Diets

The various components used in the formulation of purified diets can contain residual zinc introduced during processing of the material. Of the protein sources, egg white is the best protein source for the formulations of zinc deficient diets as it is naturally low in zinc and is a protein of good biological value (contains 0.8 to 1.5 mg Zn/kg). Casein may contain higher residues of zinc (25 to 80 mg Zn/kg) and should not be used for the formulation of zinc deficient diets, but may be used for low zinc diets. The concentration of zinc in casein, though, should be factored into the calculation of the level of zinc that is to be added to the experimental diets. Plant protein isolates are not to be used because these proteins are deficient in methionine, have low biological value, and have higher zinc residues. Often certain isolated protein sources such as casein, lactalbumin, or soy proteins are treated with chelators to remove zinc for use in zinc deficient diets. When chelating agents are used to prepare low zinc protein sources, complete removal of the chelator is important.[41] Starches and dextrin, the carbohydrate sources in the diet, can also contain zinc, and the contribution of these sources to dietary zinc should be considered when adding extra zinc to the diet. Sucrose is not recommended as the only source of carbohydrate in the diet since it can lead to the development of fatty livers as simple sugars have lipogenic properties. There is some evidence that certain dietary carbohydrates may influence zinc status, but this effect has not been confirmed by other researchers.[41]

Batch and brand variation in zinc content of diet components should be considered when formulating diets. Even if the experimental diets are obtained from commercial vendors, these precautions are critical. If diets are not purchased, but are prepared in the laboratory, prior analysis of dietary components for zinc content is essential before they are used for diet preparation so that the amount of zinc contributed by each dietary constituent is considered while formulating diets.

Although zinc carbonate is the most commonly used salt in AIN diets, zinc sulfate or zinc acetate are better sources of zinc in purified diets as the bioavailability of these salts is almost 100%.[31] While zinc carbonate ($ZnCO_3$) contains 52.1% zinc, zinc sulfate ($ZnSO_4$) · 7 H_2O and zinc acetate $Zn(C_2H_3O_2)_2 \cdot 2\ H_2O$ provide 22.7 and 29.8% zinc, respectively. To 1 kg of mineral mixture, add either 37.7g of ($ZnSO_4$) · 7 H_2O or 28.78g of $Zn(C_2H_3O_2)_2 \cdot 2\ H_2O$; this will provide 30 mg Zn/kg diet when the mineral mixture is added to the diet at 35 g/kg. Add zinc sulfate or zinc acetate to the diet and not to the salt mixture after premixing the salt with a known amount of sucrose and reduce by equal amount the quantity of sucrose that will be added to the diet. We have not encountered any problems in diets where zinc sulfate has been substituted for zinc carbonate. After preparation, diets should be analyzed for zinc concentration to ensure that they contain the desired levels of zinc; if not, the diet should be modified to obtain the desired zinc levels.

Solka Floc is the recommended cellulose source for fiber as it is low in zinc. Liquid diets are not recommended for zinc studies because of different physical characteristics. But they are recommended for immunological studies.[45] If liquid diets are used, care should be taken to ensure that adequate amounts of dietary fiber are added to the liquid diets since some commercial preparations may not contain fiber.

C. Formulation of High Zinc Diets

When designing or formulating high zinc diets, the precautions needed for zinc deficient diets are not very essential, but if high and low zinc groups are included in the same experiment, it would be advisable to take similar precautions for all diets/groups. Zinc levels in the diet can be increased by the addition of zinc sulfate or acetate. The addition of large amounts of zinc salts to drinking water is not recommended because of the undesirable taste imparted by the zinc salt (noticeable at the level of 15 mg/L), which will decrease water consumption and lead to dehydration. High levels of zinc in the diet can also reduce the palatability of the diet and hence food consumption.[18] A wide range of high zinc levels, from 100 mg/kg diet to 1000 mg/kg diet or higher, have been used in high zinc studies. High zinc levels of above 10 times normal requirements may be pharmacological doses and are not advisable for nutritional studies unless the research objective is specifically to study zinc toxicity. Moreover, very high dietary zinc levels could jeopardize experiments that have been designed to evaluate zinc and trace element interaction, zinc bioavailability, or the assessment of zinc status.

D. Precautions

1. Appropriate diet nomenclature has been suggested by the AIN ad hoc committee on standards for nutritional studies: **Cereal based or unrefined diets** are formulated mainly with plant and animal materials; **Purified diets** are made with refined ingredients; **Chemically defined diets** are those prepared with chemically pure sources of carbohydrates, nitrogen, fats, minerals, and vitamins. It is suggested that this nomenclature be used consistently when reporting diet composition. Purified diets can be unpalatable so the addition of a small percentage of sucrose or glucose is recommended to improve palatability. In general, for zinc studies, purified diets are most suitable as the composition of the diet can be controlled to minimize contamination. The composition of the AIN 93 G and M diets, analysis, and testing of the diets in rats and mice are described in detail by Reeves et al.[46] Errors in calculating diet composition will seriously affect the outcome of any experiment; hence, every effort should be made to avoid mistakes during formulation and preparation. Analyzing the diet for zinc, the key trace element under discussion here, or other trace elements before the diet is used for animal experiments can minimize mistakes and grave experimental errors. A well-mixed diet with a uniform distribution of nutrients is absolutely necessary for nutritional studies.
2. As discussed earlier, egg white solids are used in the preparation of low or zinc deficient diets. Researchers should be aware that egg white solids also contain substantial amounts of sodium, potassium, and chloride. The level of sodium chloride and potassium should be reduced and that of phosphorus increased in the mineral mixture used in diets containing egg white so that the recommended allowances for these minerals are met and not exceeded. If this is not done, the diet will be imbalanced with regard to these elements. Since egg white protein is a nutritionally complete protein, L-cystine should be omitted from the diet.

3. If an experimental design includes both modified dietary zinc concentrations as well as dietary energy restriction in long or short term studies, it is important to increase the levels of components that supply essential nutrients; for example, if the diet is restricted by 20% of control or *ad lib* intake then the diets should contain 20% more of protein, fat, fiber, mineral, and vitamin mixtures, taking care to ensure that zinc concentrations are at the desired levels but 20% higher in the restricted diet. This modification of dietary components will ensure that the restricted diet will provide normal amounts of all nutrients but 20% less total energy (Kcal) when fed at a 20% lower level than the control.[30] If such adjustments are not made in the purified restricted diet, it may not provide the recommended allowances of the essential nutrients and could introduce design errors in the experiment. Very often, such adjustments to restricted diets are ignored, but should not be as a restricted diet (dietary restriction) should supply essential nutrients at the same level as the control diet in zinc studies. Some investigators suggest that dietary restriction should include restriction of all nutrients.[31]
4. It must be noted that the AIN vitamin mixtures have been designed with a minimal margin of safety relative to requirements. The pelleting of diets can result in the loss of heat labile nutrients such as certain vitamins and oxidation of fatty acids. This loss can also occur during prolonged mixing of diets as well because of heat generation. Hence, great care should be taken not to over mix diets for prolonged periods of time and not to use high temperatures when pelleting diets. Vitamin formulations containing higher levels of vitamins are available and can be used to ensure adequate vitamin supply in pelleted diets.
5. To prevent the development of rancidity and the autoxidation of fatty acids, diets should be stored at 4°C during the course of an experiment for no more than three months or at –20°C in plastic containers for long term storage. Freezing of diets is especially important and recommended when experimental protocol forbids the use of antioxidants.
6. Certain procedures such as gelling of powdered diets using agar or crumbling of powder diets with methyl cellulose are not recommended for zinc diets. Such processing may add extra zinc and contaminate zinc deficient diets. The gel form of diet needs refrigeration and can readily be infected by bacteria and molds if stored for long even at 4°C because of higher moisture content.[45]
7. If experimental protocol calls for specified pathogen free (SPF) animals, the diets may have to be sterilized by autoclaving, fumigation, or ionizing radiation.[47] But each of the methods of sterilization have problems: heat sterilization of purified diets leads to the development of Maillard reaction products which can reduce the retention of calcium, phosphorus, magnesium, and copper;[48] irradiation can enhance oxidative damage and loss of thiamin, folic acid, and pyridoxine; chemical sterilization with ethylene oxide can leave residues in the diet.[47] Therefore, the sterilization process should be carefully examined for suitability and by prior testing in animals before such processing is employed for feeding SPF animals. However, SPF rats have seldom been used in zinc studies.

8. Most dietary variables can be eliminated by using purified diets; nonetheless, care should be taken to ensure that the ingredients are not contaminated with zinc or other trace elements and the quality is consistent. This is important irrespective of whether diets are made in the laboratory and/or purchased commercially. The major ingredients should be analyzed for zinc to ensure that the zinc contribution from the dietary components is minimal or accounted and compensated for by reducing the amount of zinc supplied by the mineral mixture. This is the most critical part of preparing diets for zinc studies and bears repetition. Use stainless steel equipment when mixing diets for zinc studies to minimize contamination from such sources.

IV. ZINC DEFICIENCY AND EXCESS

Different types of zinc deficiency can be produced in rats. Below is a description of how to produce various zinc deficiencies.

A. Severe Zinc Deficiency

For the development of relatively severe zinc deficiency, rats 28 days old, weighing 80 to 100 g, should be fed zinc deficient diets (<1 mg Zn/kg diet) for 6 to 8 weeks. This duration of feeding will often produce symptoms of deficiency such as skin lesions on neck and paws, thinning of hair, and abnormal gait in addition to significant weight loss; total cessation of growth may or may not be observed in all rats. If older animals, 45 days old, weighing 160 to 180 g are used, then they should be fed zinc deficient diets for at least 10 to 12 weeks to produce zinc deficiency.[49] Rats older than 45 to 50 days of age may not develop severe zinc deficiency because of larger bone zinc stores. All precautions discussed in the previous section should be observed to minimize zinc contamination of the environment and extraneous infection and stress in the animals. Animals should be weighed weekly and fed daily; a pair-fed group should be included for comparative purposes. The technique of pair-feeding has been described above.

B. Moderate/Marginal Zinc Deficiency

This type of zinc deficiency can be produced in weanling, young, and adult rats. The marginal zinc diets containing 1.5 to 2 mg Zn/kg can be used for the development of marginal zinc status. Feed marginal zinc diet to weanling rats, 28 days old, for at least 4 to 6 weeks, young adult rats, 45 days old, for 8 to 10 weeks, and adult rats, 55 to 60 days old for 12 to 14 weeks to produce rats with low or marginal zinc status. The duration of feeding should be adjusted by the researcher to meet the requirements of an experiment. Unlike zinc deficiency, where physical symptoms can be used to determine zinc deficiency, marginal status cannot be identified by such physical symptoms. The decline in plasma zinc levels is one indicator of low zinc status. We have found that feeding of a diet containing more than 2 mg Zn/kg diet for 4 weeks does not produce a significant decrease in plasma zinc levels,[20] so longer feeding durations are needed to lower zinc status or produce marginal zinc status in rats when diets contain more than 2 mg Zn/kg diet. Long term feeding of low zinc diets can induce cyclical feeding and reduce intake, so food intake should be monitored daily and a pair-fed group included if and when intake begins to decline. Give all rats deionized water and follow the same precautions suggested for the development for zinc deficiency.

C. Excess/High Zinc Status

Any age group of rats can be used to produce high zinc status. Feed rats diets containing high zinc levels, anywhere from 100 to 400 mg Zn/kg diet for a maximum of 1 week if high intestinal, plasma, and tissue zinc levels are desired, as beyond a period of 2 weeks zinc concentrations in all tissues start to decline and are similar to control levels by the end of five weeks. This has been clearly demonstrated by Reeves[21] in a recent study; abrupt elevation of zinc in the diet enhanced intestinal metallothionein, plasma, and tissue zinc levels for 2 weeks thereafter. Within five weeks, concentrations of metallothionein in the intestines and zinc in tissues declined to normal levels indicating that an adaptive mechanism maintains zinc homeostasis in the animals when high zinc diets are fed for a long period. Therefore, the feeding duration for high zinc diets will depend on the type of study: use 1 week feeding to obtain animals that have elevated concentrations of zinc in tissues or a period longer than five weeks when animals adapted to the high zinc diet are needed. Normal precautions should be followed for the maintenance of rats. It is advisable to give rats deionized water.

V. LIMITATIONS

Most of the limitations related to zinc deficiency, excess, diet formulation, and experimentation have been discussed above in the appropriate sections. However, here we would like to emphasize some important aspects of zinc interaction and metabolism as they relate to animal experiments. Conditions such as infection or physical or inflammatory stress can cause major disturbances in systemic zinc metabolism. Under such situations, plasma zinc declines while liver zinc increases and induces the synthesis of metallothionein. These alterations are mediated by hormones, glucocorticoids, glucagon, and catecholamines, and other factors such as interleukin-I, cyclic AMP, and interferon.[12] These modifications of zinc metabolism by environmental and other extraneous factors are important aspects of animal experimentation. Therefore, it is important that animals are free of infection and that stressful conditions are minimized in zinc studies. It has been suggested that the decline in plasma retinol along with plasma zinc levels in zinc deficient animals are a secondary consequence of reduced food intake. But a retinol supplement does not restore plasma retinol levels probably because of low levels of retinol binding protein in the intestines.[50]

The determination of marginal or suboptimal zinc status using biochemical criteria is difficult and often inconclusive because of several limitations. Plasma zinc level is most frequently used to assess zinc status, but plasma zinc responds to extraneous factors, and therefore, may not be an accurate indicator of zinc status. Moreover, plasma zinc can increase during tissue catabolism and decline during growth, which can happen in growing rats fed zinc deficient diet or even after zinc supplementation that spurs growth.[51] Also, as indicated previously, plasma zinc levels can decline in stress, infection, or during starvation. Although several other biochemical indices have been proposed as indicators of zinc status such as hair, skin, erythrocyte and leucocyte zinc levels, activities of several zinc requiring enzymes, responses to dark adaptation, and taste acuity, none have proved useful for the determination of low or marginal zinc status.[52] More recently, it has been shown that erythrocyte metallothionein is responsive to acute dietary zinc suppression and varied dietary zinc levels in humans and may be used as a marker or indicator of low, marginal, or normal zinc status.[53,54] However, the use of erythrocyte metallothionein as a marker for marginal or low zinc status has not been proven in experimental animals.

VI. SUMMARY

When formulating zinc deficient or high zinc diets, it is important to ensure that the amounts and proportions of other trace elements in the diet are maintained to avoid adverse interaction with other trace elements, unless such interactions are under investigation.

Rats are the most commonly used experimental animals in zinc studies because of their versatility, availability, adaptability, ease of handling, minimal space requirements, and short life cycle. Rat species used for nutritional studies are: Sprague Dawley or to a lesser extent Wistar for growth, deficiencies, and absorption studies, and Fischer 344 for genetic, aging, and dietary restriction experiments. After purchasing rats, acclimate them to their new surroundings for 2 to 3 days but no more than 7 days. House them individually in polypropylene and stainless steel cages with stainless steel wire screen floors or with a stainless steel wire screen rack and use low trace element fiber board as cage liners to minimize contamination. Feed the rats a control diet and deionized or distilled drinking water in plastic or glass bottles throughout the duration of the experiment. Minimize stress in animals, ensure that there is no zinc or trace metal contamination, and that animal housing meets the criteria set down by the U.S. Department of Health and Human Services (National Institute of Health, Publication # 86-23, 1985). Before initiating any animal experiments, use proper randomization procedure to group rats, and decide on an appropriate experimental design that includes all variables for statistical analysis.

For zinc experiments, use purified, powdered, or pelleted diets as diet composition can be controlled and contamination minimized; powder diets are recommended. Include a pair-fed group when a zinc deficient group is a part of the experiment. Pair feeding is the most commonly used method to compensate for the decline in food intake in rats fed zinc deficient diets and is accomplished by feeding a weight matched group the control diet at the average daily intake of the zinc deficient group given deficient diet; all other groups should be given free access (*ad libitum*) to their diets. For short term zinc deficiency studies, the inclusion of a force fed group has been suggested instead of a pair-fed group, but this procedure is not common. Do not fast rats for any duration prior to termination as fasting alters zinc metabolism because of redistribution of zinc to the liver and metallothionein synthesis which will affect the results of the experiment.

When formulating zinc diets, it is preferable to modify zinc concentrations of the diet by adding zinc salts to a basal zinc deficient diet which contains the correct amounts and proportions of all other nutrients and trace elements. Base the formulation of zinc deficient diets on the AIN 93 diets. The following precautions should be taken when preparing diets:

1. The protein source should have high biological value and be naturally low in zinc; egg white (solids) is the best protein source.
2. Use starch or dextrinized starch in the diet instead of sucrose; the new AIN 93 diet formulations have an appropriate balance of starch and sucrose.
3. Analyze all dietary components for zinc content and account for the zinc contributed to the diet by these components when adding zinc salts to modify zinc concentrations. Also analyze prepared diets before starting an animal experiment to ensure that zinc levels are within the desired range.
4. Zinc sulfate ($ZnSO_4 \cdot 7\ H_2O$) or zinc acetate ($Zn(C_2H_3O_2) \cdot 2H_2O$) are the preferred sources of zinc in the diet because of their higher bioavailability than the commonly used zinc carbonate and are added directly to the diet and not to the salt mixture.
5. Stringent precautions are needed to prevent contamination of diets during preparation; avoid high temperatures while pelleting diets and prolonged mixing as these will result in the loss of certain heat labile nutrients.

6. When preparing excess or high zinc diets, do not use exceedingly high levels as it will reduce the palatability of the diets.
7. Store diets at low temperatures (4°C or –20°C) to prevent the development of rancidity.

Severe zinc deficiency can be produced in weanling rats (28 to 30 days old) by feeding them zinc deficient diets for 6 to 8 weeks or in young adult rats (45 days old) by feeding them zinc deficient diets for 10 to 12 weeks. The physical symptoms of zinc deficiency are weight loss or cessation of growth (especially in weanling rats), hair loss, and lesions in the paws and skin around the neck. Plasma zinc levels also decline in zinc deficiency. To develop high zinc status, feed rats diets containing 100 to 400 mg Zn/kg diet for one week so that rat tissues will contain the highest amount of zinc and no more than 2 weeks because adaptive homeostatic mechanisms reduce tissue zinc to control levels by five weeks.

Some of the important aspects of zinc metabolism as they relate to animal experiments include:

1. Infection or inflammatory or physical stress can result in major disturbances in systemic zinc metabolism which are mediated by glucocorticoid hormone, glucagon, interleukin 1, cyclic AMP and interferon.
2. Plasma retinol also declines with plasma zinc levels in zinc deficiency and it is not amenable to retinol supplementation.
3. Plasma zinc levels can also decline during growth and increase during tissue catabolism.
4. Several biochemical indices have been proposed as indicators of zinc status and include plasma and blood cell zinc levels, activities of zinc requiring enzymes and taste acuity.

ACKNOWLEDGMENT

The authors sincerely thank Dr. Ronald J. Rose of Harlan Teklad, Madison, WI for providing important information on diet formulation and precautions.

REFERENCES

1. Kellin, D. and Mann, T., Carbonic Anhydrase. Purification and nature of the enzyme. *Biochem. J.*, 34, 1163, 1940.
2. Vallee, B.L. and Auld, D.S., Zinc coordination, function, and structure of zinc enzymes and other proteins. *Biochemistry*, 29, 5647, 1990.

3. Bremner, I. and Beattie, J.H., Metallothionein and trace minerals. *Ann. Rev. Nutr.*, 10, 63, 1990.
4. Berg, J.W., Zinc finger domains: hypotheses and current knowledge. *Ann. Rev. Biophys. Biochem.*, 19, 405, 1990.
5. Walsh, C.T., Sandstead, H.H., Prasad, A.S., Newberne, P.M., and Fraker, P.J., Zinc: health effects and research priorities for the 1990's. *Environ. Hlth. Perspect.*, 102 (Supplement 2) 5, 1994.
6. Keen, C.L., Clegg, M.S., and Hurley, L.S., The role of zinc in prenatal and postnatal development. In: Prasad, A.S., Ed., *Essential and Toxic Trace Elements in Nutrition*, New York: Alan R. Liss. 1988, 203.
7. Sandstead, H.H., Wallwork, J.C., Halas, E.S., Tucker, D.M., Overgsten, C.L., and Strobel, D.A., Zinc and the central nervous functions. In: Sarkar, Ed., *Biological Aspects of Metals and Metal Related Diseases*, New York: Raven Press, 1983, 225.
8. Wapnir, R.A., Protein nutrition and mineral absorption. CRC Press, Boca Raton, FL, 1990, 131.
9. Sandstrom, B. and Lonnerdal, B., Promoters and antagonists of zinc absorption. In: Mills, C.F., Ed., *Zinc in Human Biology.* London: Springer Verlag, 1989, 57.
10. Panemangalore, M., Interaction among zinc, copper and cadmium in rats: Effects of low zinc and copper diets. *J. Trace Elem. Exp. Med.*, 6, 125, 1993.
11. Reeves, P.G., Rossow, K.L., and Bobilya, D.J., Zinc induced metallothionein in the intestinal mucosa, liver, and kidney of rats. *Nutr. Res.*, 14, 897, 1994.
12. Bremner, I. and May, P.M., Systemic interactions of zinc. In: Mills, C.F., Ed., *Zinc in Human Biology.* London: Springer Verlag; 1989, 95.
13. Flanagan, P.R., Haist, J., and Valberg, L.S., Zinc absorption, intraluminal zinc and intestinal metallothionein in zinc deficient and replete rats. *J. Nutr.*, 113, 962, 1983.
14. King, J.C., Assessment of zinc status. *J. Nutr.*, 120, 1474, 1990.
15. Golden, M.H.N., The diagnosis of zinc deficiency. In: Mills, C.F., Ed., *Zinc in Human Biology,* London: Springer Verlag, 1989, 323.
16. Shay, N.F., Beverly, J.L., Rains, T.M., Jr., Paul, G.L., and Randall, A.C., Central administration of neuropeptide Y restores intake to normal in zinc deficient rats. *FASEB J.*, 9, A867, 1995.
17. Rains, T.M. and Shay, N.F., Zinc status specifically changes preference for carbohydrate and protein in rats selecting from separate carbohydrate-protein-and fat-containing diets. *J. Nutr.*, 125, 2874, 1995.
18. Fox, M.R.S., Zinc Excess. In: Mills, C.F., Ed., *Zinc in Human Biology.* London: Springer Verlag, 1989, 366.
19. Fosmire, G., Zinc toxicity. *Am. J. Clin. Nutr.*, 51, 225, 1990.
20. Bebe, F.N. and Panemangalore, M., Modulation of tissue trace metal concentrations in weanling rats fed different levels of zinc and exposed to oral lead and cadmium. *Nutr. Res.* (in press).
21. Reeves, P.G., Adaptation responses in rats to long-term feeding of high zinc diets: emphasison metallothionein. *J. Nutr. Biochem.*, 6, 48, 1995.
22. Derelanko, M. and Mollinger, M.A., Eds. *CRC Handbook of Toxicology.* CRC Press, Boca Raton, FL, 1995.
23. Kennedy, M.L. and Failla, M.L., Zinc metabolism in genetically obese (ob/ob) mice. *J. Nutr.*, 117, 886, 1987.
24. Reis, S.L., Keen, C.L., Lonnerdal, B., and Hurley, L.S., Mineral composition and zinc metabolism in female mice of varying age and reproductive status. *J. Nutr.*, 118, 349, 1988.
25. Woodward, W.D., Filveau, S.M., and Allen, O.B., Decline in serum zinc levels throughout the life span of the laboratory mouse. *J. Gerontol.*, 39, 521, 1984.
26. Krause, B.R. and Hartman, A.D., Accumulation of adipocyte cholesterol during hypolipidemic drug treatment in cholesterol fed rats. *Biochem. Biophys. Acta*, 713, 485, 1982.

27. Bentley, P.J. and Grubb, B.R., Effect of zinc deficient diet on tissue zinc concentrations in rabbits. *J. Anim. Sci.*, 69, 4876, 1991.
28. Quarterman, J. and Humphries, W.T., The production of zinc deficiency in the guinea pig. *J. Comp. Pathol.*, 93, 261, 1983.
29. U.S. Department of Health and Human Services. Guide for the Care and Use of Laboratory Animals. 1985, NIH Publication No. 86-23.
30. Panemangalore, M., Lee, C.J., and Wilson, K., Comparative effects of dietary energy restriction in young adult and aged rats on body weight, adipose mass, and lipid metabolism. *Nutr. Res.*, 6, 981, 1986.
31. Hart, R.W., Neuman, D.A., and Robertson, R.T., Eds. *Dietary Restriction: Implications for Design and Interpretation of Toxicity and Carcinogenicity.* Washington DC: ILSI Press, 1995.
32. Steel, R.G.D. and Torrie, J.H., *Principles and Procedures of Statistics*. McGraw-Hill, New York, NY, 1980.
33. O'Dell, B.L. and Reeves, P.G., Zinc status and food intake. In: Zinc in Human Biology, Mills, C.F., Ed., Springer Verlag, London, 1989, 173.
34. Flanagan, P.R., A model to produce pure zinc deficiency in rats and its use to demonstrate that dietary phytate increases the excretion of endogenous zinc. *J. Nutr.*, 113, 447, 1983.
35. Park, J.H.Y., Grandjean, C.J., Hart, M.H., Erdman, S.H., Pom, P., and Vanderhoof, J.A., Effect of pure zinc deficiency on glucose tolerance and glucagon levels. *Am. J. Physiol.*, 251, E273, 1986.
36. Quarterman, J., Williams, R.B., and Humphries, W.R., An apparatus for the regulation of food supply to rats. *Br. J. Nutr.*, 24, 1049, 1970.
37. Reeves, P.G. and O'Dell, B.L., The effect of zinc deficiency on glucose metabolism in meal fed rats. *Br. J. Nutr.*, 49, 441, 1983.
38. Baker, D.H. and Ammerman, C.B., Zinc bioavailability. In: Ammerman, C., Baker, D.H., and Lewis, A.J., Eds. *Bioavailability of Nutrients for Animals*. Academic Press, San Diego, 1995, 367.
39. Masoro, E.J., Food restriction in rodents: an evaluation of its role in the study of aging. *J. Gerontol. (Biol. Sci.)*, 43, 859, 1988.
40. Coates, M.E., Nutritional considerations in the production of rodents for aging studies. *Neurobiol. Aging*, 12, 672, 1991.
41. Smith, J.C., Methods in trace element research. In: Mertz, W., Ed. *Trace Elements in Human Nutrition*, Vol 1. Academic Press, San Diego, 1987, 21.
42. National Academy of Sciences, Nutrient requirements of laboratory animals. Washington, DC, NAS-NRC: Number 10. 1972.
43. Ad hoc Committee Report of the American Institute of Nutrition Ad Hoc Committee on Standards. *J. Nutr.*, 107, 1340, 1977.
44. Reeves, P.G., Nielsen, F.H., and Fahey, Jr., G.C., AIN 93 purified diets for laboratory rodents: final report of the American Institute of Nutrition ad hoc writing committee on reformulation of AIN 76 rodent diet. *J. Nutr.*, 123, 1939, 1993.
45. Coates, M.E., Feeding and watering. In: Tuffery, A.A., Ed. *Laboratory Animals*. Chichester: John Wiley, 1995, 107.
46. Reeves, P.G., Rossow, K.L., and Lindauf, J., Development and testing of AIN purified diets for rodents: results on growth, kidney calcification and bone mineralization in rats and mice. *J. Nutr.*, 123, 1923, 1993.
47. Coates, M.E., Diets for germfree animals. Part 1. Sterilization of diets. In: Coates, M.E. and Gustaffson, B.E., Eds. *The Germfree Animal in Biomedical Research*. London: Laboratory Animals Ltd. 1984, 85.
48. Andrieux, C., Sacquet, E., and Gueguer, L., Interaction between Maillard's reaction products, microflora of the digestive tract and mineral metabolism. *Reprod. Nutr. Develop.*, 20, 1061, 1980.
49. Panemangalore, M. and Brady, F.O., The influence of zinc status on the levels metallothionein in isolated perfused rat liver. *J. Nutr.*, 109, 1825, 1979.

50. Smith, J.C., The vitamin A-zinc connection. A review. *Ann. N.Y. Acad. Sci.*, 355, 62, 1980.
51. Prasad, A.S., Clinical manifestations of zinc deficiency. *Ann. Rev. Nutr.*, 5, 341, 1985.
52. Ruz, M., Cavan, K., Bettger, W.J., Thompson, L., Berry, M., and Gibson, R.S., Development of a dietary model for the study of zinc deficiency in humans and evaluation of some biochemical and functional indices of zinc status. *Am. J. Clin. Nutr.*, 53, 1292, 1991.
53. Grider, A., Bailey, L.B., and Cousins, R.J., Erythrocyte metallothionein as an index of zinc status in humans. *Proc. Natl. Acad. Sci.*, 86, 1259, 1990.
54. Thomas, E.A., Bailey, L.B., Kauwell, G.A., Lee, D-Y., and Cousins, R.J., Erythrocyte metallothionein response to dietary zinc in humans. *J. Nutr.*, 122, 2408, 1992.

Chapter 13

ZINC RADIOTRACER IN THE STUDY OF THE MECHANISM OF ZINC HOMEOSTASIS

Donald Oberleas
Department of Food and Nutrition
Texas Tech University
Lubbock, Texas

CONTENTS

0-8493-9611-5/97/$0.00+$.50

I. HISTORY

Zinc deficiency was recognized as a practical problem in domestic animals in the early 1950s. This was associated with isolation, purification, identification, and industrial synthesis of vitamin B_{12}. Several events in agricultural evolution played a significant role in producing a nutritional zinc deficiency syndrome described as "parakeratosis." This deficiency syndrome, characterized by decreased growth rate, anorexia, hyperkeratosis, and parakeratosis, had escaped detection as a practical problem for many centuries.

The presence of phytate (*myo*-inositol 1,2,3,5/4,6-hexakis [dihydrogen phosphate]) as a storage form of phosphate in plant seeds, roots, and tubers has been known for more than a century. It had been studied extensively in relation to calcium and phosphate metabolism with some interest in iron and magnesium. It was not until the late 1950s, when scientists began to notice that zinc was less available from diets containing plant proteins than from those containing animal proteins, that phytate became of interest.

II. RADIOACTIVE ZINC TRACERS

Radioactive tracers are identical in all respects to normal matter, except for one property: isotopic tracer atoms have a tag or label that makes them instantly recognizable among a population of normal atoms. The label is radioactivity. This tag enables detection and thus reveals the physiological or biochemical reactions. The beginning of tracer methods date to the early 1920s when Hevesy and coworkers used naturally occurring isotopes (lead, bismuth, and thorium) and studied their distribution and metabolism in plants and animals. In the early 1930s, study of the distribution of artificially labeled phosphate in animals was begun. This was the true beginning of the use of radiotracers experimentally in animals.[1]

Radioactivity is released as the nucleus becomes transformed to a stable configuration. The random nature of decay and the rapidity of disintegration determines the usefulness of an isotope for biological work. Isotopes with long half-lives are essential. With zinc, ten unstable isotopes are recognized. These range from ^{61}zinc with a half-life of 88.1 seconds to ^{65}zinc at 243.6 days. The latter emits a strong gamma ray with 1.115 Mev of energy and a positron with energy of 0.325 Mev. Only ^{65}Zn has the ideal characteristics for biological research.

III. THE ESSENTIALITY OF ZINC

The practical significance of zinc was not realized until peanut meal was used as the protein source for swine diets and showed that supplemental zinc was required to prevent or cure parakeratosis.[2] Since these diets contained 34 to 44 mg/kg of native zinc, this condition could not be considered a simple case of inadequate dietary zinc.

The first demonstration that zinc was less available to the chick from plant protein sources than from animal protein sources was confirmed when phytate was added to casein-gelatin-based diets and showed decreased bioavailability of zinc to an extent similar to that when isolated soybean protein was fed.[3,4,5] The effect of phytate was later confirmed in swine.[6] Rats were first utilized as a model animal for the phytate/zinc phenomenon in an extensive series of studies.[7,8,9]

IV. DESCRIPTION OF THE DIETS

Several dietary models are available for the study of zinc deficiency. Typical diets are shown in Table 1 as no single diet should be considered absolute in these types of studies. A variety of dietary variables may be incorporated. The most practical model utilizes isolated soybean protein or other oil-extracted legume seed proteins as the protein source. These protein sources have variable but quantifiable amounts of phytate. A dietary protein content of 200 g/kg of isolated soybean protein or the equivalent of 18% protein provides optimal growth.

Other dietary ingredients include a carbohydrate source such as glucose monohydrate. Combinations of corn starch with sucrose may also be used. The water-soluble vitamin premix and zinc premix are prepared in glucose monohydrate and if used will contribute some carbohydrate. The vegetable oil may be soybean, corn, cottonseed, or other suitable vegetable oil. The vegetable oil serves two purposes: (1) as a source of energy and essential fatty acids and (2) the diet tends to be very dusty during mixing which is controlled by the vegetable oil. About 50 g of vegetable oil per kg diet (part from the fat-soluble vitamin premix) will control the dust; beyond 100 g/kg the diet becomes oily and more difficult to handle. This diet, with zinc supplementation at 20 g zinc premix/kg diet, regardless of dietary protein, should provide at least a 45 g/week growth rate for 4 weeks starting with a 70 g male rat.

In rodents, as in other species, the severity of zinc deficiency is relative to dietary factors other than zinc, namely, phytate and calcium.[10] Calcium is frequently a variable and may be added to achieve the calcium level desired. A level of 16 g/kg of dietary calcium maximizes the phytate effect in each diet but is not an essential factor in studies with phytate and zinc.

V. EXPRESSION OF ZINC ADEQUACY AS MOLAR RATIOS

Zinc deficiency has an early and remarkable effect on growth. Thus, with an otherwise balanced diet, growth rate may be used as a sole

TABLE 1.

Typical Diet Models for the Generation of Zinc Deficiency in Rats

Ingredient	g/kg
Glucose monohydrate	600–650
Vegetable oil	40–90
Protein (isolated soybean protein or equivalent of extracted legume protein or casein as needed)	200–400
Cellulose[a]	30–50
Mineral premix	35–50
Fat-soluble vitamin premix	10
Water-soluble vitamin premix	10
Calcium carbonate[a]	20
Phytic acid or Sodium phytate[a]	10
Zinc premix[a]	10–20

[a] When used, added at the expense of glucose. Mineral premix contained (g/kg): calcium carbonate, 214; dicalcium phosphate · $2H_2O$, 333; magnesium carbonate, 20; magnesium sulfate · $7H_2O$, 24; sodium chloride, 102; potassium chloride, 17; ferric phosphate (soluble), 16; monobasic potassium phosphate, 252; manganese sulfate · H_2O, 15; cupric sulfate · $5H_2O$, 1.3; aluminum potassium sulfate · $12H_2$, 0.19; potassium iodide 0.5; cobaltous chloride · $6H_2O$, 0.05; sodium fluoride, 0.8. Fat-soluble vitamin premix (g/kg): vitamin A acetate (500,000 IU/g), 4.0; vitamin D (150,000 ICU/g), 1.9; menadione, 1.0; alpha tocopherol, 3.0; antioxidant, 10.0; vegetable oil, 980.1. Water-soluble vitamin premix (g/kg): thiamine HCl, 1.6; riboflavin, 1.6; niacin, 8.0; pyridoxine HCl, 1.6; calcium pantothenate, 4.0; biotin, 0.02; folic acid, 0.5; cyanocobalamin, 0.005; glucose 990.675. Methionine was added at 1 or 3 g/kg diet. Choline (choline chloride) added at 1 g/kg diet. Zinc premix (g/kg): zinc carbonate, 10; glucose, 990.

objective criterion for evaluation of zinc status and correlates with phytate to zinc molar ratio. Optimal growth rate for young rats has been achieved by dietary phytate to zinc molar ratio between 3 and 6. At a lower level of dietary calcium, a larger ratio has since been shown to provide adequate zinc.[11]

It was clear very early that the phytate complexed zinc in the gastrointestinal tract and decreased zinc bioavailability.[7] Beyond the minimum dietary zinc content of about 6 mg/kg, it is only the concentration of zinc relative to the concentration of phytate that is important. Complementary to rat studies, *in vitro* experiments have indicated that zinc phytate was less soluble than calcium phytate. The calcium to zinc to phytate complex shows the calcium to phytate interaction as a kinetic synergism.[7,8,9] Later *in vitro* studies further showed other minerals including magnesium, calcium, copper, zinc, and lead complex inversely to their atomic mass. In this work, calcium was shown to be a coincidental tertiary element. Any of the elements studied when included with any other element in solution with phytate provided the same kinetic synergism.[12] The complexations described occurred maximally at about pH 6 which is the approximate pH of the duodenum and upper jejunum where zinc and calcium are preferentially absorbed.

In vivo and *in vitro* evidence has indicated that a chemical complexation between phytate and minerals should occur in the small intestine. By converting mass units to molar equivalents, a molar ratio of phytate to zinc [phytate to zinc] or [calcium][phytate] to [zinc] represents the best means of expressing the net effect of phytate.[10,13] In neither expression has a precise critical molar ratio relative to the dietary adequacy been determined.

VI. MECHANISM OF PHYTATE EFFECT

The general conclusion has been that dietary phytate complexes dietary zinc within the small intestine and prevents the absorption of dietary zinc resulting in zinc deficiency. However, it has been well documented that endogenous zinc is recycled via the pancreas, up to 3 to 4 times the amount of zinc consumed during the period of an average day.[14,15,16] The gastrointestinal tract is the normal excretory route for zinc.

Based on the variability of zinc depletion described above, it was shown that regardless of the phytate to zinc molar ratio, approximately 45% of dietary zinc was absorbed by rats.[17] It was clear that more than an effect on dietary zinc was involved in the phytate effect on zinc metabolism. Though the presence of endogenous zinc was well established, little was known about the nature or vulnerability of the pancreatic zinc secretion toward phytate complexation. It was obvious that the magnitude of the endogenous zinc secretion required an efficient mechanism of reabsorption to maintain homeostasis. If endogenous zinc were complexed by phytate, then a larger pool of zinc was available for complexation than was reflected by the dietary zinc content.

A rat model has been developed that allows rapid equilibrium labeling of the endogenous zinc pools. This model also allows the study of the effect of phytate on the endogenous zinc pools. In this model, young rats were depleted with a phytate containing, low zinc diet for 1 to 4 weeks. Without depletion, the pancreatic cells will not be uniformly labeled and no differential in fecal zinc can be demonstrated. The pre-collection depletion allowed sufficient depletion of the pancreatic acinar cells that allow an equilibrium within the body, particularly within the pancreas. With an equilibrium within the pancreas, rats are divided between phytate-containing and phytate-free diets and placed in metabolism cages. The rats are injected with 10 μCi ^{65}zinc intraperitoneally or intravenously at this stage. There seems to be little difference in the equilibrium process between the mode of radioactive zinc administration. The ^{65}zinc in feces collected over the following several days and analyzed will clearly demonstrate the phytate effect on the endogenous zinc secreted via the pancreas.[18] It is essential that the collections be made at nearly the same time every day for the extent of the collection period. It is also essential that the collection period be at least 10 to 14 days. The first 1 to 3 days of collection may be artifactual because homeostasis requires 7 to 10 days. The pancreas contains two distinct zinc pools. The zinc which is incorporated into carboxypeptidase is tightly bound and therefore not vulnerable to complexation by phytate. Therefore, the zinc affected by phytate represents a second loosely complexed, labile pool of zinc. This can be demonstrated only with the most severe depletion of the pancreas prior to the injection of radioactive zinc.

The absorption/reabsorption of the combined dietary and endogenous pancreatic pools of zinc is essential for the maintenance of zinc homeostasis. Under these circumstances, the dietary intake is necessary for replacement of obligatory excretory losses. A mathematical model can be constructed from the above evidence. Figure 1 shows a theoretical model based on the experimental evidence.[18] This mathematical model, based on relative relationships, assumes that the zinc ingested from the diet is taken as unity, with a value of "1". Pancreatic zinc secretion is variable throughout a day, but is equivalent over a 24 hour period to 2 to 4 times the size of the dietary contribution. The duodenal pool representing the combination of the dietary and pancreatic pools becomes the critical mass of zinc available for absorption or complexation by phytate and excretion. Fractional absorption of zinc from the dietary pool has been shown to be from 10 to 50%.[19] If this same absorption factor is applied to the duodenal pool, a considerably larger amount of zinc is available to sustain homeostasis. An initial estimate indicates that the zinc contained in stable complexes such as carboxypeptidases, not affected by phytate, represents 20 to 30% of the pancreatic pool and is always excreted. This allows the reabsorption of 70 to 80% of the duodenal pool which is necessary to sustain homeostasis.[20]

To apply the same mathematical model to a chemically zinc deficient, nonphytate containing diet, one must assume that the dietary pool is considerably smaller than 1, e.g., 0.1. If the dietary zinc intake was reduced, e.g., by feeding an egg albumin protein source, fecal losses would emanate primarily from the stable complexes contained in the pancreatic secretion. Under these circumstances, the stable complexes in the pancreatic pool become the major source of excreted zinc and when combined with the dietary zinc pool provides too little zinc for absorption/reabsorption to maintain zinc homeostasis. Though the onset of symptoms of zinc deficiency is slower with a phytate-free, low zinc diet, the ultimate outcome remains the same.

VII. ANALYTICAL PROCEDURES

The analytical procedures needed to complete this research include the measurement of radioactivity and quantitative zinc and phytate

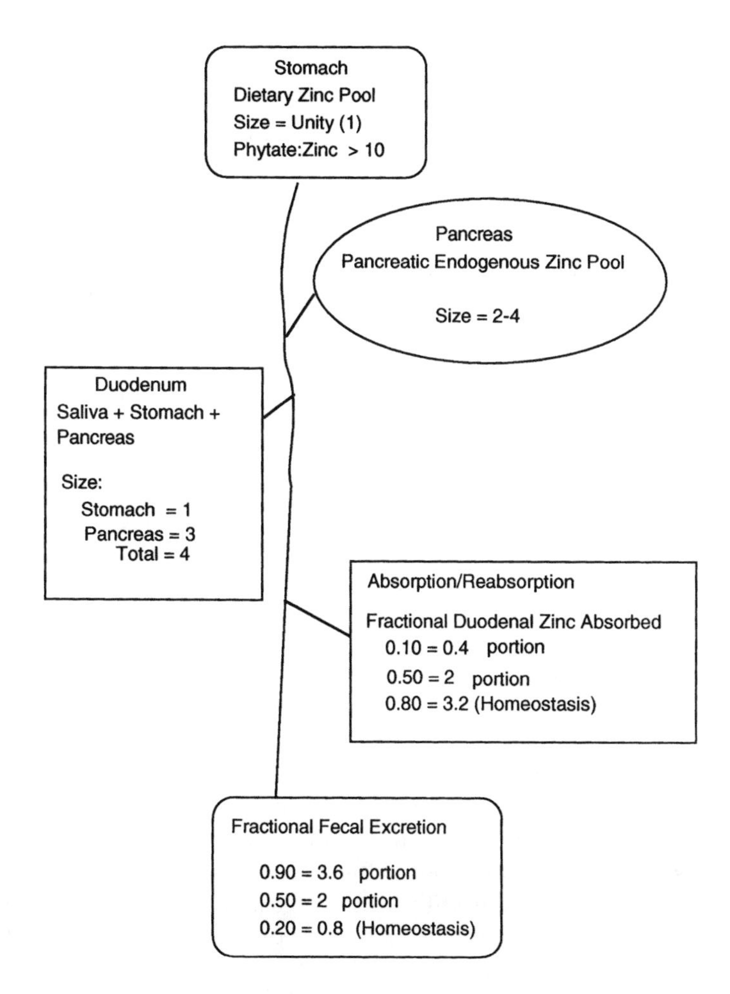

Figure 1
Shows the mechanism of zinc homeostasis. This model assumes a daily dietary intake expressed as unity (1). All other values are relative to the dietary intake. (From Oberleas, D., *J. Inorg. Biochem.*, 62, 1996. With permission.)

analyses. Radioactive ^{65}Zn emits a strong gamma ray with 1.115 MEV of energy. An excellent gamma scintillation counter capable of counting at the appropriate energy and providing good counting efficiency is needed. If the instrument is stable and the counting background consistent, the fecal samples may be expressed in net counts per minute. The comparison to reflect the effect of phytate is the ratio of activity from animals consuming a phytate containing diet to that of a nonphytate containing diet. So long as the radioactivity is expressed in the same units, the ratios represent the differential of activity.

Counting the radioactivity in pancreatic fluid is a more tedious task. Depending on the size of the rat at the time of collection, the flow rate via the pancreatic duct may be as little as 0.5 mL/h up to about 1.5 mL/h.

The limitation for collection ranges from 1 to 4 hours until the animal becomes totally exhausted and succumbs to the procedure. The success rate for intubating the common bile duct may be as low as 80%.

To obtain good counting geometry, it is necessary to pipet identical volumes of pancreatic/biliary fluid into the tubes for counting. There are several ways in which the radioactivity may be expressed. The first method is to calculate the ratio of radioactivity from animals fed phytate vs. nonphytate diets, per unit volume of fluid, per unit concentration of protein, compared with total zinc in the fluid, specific activity compared to plasma, etc.

Phytate analysis has gone through several magnitudes of change in this century. The earliest method was based on the insolubility of ferric phytate in dilute acid solution, either titrating to an endpoint with an iron sensitive indicator or adding a standard concentration of iron and back titrating the excess to determine the concentration of phytate by difference.[21] More recently, phytate methods have incorporated precipitation of phytate with ferric iron, washing the precipitate with dilute acid, wet ashing the precipitate, and calculating a phytate equivalent from quantitative phosphate analysis.[22] These methods were cumbersome and time consuming but gave analytical results that were adequate for the expression of phytate to zinc molar ratios for most rat diets. Molar ratios could easily be calculated with sufficient precision to differentiate between zinc deficiency and zinc adequacy. A magnitude improvement in sensitivity was made by first purifying the phytate on a small amount of strong anion exchange resin prior to ashing and phosphate analysis, but this method was still time consuming.[23] HPLC technology has allowed two very recent modifications of the phytate methodology that create a new magnitude in sensitivity and precision with a more reasonable analytical time requirement. These methods include a reverse phase isocratic method[24] and a gradient phase anion exchange method that incorporates a post column iron/sulfosalicylic acid reaction with visible absorption output.[25] These HPLC methods separate the extracted phytate-containing fraction into the various inositol phosphate esters.

Though the precision of the analytical procedure is improved by HPLC, it may not improve the usefulness of the resulting data. The important statistic is the [phytate] to [zinc] molar ratio or more precisely the [calcium][phytate] to [zinc] molar ratio. Estimates on the basis of phytate to zinc molar ratio for the division between zinc adequacy and zinc deficiency has ranged between 10 and 15.[11] It should be obvious that 10 to 15 molecules of phytate are not required to complex 1 atom of zinc. With a molecular weight of 660 for undissociated inositol hexaphosphate, a variation of 660 mg/kg in phytate analysis would vary the phytate to zinc molar ratio by only 1. With these methods sensitive to less than 1 μg/g, the analytical precision available today far exceeds that needed for the expression of zinc adequacy by this method.

VIII. POTENTIAL PROBLEMS

There are many potential problems in this rather involved procedure. The first is the development of appropriate dietary models. The protocol requires two dietary proteins, namely, one containing native phytate and a comparable one with a nonphytate protein source. The nonphytate protein sources are somewhat limited to casein and lactalbumin or ovalbumin. The phytate containing protein sources are numerous, but may have variable concentrations of protein which need to be analyzed and equalized within the diets. The diets may be formulated with a nonphytate protein with added sodium phytate. All other nutrients in the diets need to be identical so that only the intended variables are being studied. These should be limited to phytate, zinc, and calcium or other tertiary divalent elements. The diets should be analyzed for phytate, zinc, and calcium if calcium is a teritary variable. A phytate to zinc molar ratio should be calculated and compared to weight gain in a young animal. The latter comparison may be used for comparison of the relative success of the depletion phase.

The need to deplete the animals prior to the injection of ^{65}zinc is imperative. The injection of a trace dose of a trace mineral into an animal that is otherwise in homeostasis does not provide sufficient equilibrium to observe the differential effect of phytate. Injection may be intravenously, which is very tedious and is usually done via the tail vein, or intraperitoneally, which is easier and provides concomitant equilibrium. In depleted animals, a large excretion of radioactive zinc is observed regardless of the dietary conditions, creating a large artifact in the collection data for 1 to 3 days. True equilibrium does not occur until about the 7th day post injection. Therefore, a 10 to 14 day fecal collection is necessary to observe the effect of phytate on zinc homeostasis.

Phytate analyses have exhibited some problems with the improvement of analytical technology. With the early methods, the precipitate or iron complexes were assumed to be only inositol hexaphosphate. As technology has improved, it has been learned that when cereals and legumes are processed, some of the phytate is hydrolyzed to lower esters. This mixture of esters creates a small variation in the concentration of some foodstuffs.

With the advent of anion exchange analysis and subsequent HPLC methodologies, a new variable has been introduced that has not been addressed in the literature. A compound, as yet unidentified, has been discovered that has an anion character, contains phosphate, complexes iron, has an absorption spectrum in the 257 to 260 nm range, and elutes with the inositol quadriphosphate fraction. With these characteristics, it becomes a contaminating factor in the most recent phytate analytical procedures. It is also present in casein and thus can be a contaminating factor when casein-based diets to which phytate has been added are analyzed. For the casein-based diets, this is easily corrected by analyzing a similar nonphytate containing casein-based diet and subtracting the difference. This correction is not available for other foodstuffs or for mixtures of foods. Therefore, a phytate analytical method should be chosen which allows the quantitation of each inositol phosphate ester.

IX. BRIEF SUMMARY

Zinc deficiency produced by use of a nonphytate containing protein source provides a model suitable as an academic curiosity for the study of zinc metabolism. On a practical level, the discovery of the effect of phytate in the diet on zinc homeostasis thrusts zinc deficiency toward a practical problem. The rat is an ideal model for the study of zinc homeostasis. The results expressed on a molar basis are equally applicable to other monogastric species including man. The dietary model requires a plant seed protein source and the outcome can be evaluated by use of a phytate to zinc or calcium × phytate to zinc molar ratio in the absence of overt symptoms. The symptoms are few and subtle, particularly in the adult of any species. The mechanism demonstrates that the effect of phytate is primarily on the labile endogenous zinc secreted via the pancreas. Thus, the major concern in supplying dietary zinc is to replace that which is lost via the gastrointestinal tract as both stable zinc complexes and that excreted as phytate complexes. Calcium is not essential for phytate complexation of zinc, but is the mineral in the highest concentration in most diets. Thus, calcium has been shown to be the primary synergistic agent affecting the homeostatic change. A mathematical model has been developed which fits all nutritional models for zinc homeostasis. Radioactive ^{65}zinc is a valuable tool in the study of this phenomenon.

ACKNOWLEDGMENT

I should like to thank Drs. Boyd L. O'Dell and Barbara F. Harland for reviewing this manuscript and making many helpful suggestions.

REFERENCES

1. Wolf, G., *Isotopes in Biology*, Academic Press, New York, 1996.
2. Tucker, H.F. and Salmon, W.D., Parakeratosis or zinc deficiency disease in the pig. *Proc. Soc. Exptl. Biol. Med.*, 88, 613, 1955.
3. O'Dell, B.L. and Savage, J.E., Symptoms of zinc deficiency in the chick. *Fed. Proc.*, 16, 394, (abstract) 1957a.
4. O'Dell, B.L. and Savage, J.E., Potassium, zinc and distillers dried solubles as supplements to a purified diet. *Poul. Sci.*, 34, 459, 1957b.

5. O'Dell, B.L. and Savage, J.E., Effect of phytic acid on zinc availability. *Proc. Soc. Exptl. Biol. Med.*, 103, 304, 1960.
6. Oberleas, D., Muhrer, M.E., and O'Dell, B.L., Effects of phytic acid on zinc availability and parakeratosis in swine. *J. Animal Sci.*, 21, 57, 1962.
7. Oberleas, D., *Dietary Factors Affecting Zinc Availability*, Ph.D. Dissertation, University of Missouri, Columbia, 1964.
8. Oberleas, D., Muhrer, M.E., and O'Dell, B.L., Dietary metal-complexing agents and zinc availability in the rat. *J. Nutr.*, 90, 56, 1966a.
9. Oberleas, D., Muhrer, M.E., and O'Dell, B.L., The availability of zinc from foodstuffs. In: *Zinc Metabolism*, A.S. Prasad, Ed., C.C. Thomas, Springfield, 1966b, 225.
10. Oberleas, D., Factors influencing availability of minerals. In: *Proceedings of the Western Hemisphere Nutrition Congress IV*, P.L. White and N. Selvey, Eds., Publishing Sciences Group, Acton, 1975, 156.
11. Morris, E.R., Phytate and dietary mineral bioavailability. In: *Phytic Acid: Chemistry and Applications*, E. Graf, Ed., Minneapolis, Pilatus Press, 1986, 57.
12. Chan, H.C., *Phytate and Cation Binding Activity.* M.S. Thesis, Texas Tech University, Lubbock, TX, 1988.
13. Davies, N.T., Carswell, A.J.P., and Mills, C.F., The effect of variation in dietary calcium intake on the phytate-zinc interaction in rats. In: *Trace Elements in Man and Animals-V, (TEMA-5)*, C.F. Mills, I. Bremner, and J.K. Chesters, Eds., Commonwealth Agricultural Bureaux, Farnham Royal, U.K., 1985, 456.
14. Birnstingl, M., Stone, B., and Richards, V., Excretion of radioactive zinc (Zn-65) in bile, pancreatic and duodenal secretions of the dog. *Am. J. Physiol.*, 186, 377, 1956.
15. Pekas, J.C., Zinc 65 metabolism: gastrointestinal secretion by the pig. *Am. J. Physiol.*, 211, 407, 1966.
16. Cragle, R.G., Dynamics of mineral elements in the digestive tract of ruminants. *Fed. Proc.*, 32(8), 1910, 1973.
17. Lo, G.S., Steinke, F.H., Ting, B.T.G., Janghorbani, M., and Young, V.R., Comparative measurement of zinc absorption in rats with stable isotope 70-Zn and radioisotope 65-Zn. *J. Nutr.*, 111, 2236, 1981.
18. Oberleas, D., Mechanism of zinc homeostasis. *J. Inorg. Biochem.*, 62, 231, 1996.
19. Mills, C.F., Reassessment of human trace element requirements by a WHO/FAO/IAEA expert committee. In: *Proceedings of the Eighth International Symposium on Trace Elements in Man and Animals (TEMA-8)*, M. Anke, D. Meissner, and C.F. Mills, Eds., Verlag Media Touristik, Gersdorf, 1993, 29.
20. Kwun, I.S., *Effect of Phytate Added to a Casein-Based Diet on Endogenous Zinc and Study of Pancreatic/Biliary Fluid Fractions in Rats.* Ph.D. Dissertation, Texas Tech University, Lubbock, TX, 1995.
21. Heubner, W. and Stadler, H., Uber eine Titrationmethode zur Bestimmung des Phytins. *Biochim. Z.*, 64, 422, 1914.
22. Oberleas, D., The determination of phytate and inositol phosphates. In: *Methods of Biochemical Analysis*, New York, Wiley, 20, 87, 1971.
23. Harland, B. and Oberleas, D., Anion-exchange method for determination of phytate in foods: Collaborative study. *J. Assoc. Off. Anal. Chem.*, 69, 667, 1986.
24. Lehrfeld, J., High-performance liquid chromatography analysis of phytic acid on a pH-stable, macroporous polymer column. *Cereal Chem.*, 66, 510, 1989.
25. Rounds, M.A. and Nielsen, S.S., Anion-exchange high-performance liquid chromatography with post-column detection for the analysis of phytic acid and other inositol phosphates. *J. Chromat.*, 653, 148, 1993.

OTHER TRACE ELEMENTS

Chapter 14

Dietary Boron Deficiency and Supplementation

Curtiss D. Hunt
*U.S. Department of Agriculture/ARS
Grand Forks Human Nutrition
Research Center
Grand Forks, North Dakota*

CONTENTS

0-8493-9611-5/97/$0.00+$.50

I. OVERVIEW OF BORON ESSENTIALITY AND FUNCTION

There is universal agreement that vascular plants, diatoms, and some species of marine algal flagellates have acquired an absolute requirement for boron.[1,2] Although a specific biochemical role for this element in the metabolism of higher plants remains to be elucidated, the unambiguous characteristics of boron deficiency are sufficient to define precisely the boron requirement of many plant species.[3] Findings from numerous studies indicate that diets low in boron (≤ 0.3 μg [0.028 μmol]/g) supplemented with inorganic boron, in amounts (~ 2 μg [~0.185 μmol]/g) equivalent to that found in diets comprised of mainly fruits and vegetables, are sufficient to affect several aspects of animal and human physiology. The physiological influence of dietary boron is typically more pronounced during concurrent nutritional insult, especially vitamin D deficiency.

A. Vitamin D, Mineral, and Bone Metabolism

Dietary boron deprivation (≤ 0.3 μg [0.028 μmol]/g) increases gross bone abnormalities in the vitamin D-deficient chick[4] and chick embryo.[5] At the microscopic level, dietary boron deprivation exacerbates distortion of the marrow sprouts, a characteristic of vitamin D deficiency.[4] In humans, boron supplementation (3 mg [0.277 mmol] B/day) of a boron-low (0.25 mg [0.023 mmol] B/2000 kcal) diet, increased serum 25-hydroxycholecalciferol.[6,7]

Dietary boron influences mineral metabolism through uncharacterized mechanisms. In the vitamin D-deficient rat, low dietary boron (0.158 μg [0.015 μmol] B/g) decreases the apparent absorption and retention of calcium and phosphorus and increases femur magnesium concentrations.[8] Low dietary boron (0.180 μg [0.017 μmol] B/g) decreases femoral calcium and phosphorus concentrations in vitamin D adequate, but not inadequate chicks.[9] Dietary boron also influences the concentrations of several minerals in the hearts of vitamin D-deprived rats.[9] In humans, dietary boron influences serum calcitonin and ionized calcium concentrations[6] as well as the urinary excretion of calcium.[10,11] The mechanism(s) through which boron modulates mineral metabolism in a variety of tissues, especially during concurrent vitamin D deficiency, is under active investigation.

B. Energy Substrate Utilization

Dietary boron deficiency exacerbates vitamin D deficiency-induced perturbations in energy substrate utilization. For example, in the vitamin D-deficient, boron-deprived (0.465 μg [0.043 μmol] B/g) chick, the abnormally elevated concentrations of plasma glucose is decreased by supplemental boron.[4] In boron-deprived (0.23 mg [0.021 mmol] B/2000 kcal) postmenopausal women fed a low-magnesium diet,[12] a dietary boron supplement (3.23 mg [0.30 mmol]/day) decreased serum glucose concentrations (in the normal range) by approximately 6%. Dietary boron deprivation markedly increases plasma insulin concentrations in the vitamin D-deprived rat[13] and also increases peak pancreatic insulin secretion from isolated, perfused chick pancreata by nearly 75%.[14] Plasma triglyceride concentrations, depressed in the vitamin D-deficient, boron-deprived chick, are increased by supplemental boron.[15] Also, in the same community-based study mentioned above,[12] boron supplementation of a boron-low diet increased serum triglyceride concentrations. The relationships among boron, insulin secretion, and plasma glucose concentrations are under active investigation.

Many boron nutrition experimental protocols developed within the last 12 years reflect fuller appreciation for characteristics of boron geochemistry, boron inorganic chemistry, boron metabolism in plants and animals, and boron anthropomorphic sources. As described below, those characteristics are manipulated to reduce environmental boron contamination, to obtain boron-low drinking water, and to construct boron-low diets (< 0.1 μg [0.009 μmol]/g).

II. WATER

A. Boron Content

Weathering of clay-rich sedimentary rock is the major source of total boron mobilized into the aquatic environment.[16] Many clays have a large boron adsorption capacity because of the substantial quantity of amorphous iron and aluminum hydroxy oxides.[17] Sewage effluent containing as much as 5 μg (0.467 μmol) boron per mL can be an important source of boron, especially in areas where boron compounds are major constituents of detergent products.[16] Because soil types and human population densities vary, the concentration of boron in surface water[16] may vary substantially between geographic locations and water sources [μg (μmol)/mL]; examples of this include: U.K., 0.046 to 0.822 (0.004 to 0.077); Italy, 0.400 to 1.00 (0.037 to 0.093); Sweden, 0.001 to 1.046 (0.00009 to 0.098); Germany, 0.100 to 2.00 (0.009 to 0.187); and U.S., 0.001 to 5 (0.00009 to 0.467). Within the state of California, the variance of boron [μg (μmol)/mL] in water is large: Susan Bay — lower San Joaquin River water samples, 0.65 (0.060); spring and well water surrounding the Salton Sea, 2.60 (0.241); and various water samples from Casa Diablo including hot and cold springs, lake waters, and streams from area of volcanic activity, 8.41(0.779).[18] Shallow groundwater in the San Joaquin Valley region of California may have boron concentrations as high as 120 μg (11.2 μmol) per mL.[16] The boron content of any of those waters from California is well above that of municipal purified river water in Grand Forks, North Dakota (≤ 0.015 μg [0.001 μmol]/mL). Furthermore, wells in the same locale but at different depths can contain water with differing boron concentrations. For example, water from a shallow well located 2 miles west of Calistoga, CA, was derived mainly from runoff water and contained 0.04 μg (0.004 μmol)/mL. Water from an adjacent deep well contained concentrations of boron (6.79 μg [0.629 μmol]/mL) that were toxic to grape plants able to thrive in local soils.[18]

B. Bioavailability

Boron is normally present in water as undissociated boric acid [$B(OH)_3$][16] although organic complexes of boron may be present in fresh waters.[19] The bioavailability of boron from water, as boric acid, is probably very high in most mammals including rodents. As described below, inorganic boron supplements that quickly transform into boric acid in water solutions are completely absorbed from the intestinal tract. Furthermore, in plants, boron is not available for uptake until it is mineralized from decomposed organic matter, or artificially added to soils (irrigation or fertilization), and subsequently distributed between the solid and liquid phases of the soil.[17,20]

C. Water Usage and Storage

The generally high bioavailability of boron in water requires that drinking water provided to rodents in boron nutrition studies contain negligible amounts of boron (≤ 0.015 μg [≤ 0.001 μmol]/mL). Studies of *in vitro* boron nutrition probably require extremely high-quality water. Preferably, even potable water low in boron should be processed to remove other elements that modify the effects of boron in the model system.

From the examples of boron content of water given above, it is clear that local municipal water supplies may be sufficiently low in boron to use directly in such studies without further purification. However, boron usually must be removed from drinking water, a process commonly accomplished by normal demineralization (non-glass [for reasons described below] still) or reverse osmosis* which provides water of relatively high purity (~15.0 MΩ·cm). In my laboratory, the process of reverse osmosis is followed by further purification and deionization steps to achieve nearly "pure" water** as measured by resistivity (~18.0 MΩ·cm). These steps, in sequence are *mixed* bed ion [cation and anion] exchange, ultraviolet light, activated carbon, second mixed bed ion exchange, and 0.2 μ filter. Boron-specific resins also effectively remove boron from solution[21] although one such resin that contains a hydrophobic styrene backbone and a tertiary amine group will only absorb boron in the alkaline form [$B(OH)_4^-$].[22] A quartz sub-boiling point still containing mannitol (6 mg [0.033 mmol]/mL water) can be used to reduce the boron content of water.[23] Regardless of the method used to purify water, the water should be tested on a regular basis by a sensitive assay method to verify boron content.

Boron, in the form of borax, has been used since Egyptian and Roman times for the preparation of hard (borosilicate) glasses. Therefore, boron is either a constituent or contaminate of most laboratory glassware.[24] Boron leaches readily from most glassware, especially under conditions of alkalinity.[25] Even the brief time taken to quantify volumes of demineralized, deionized*** water with a standard glass volumetric pipette is sufficient exposure to glass to increase the boron content of the water sample from background concentrations[26] to half that found in rat plasma (0.050 μg [0.004 μmol]/mL).[27] Therefore, neither drinking water nor reagent solutions used in boron nutrition research should come in contact with glassware even for short periods of time. Because boron is also used in the manufacture of many detergents, soaps, and cleaners,[16] special care should be taken to rinse plasticware of all cleaning agents as described in the accompanying chapter on boron measurements in rodent diets and tissues.

---

* Feed water for the reverse osmosis (RO) process that contains chlorine or water-hardening cations (calcium, magnesium, and iron ions) should be processed through activated charcoal and an ion-exchange system (typical water-softening) to prolong the life of the RO filter.

** Pure water has a theoretical specific resistance of 18.2 MΩ·cm at 25°C. Analysis of water purity by the measure of resistivity is based on ionization of materials in solution and does not take into account uncharged molecules, for example, many organic molecules. In potable water, inorganic boron is essentially present only as the mononuclear species $B(OH)_3$ and $B(OH)_4^-$. The uncharged $B(OH)_3$ is the dominant species in most water (pH 1, ~100%; pH 7.4, >98%; pH 9.25, 50%; pH 11, ~0%).[17] Therefore, determination of water purity by typical resistivity measurements may not provide an accurate indication of boron concentrations in water.

*** The term "demineralized" is used instead of "distilled" to denote the removal of minerals from aqueous solutions by all methods including distillation by volatilization and subsequent condensation. The term "deionized" is used to denote the further removal of trace amounts of both cations and anions from aqueous solutions.

III. DIET FORMULATION

A. Boron Sources and Distribution

To observe consequences of boron deprivation in rodents, the basal diet should not contain more than 0.20 μg [0.018 μmol] B/g of air-dried diet. The concentration of boron in some finished commercial diets may be approximately two to five times this amount. Therefore, dietary boron supplemented in the few μg/g range to these commercial diets is not likely to have much effect on general metabolism. For example, the concentration of boron in alfalfa has been reported to be as high as 42 μg [3.88 μmol] boron/g dry material.[28] In a 10% alfalfa ration, the amount of boron contributed by the alfalfa component alone would increase the total boron concentration to 4.2 μg [0.388 μmol] boron/g dry ration. Samples of Ralston Rodent Laboratory Chow #5001,* a common diet for the laboratory rat, were found to contain 12.1 to 13.7 μg [1.12 to 1.27 μmol] boron/g.[29] Because boron is present in all plant food sources commonly consumed by laboratory rodents, "background" boron concentrations occur in rodent blood and tissue. On the other hand, balanced rodent diets low in boron content (≤0.2 μg [0.018 μmol]/g) are not difficult to formulate.

Boron is unequally distributed within Angiospermae, the class of plants most often utilized in animal or human diets. For example, most species within the subclass Dicotyledoneae, which includes nuts, fruits, vegetables, tubers, and legumes, have a much higher concentration of boron than do species in the subclass Monocotyledoneae, which includes the gramineous grain crops (corn, rice, wheat, barley).[18] The different content of boron in broccoli flowers (1.852 μg [0.171 μmol]/g fresh weight) and stalks (0.889 μg [0.082 μmol]/g fresh weight) illustrates the compartmentalization of the element even within the same plant.[18] Finally, the geographical source of plant source foodstuffs is important when calculating usual individual boron intake because the concentration of boron in plants varies with soil type, length of exposure, rate of transpiration, and different agricultural practices.[3,18,30]

Compared to many foods of plant origin, the concentration of elemental boron in animal muscle and milk is very low. However, the total amount of animal based-foods consumed must be considered when estimating total daily boron intake. For example, human toddlers receive approximately 13% of their total daily boron intake from milk, a food with a relatively low concentration of boron, because of the relatively large volume of milk consumed.[31]

---

* Mention of a trademark or proprietary product does not constitute a guarantee or warranty of the product by the U.S. Department of Agriculture, and does not imply its approval to the exclusion of other products that may also be suitable.

B. Carbohydrate, Protein, and Fat

Rodent diets low in boron reflect careful selection of diet protein, carbohydrate, and fat sources. For such diets, protein from animal-based products (casein, egg white, egg yolk, lactalbumin, lactose, gelatin, fish meal, blood fibrin, skim milk powder) and monocotyledon-based products (wheat gluten) is more appropriate than protein from dicotyledon-based products (soy protein, peanut meal). High protein casein is (Table 1) used as the protein source in this laboratory in lieu of more reliable, attainable protein sources. Products from monocotyledon plants (arrowroot, corn, corn starch, rice, sago palm, wheat starch) are more appropriate sources of carbohydrate than products from dicotyledon plants (beans, potato starch). Sucrose (and possibly fructose and dextrose), a highly refined product from the sugar beet (Dicotyledoneae), contains low amounts of boron (0.10 μg [0.009 μmol]/g).[18]

TABLE 1.

Composition of Low-Boron Diet for Rats[a,b,c]

Ingredient	Amount Supplemented to Diet	Boron Content[d]	Boron Contribution to Diet
	g/kg	*mg/kg*	*μg/kg*
Corn, ground, acid washed[e]	717.64	0.027	19.37
Casein, high protein[f]	160.00	0.043	6.88
Corn oil[e]	75.00	NA[g]	
Mineral mix[h]	24.84	>0.386	>9.59
$CaHPO_4$[i]	17.00	2.360	40.12
Vitamin mix[j]	4.57	0.000	0.00
Choline chloride[k]	0.75	NA	
DL-alpha-tocopherol[e]	0.20	NA	

[a] Basal diet typically contains ≤0.100 mg B/kg diet on an air-dried basis by analysis.[52] Each batch should be analyzed for boron content.

[b] Supplemented with orthoboric acid (H_3BO_3, Puratronic; Johnson Matthey Chemicals Ltd., Aesar, Seabrook, NH) in a boron pre-mix (for example, 0.0046 g H_3BO_3 [800 μg B]/g anhydrous dextrose [ICN Biochemicals, Cleveland, Ohio]) as needed.

[c] To meet revised amino acid requirements[35], supplements of cysteine (1.7 g/kg diet) and methionine (1.7 g/kg diet) should be added to the diet.

[d] Analysis[26] of a representative lot of a given ingredient and therefore not necessarily the average boron content of the ingredient.

[e] ICN Pharmaceuticals. Corn was acid-washed with HCl as described in Diet Preparation.

[f] Teklad, Division of Harlan Industries, Inc. (Madison, WI).

[g] No analysis available.

[h] See Table 3.

[i] 'Baker Analyzed', J.T. Baker, Phillipsburg, NJ.

[j] See Table 2.

[k] GIBCO, Grand Island, NY.

Monocotyledon plant products (wheat bran, oat bran) are more appropriate sources of fiber than dicotyledon plant-based products (guar gum, gum arabic, gum tragacanth, locust bean gum, xanthan gum). Fiber sources including cellulose, hemicellulose, lignin, and pectin must be identified because those products can be derived from plant species in either subclass of Angiospermae. Preferably, the carbohydrate source should have sufficient fiber content to eliminate the use of other fiber sources. Few fat or oil sources have been assayed for boron content. However, based on the source of the fat or oil, it is reasonable to assume that, in terms of boron content, animal-based fats and oils (beef tallow, butter fat, cod liver oil, lard, menhaden oil, shark liver oil) and monocotolyden-based fats and oils (coconut oil, corn oil, palm oil, wheat germ oil) are more appropriate sources of fat than products from dicotyledons (cottonseed oil, linseed oil, olive oil, peanut oil, safflower oil, soy bean oil, or sunflower oil).

In physiological concentrations, inorganic boron is essentially present in biological fluids only as the mononuclear species $B(OH)_3$ and $B(OH)_4^-$. Within the normal pH range of the gut, $B(OH)_3$ would prevail as the dominant species (pH 1, ~100% $B(OH)_3$; pH 7.4, > 98%; pH 9.25, 50%; pH 11, ~0%).[32] The uncharged $B(OH)_3$ is capable of forming complexes with several biologically important ligands that typically contain adjacent hydroxyl groups in the *cis* position. The reactivity of boric acid with the ligand generally increases as the number of adjacent hydroxyl groups increases. Riboflavin and the pyridine nucleotides are examples of ligands that bind boron. The exact form of the complex varies according to pH and ligand:boron ratio. The relevant *cisoid* diol conformations are also present in several biologically important sugars and their derivatives (sugar alcohols, -onic, and -uronic acids), such as mannose, ribose, galactose and fructose.[33]

C. Vitamins

Vitamin supplements for animal diets typically contribute inconsequential amounts of boron to the total diet. For example, in-house analysis of the standard AIN-76A vitamin mix and a vitamin mix used in a diet for rats in studies of boron nutrition (Table 2) revealed undetectable amounts of boron (<0.015 μg [0.001 μmol]/g).[26] However, chemical coordination of boron with certain vitamins or their carriers may be of concern. For example, mannitol, a polyhydroxy compound and the usual carrier for vitamin B_{12} dietary supplements, reacts with boric acid to form stable mannitol-boron complexes even in the presence of 6 mol/L hydrochloric acid solutions.[34] Depending on the mannitol/$B(OH)_3$ molar ratio, the boron-mannitol complexes contain one or two boron atoms per mannitol molecule. The typical molar concentration of boron in boron-supplemented research diets for rats (0.280 μmol/g; 3 μg/g) approaches the molar

TABLE 2.

Composition of Vitamin Mix for Low-Boron Diet for Rats[a]

Ingredient	g/kg mix[b]
D-dextrose, anhydrous[c]	943.6565
i-Inositol[d]	10.9409
Vitamin B_{12} (0.1% in mannitol)[c,d]	10.9409
DL-Pantothenic acid, calcium salt[c]	10.5011
Nicotinic acid[e]	6.5646
Riboflavin[e]	5.9081
Retinyl palmitate (500,000 IU/g)[c]	3.5011
Pyridoxine HCl[e]	3.2823
Thiamine HCl[e]	2.1882
p-Aminobenzoic acid[e]	1.0941
Vitamin D_3 powder in endosperm carrier (400,000 IU/g)[f]	0.5470
Folic acid[c]	0.4376
Biotin[c]	0.2188
Menadione[e,g]	0.2188

[a] Vitamin mix contains undetectable amounts of boron (≤0.015 µg B/g)[26] by analysis.[52]

[b] Vitamin mix, fed at 0.457% of diet, provides (g/kg of diet): *i*-inositol, 0.050; Ca *d*-pantothenic acid, 0.048; retinyl palmitate (500,000 IU/g), 0.016; niacin, 0.030; riboflavin, 0.027; vitamin B_{12}, 0.000050; pyridoxine, 0.015; thiamine, 0.010; *p*-aminobenzoic acid, 0.005; cholecalciferol (400,000 IU/g), 0.0025; folic acid, 0.002; biotin, 0.001; menadione, 0.001. To ensure adequacy, water soluble vitamins were supplemented in excess of the requirement.[35]

[c] ICN Pharmaceuticals, Inc., Life Sciences Group (Cleveland, OH).

[d] Commercial sources of cystalline vitamin B_{12} (without mannitol) are available and may be more suitable for titrating a boron requirement.

[e] GIBCO (Grand Island, NY).

[f] Dyets Inc. (Bethlehem, PA).

[g] New evidence suggests that the requirement for vitamin K should be met by vitamin K_1 (phylloquinone; 1 mg/kg) instead of menadione because the rat actively absorbs phylloquinone and gives preference to phylloquinone metabolically.[35]

concentration of mannitol in diets (0.274 µmol/g) as the consequence of adding vitamin B_{12} to meet the new National Research Council recommendation of 50 ng vitamin B_{12}/g diet.[35] Mannitol is readily absorbed from the gut into the blood stream but does not enter intracellular spaces before elimination via the urine.[36] Although an equilibrium is established, it is conceivable that mannitol-boron complexes effectively reduce the amount of boron available to tissues and thus, enhance the signs of boron deficiency.

The vitamin pyridoxine, as a 1,-3-diol compound, reacts strongly with borate but the pyridoxine-borate (2:1) complex apparently has full activity as a vitamin in the rat; this indicates that it is completely dissociated in the animal to give the unchanged vitamin.[37] However, because the teratogenic effects of high doses of boric acid in the chick embryo can be mitigated by *in ovo* injections of pyridoxine hydrochloride,[38] it is reasonable to assume that pyridoxine-borate complexes have reduced metabolic activity. The ribityl side chain of riboflavin forms a water-soluble complex with boron and the riboflavin-borate complexes apparently have full biological activity.[39] However, the teratogenic effects of high doses of boric acid in the chick embryo can be mitigated by *in ovo* injections of riboflavin.[40] *In vitro* binding of riboflavin to serum proteins is reduced by borate at concentrations of 1.0 μmol/mL,[41] which is much higher than those found normally in rat plasma (0.005 μmol/mL).[27]

D. Minerals

Individual minerals used in diets for rats in studies of boron nutrition can vary markedly in boron content compared to other minerals, compared to different forms of the same mineral (i.e., $CaCO_3$ vs. $CaHPO_4$), or compared to different commercial supplies of the same form. For example, the diet described in Table 1 relies on ground corn to supply any possible silicon requirement because in-house analysis[26] of inorganic sources of silicon indicates unacceptable amounts of contaminant boron (7.15 μg [0.661 μmol] B/g $Na_2SiO_3{\cdot}9H_2O$). A relatively pure, economical, and available source of $CaHPO_4$ added (17.0 g/kg diet) to meet the calcium and phosphorus requirement contributes twice as much boron (~40 ng [3.70 nmol] B/g diet) as does 717.0 g of acid-washed ground corn (~19 ng [1.76 nmol] B/g diet). The amount of $Mn(C_2H_3O_2)_2{\cdot}4H_2O$ (Table 3) added (0.225 mg/g diet) to supply the older[42] manganese requirement (50 μg/g diet) contributes more boron than does a relatively much larger amount of KCl (7.000 mg/g diet) to supply the potassium requirement (4.17 ng [0.386 nmol] vs. 5.96 [0.551 nmol] B/g diet). The amount of boron in similar grades of $CaCO_3$ (not used in the diet described in Table 1) can vary as much as seven-fold between suppliers.[26] Therefore, construction of new boron diets requires analysis of each source of carbohydrate, protein, fat, and mineral. Furthermore, the calculated boron content of a given batch of finished diet should not substitute for actual boron analysis of that batch.

E. Bioavailability

The bioavailability of boron in foods is unknown but probably is high. Postmenopausal women fed a low-boron diet (0.36 mg [0.033 mmol/d) composed of common foodstuffs excreted 100% of their total daily boron intake in the urine.[10] In other words, all of the boron in the low-boron diet was absorbed and ultimately excreted in the urine.

The speciation of boron in foods has not been determined but probably is complex and dependent upon the nature of the integral ligands. If plant and animal boron absorption mechanisms are analogous, the organic forms of boron per se are probably unavailable to humans; the

TABLE 3.

Composition of Mineral Mix for Low-Boron Diet for Rats

Ingredient		Boron Content[b]	Boron Contribution to Diet
	g/kg mix[a]	*mg/kg ingredient*	*µg/kg diet*
Corn, ground, acid washed[c]	313.516	0.027	0.210
KCl[d]	279.049	0.595	4.124
Iron mix[e]	180.515	NA[f]	
Mg $(C_2H_3O_2)_2 \cdot 4H_2O$[d]	140.900	0.920	3.220
NaCl[d]	80.445	0.303	0.605
$Mn(C_2H_3O_2)_2 \cdot 4H_2O$[g]	1.796	26.500	1.182
$Zn(C_2H_3O_2)_2 \cdot 2H_2O$[h]	2.027	0.227	0.011
$CuSO_4 \cdot 5H_2O$[i]	1.202	1.537	0.046
$Na_2HA_sO_4 \cdot 7H_2O$[h]	0.168	0.396	0.002
$(NH_4)_6Mo_7O_{24} \cdot 4H_2O$, grade 1[g]	0.074	8.505	0.016
$Cr(C_2H_3O_2)_3 \cdot H_2O$, purified[j]	0.096	1.755	0.004
NaF[g]	0.089	NA	
$NiCl_2$[k]	0.089	NA	
KI, ultrapure[l]	0.016	0.227	0.000
NH_4VO_3, Puratronic[f]	0.009	NA	
Na_2SeO_3[l]	0.009	0.233	0.000

[a] Mineral mix, fed at 2.484% of diet, provides (per kg of diet): Cl, 5.16 g; K, 3.64 g; Na, 787 mg; Mg, 400 mg; Fe, 40 mg; Zn, 15.0 mg; Mn, 10.0 mg; Cu, 8.0 mg; As, 1.0 mg; Mo, 1.0 mg; Ni, 1 mg; F, 1 mg; Cr, 0.5 mg; I, 0.3 mg; V, 0.10 mg; Se, 0.10 mg.

[b] Analysis[26] by [52] of a representative lot of a given ingredient and therefore not necessarily the average boron content of the ingredient.

[c] ICN Pharmaceuticals, Inc., Life Sciences Group, Cleveland, OH. Corn was acid washed as described in Diet Preparation.

[d] Baker Analyzed, J.T. Baker, Phillipsburg, NJ.

[e] 1.61 g iron sponge, 22 mesh (Puratronic, Johnson Matthey, Aesar, Seabrook, NH) dissolved in 17.47 mL of 6M HCl ("double distilled from Vycor," GFS Chemicals, Columbus, OH) then mixed to dryness in 178.9 g acid-washed ground corn (prepared as described in Diet Preparation).

[f] No analysis available.

[g] Johnson Matthey Chemicals, Ltd., Aesar, Seabrook, NH.

[h] Certified ACS, Fisher Scientific Co., Fair Lawn, NJ.

[i] Puratronic, Johnson Matthey Chemicals, Ltd., Aesar, Seabrook, NH.

[j] Fisher Scientific Co., Fair Lawn, NJ.

[k] Johnson Matthey, Alfa Products, Morton Thiokol, Danvers, MA.

[l] Alfa Products, Morton Thiokol, Danvers, MA.

organic forms of boron in soil must be mineralized to be available to plants.[42] However, the strong association between polyhydroxyl ligands and boron is easily and rapidly reversed by dialysis, change in pH, heat, or the excess addition of another low-molecular polyhydroxyl ligand.[33] Thus, within the intestinal tract, most ingested boron is probably converted to $B(OH)_3$, the normal end product of hydrolysis of most boron compounds.[43]

High boron intakes from natural foodstuffs may stimulate some uncharacterized mechanism that limits boron absorption. For example, the cow[44] and sheep[45] excrete only 30 and 41%, respectively, of total dietary boron in the urine when fed natural foodstuffs. Compared to humans fed typical amounts of boron (0.48 mg [0.044 mmol] B/kg body weight), the boron intake on a body weight basis is considerably higher in these animals (heifer, 0.715 mg [0.066 mmol]; sheep, 0.667 mg [0.062 mmol] B/kg body weight). In another animal study,[46] female rats consuming water high in boron (100 μg [9.25 μmol]/mL) for 21 d exhibited increased plasma boron concentrations although some mechanism concurrently eliminated any excess boron from the liver and brain against their own concentration gradients. In cows, the percent of filtered boron reabsorbed by the kidneys decreased significantly with increased boron intake.[44] Therefore, the design of studies on boron bioavailability must take into account uncharacterized mechanisms that apparently limit boron absorption.

F. Boron Supplementation

Residual contamination of boron supplements by other ultratrace elements with uncharacterized effects on metabolism is an obvious concern. Therefore, to determine the physiological roles of boron, special efforts should be taken to ensure that dietary boron supplements represent the most pure, but still natural, sources of inorganic boron available. In boron nutrition studies with rodents conducted in my laboratory, boron is added to the experimental diets as orthoboric acid (H_3BO_3). This is a common inorganic form of boron of high purity (99.9995%) that is absorbed well from the gastrointestinal tract.[47]

In studies designed to "titrate" the physiological effects of boron, I have found that dietary boron increments of 0.1 μg [0.009 μmol]/g diet added to background concentrations (~0.1 μg [0.009 μmol]/g) can be detected by inductively-coupled argon plasma spectrographic (ICP) analysis.[15] By starting with background boron concentrations, adding the first boron increment based on the minimum detectable (~0.1 mg/kg), and then making subsequent additions based on natural log increments of the first, a useful "titration curve" for the effects of boron, for example, on bone composition, across the physiological range is created (0.1 [0.009], 0.2 [0.018], 0.4 [0.036], 0.8 [0.072], 1.6 [0.144], 3.2 [0.288], 6.4 [0.576], 12.8 μg [1.152 μmol] B/g diet).

G. Special Dietary Considerations

Correct selection of foodstuffs is important in boron kinetics studies that use isotopic dilution measured by mass spectroscopy.[48] This technique to determine body pool size and distribution of boron depends upon changes in the distribution of the two stable boron isotopes (^{10}B and

^{11}B) over time in body tissues and fluids after absorption of ^{10}B-enriched foodstuffs. Broccoli, with a relatively high boron requirement and thus high uptake (26.28 μg [2.43 μmol]/g dry weight), good tolerance to hydroponic culture, and rapid growth cycle, is a suitable foodstuff for studies of intrinsically labeled ^{10}B isotope distributions in tissues and fluids.

The equilibrium constants for the boron-isotope exchange reaction between the two stable boron isotopes in $B(OH)_3$ (regular triangle form) and $B(OH)_4^-$ (tetrahedron form) are shifted to the right (K = 1.0206 at 0°C and 1.0194 at 25°C).[49]

$$^{10}B(OH)_3 + {}^{11}B(OH)_4^{\ddagger} = {}^{11}B(OH)_3 + {}^{10}B(OH)_4^{\ddagger}$$

Thus, the ^{10}B isotope is preferentially complexed in solution with the net result that boron isotopic fractionation occurs in biological systems, a well established phenomenon for elements with low atomic number. Therefore, the presence of dietary constituents that complex with boron may influence apparent boron kinetics as determined by isotopic dilution. Boron fractionation occurs in plants and varies according to plant tissue.[23] Thus, special care must be taken to ensure that the test dose of the intrinsically-labeled plant material is taken from similar tissues and homogenized thoroughly.

Studies designed to determine boron bioavailability from foods probably should use common foodstuffs derived from the plant subclass Dicotyledoneae which are known[50,51] to have high boron content (ug [μmol]/g wet weight), e.g., avocados (10.9 [1.01]), peanuts (5.83 [0.539]), grapes (1.40 [0.129]), sweet cherries (2.28 [0.211]), pears (2.27 [0.210]), plums (2.18 [0.202]), black-eyed peas (2.12 [0.196]), red beans (2.06 [0.191]), and apples (1.87 [0.173]). On the other hand, cooked white rice, derived from a grain in the subclass Monocotyledoneae, contains only 0.09 μg [0.008 μmol] B/g would be a poor choice. Dried preparations of the high boron, for example, apple powder, are sometimes available from commercial sources.

IV. DIET PREPARATION

A. Environmental Boron Contamination

Background contamination from dietary components described in Tables 1, 2, and 3, contribute approximately 0.1 mg [0.009 mmol] B/kg of diet. The amount of orthoboric acid (H_3BO_3) needed to double the background boron contamination in one kilogram of diet is 0.57 mg, an amount similar to the weight of approximately 3 grain-sized particles of table salt. Therefore, full appreciation for, and complete elimination of, boron from the environment is critical for success in nutritional studies of boron.

The reverse form of the contamination phenomenon also must be fully appreciated for correct addition of the boron supplement to the diet. Ninety-seven salt grain-sized particles of orthoboric acid, each weighing the same as grains of table salt (0.174 mg) and distributed in one kilogram of finished diet, would give a theoretical boron concentration of 3 mg [0.277 mmol]/kg diet. However, failure to grind the source of the supplemental boron to an extremely fine powder before addition to boron mix (see below) causes remarkably complex sampling problems and poor dietary boron distribution to a given group of animals.

Air supplies to the diet preparation area should be filtered and all working surfaces should be damp-mopped prior to diet preparation. Mixer parts that come in contact with the diet, utensils, weigh vessels (other than new plastic weigh boats), stirring rods, diet containers, etc., should be rinsed with tap water, washed with low-sudsing soap dissolved in demineralized water, rinsed in demineralized water, soaked in a chelating agent dissolved in demineralized water, then rinsed again three times with demineralized water, then again three times with demineralized, deionized water. Glass containers should be avoided but may be used to transfer exceptionally dry ingredients including anhydrous chemicals.

B. Pre-Mixes

Direct addition of boron and other ingredients required in small amounts, and treatments to the diet should be avoided because accurate weighing, transfer, and dietary distribution of minute quantities of those materials is highly unlikely. Boron sources should be ground finely (in my laboratory, they are ground once [10 min/g], then ground again later [1 hr/g] in a glass mortar and pestle) then distributed in a preground carrier (for example, 0.0046 g H_3BO_3 [800 μg {74.0 μmol} B]/g dextrose) which is shaken (45 min) in an appropriately-sized screw-capped polypropylene bottle that contains 5 to 6 appropriate-sized burundum grinding media ("ceramic marshmallows"). To maintain relatively constant additions of sugar, larger amounts of supplemental boron should be added through the use of progressively more concentrated boron mixes (for example, between 1 and 5 μg B/g diet: 0.0246 g H_3BO_3 [4297 μg B {397 μmol}]/g dextrose). The amount of all premixes needed must be carefully calculated in advance and made in excess to avoid variations between multiple mixes.

High purity elemental iron, obtained in the form of iron sponge (Table 3), must be oxidized to the ferrous state prior to addition to the mineral mix. The iron sponge is dissolved in 6 mol HCl (10.85 mL/g iron sponge). This is best accomplished in an open Teflon® screw-cap bottle containing a rotating Teflon-coated stirring bar on a source of low heat. The solution must be removed from heat and the bottle capped immediately after the sponge has dissolved and the solution is still dark green. Oxidation to the ferric state (yellow-color) should be avoided. After cooling, the iron mixture is mixed by hand in dry acid-washed ground corn (see below) and allowed to dry overnight to avoid unwanted reactions with other mineral elements.

To minimize bulk of the mineral mix, minerals supplemented to the diet in excess of 10 g/kg diet should be added directly to the diet. Each mineral supplement (for examples, see Table 3) should be ground separately, then weighed, and place directly in an appropriate-sized mixing container. A ball-mill ceramic mixer containing 15 to 20 appropriately-sized burundum grinding media is recommended. After addition of the ground corn carrier, the ball-mill mixer should operate for 4 hours. Prolonged mixing generates unwanted heat, and insufficient drying of the iron mix, or insufficient corn carrier in the iron mix causes unwanted chemical reactions; the finished premix should not appear blackened. The finished mineral mix must be stored in a cool location (not a refrigerator with vibrating machinery) in an air-tight plastic container, and, before use, checked for evidence of settling.

Vitamin premixes for diets used in boron nutrition studies (for example, Table 2) should also be mixed in plastic containers. To ensure better distribution, dietary supplements of choline chloride (dissolved in demineralized, deionized water) and DL-alpha-tocopherol are distributed in the oil source. The finished vitamin mix must be stored in a cool location in an air-tight plastic container, and, before use, checked for evidence of settling.

C. Acid-Washing of Corn

The ground corn used in the diet described in Table 1 is acid-washed to remove mineral contamination, thereby providing consistent background mineral content among batches of corn. In brief, 2.8 L of 2.0 *N* HCl are added per kg of yellow ground corn and the slurry is mixed in a plastic container with a belt-driven polymethacrylate stirrer for 30 min. After settling for 15 min, the supernatant is poured off and 1.2 L of deionized water is added before the slurry is mixed for another 30 min. The rinsing procedure is repeated three times with rinsing time reduced to 15 min. The slurry is then poured into a Buchner funnel, and rinsed with 4.8 L of deionized water. The majority of water is removed by placing the funnel containing the corn on top of a 5 L glass vacuum jar. The damp corn is then transferred to a plastic pan and placed in a hot air (75°C) convection oven and allowed to dry for 48 hr.

D. Diet Mixing

In my laboratory, all pre-weighed dry ingredients (boron mix, acid-washed corn, casein, mineral mix, and bulk minerals), except for the

14. Bakken, N.A., *Dietary Boron Modifies the Effects of Vitamin D Nutriture on Energy Metabolism and Bone Morphology in the Chick*, University of North Dakota, Grand Forks, ND, 1995.
15. Hunt, C.D. and Herbel, J.L., Physiological amounts of dietary boron improve growth and indicators of physiological status over a 20-fold range in the vitamin D_3-deficient chick, in *Trace Element Metabolism in Man and Animals-8*, Anke, M., Meissner, D., and Mills, C., Eds., Gersdorf, Germany, 1994, 714.
16. Butterwick, L., de Oude, N., and Raymond, K., Safety assessment of boron in aquatic and terrestrial environments, *Ecotox. Environ. Safety*, 17, 339, 1989.
17. Hue, N., Hirunburana, N., and Fox, R., Boron status of Hawaiian soils as measured by B sorption and plant uptake, *Commun. in Soil Sci. Plant Ana.*, 19, 517, 1988.
18. Hunt, C.D., Shuler, T., and Mullen, L.M., Concentration of boron and other elements in human foods and personal-care products, *J. Am. Diet. Assoc.*, 91, 558, 1991.
19. Krauskopf, K.B., Geochemistry of Micronutrients, in *Micronutrients in Agriculture*, Mortvedt, J., Giordano, P., and Lindsay, W., Eds., Soil Science Society of America, Inc., Madison, WI, 1972, 7.
20. Gupta, U.C., James, Y.W., Campbell, C.A., Leyshon, A.J., and Nicholaichuk, W., Boron toxicity and deficiency: a review, *Can. J. Soil Sci.*, 65, 381, 1985.
21. Mair, J.W. and Day, H.G., Curcumin method for spectrophotometric determination of boron extracted from radiofrequency ashed animal tissues using 2-ethyl-1,3-hexanediol, *Anal. Chem.*, 44, 2015, 1972.
22. Aggarwal, J.K. and Palmer, M.R., Boron isotope analysis. A review, *Analyst*, 120, 1301, 1995.
23. Vanderpool, R.A. and Johnson, P.E., Boron isotope ratios in commercial produce and boron-10 foliar and hydroponic enriched plants, *Agric. Food Chem.*, 40, 462, 1992.
24. Burdo, R.A. and Snyder, M.L., Determination of boron in glass by direct current plasma emission spectrometry, *Anal. Chem.*, 51, 1502, 1979.
25. Green, G.H., Blincoe, C., and Weeth, H.J., Boron contamination from borosilicate glass, *J. Agric. Food Chem.*, 24, 1245, 1976.
26. Hunt, C.D, unpublished data, 1995.
27. Nielsen, F.H, Shuler, T.R., and Uthus, E.O., Dietary arginine and methionine effects, and their modification by dietary boron and potassium, on the mineral element composition of plasma and bone in the rat, *J. Trace Elem. Exp. Med.*, 5, 247, 1992.
28. Pau, J.C.-M., Pickett, E.E., and Koirtyohann, S.R., Determination of boron in plants by emission spectroscopy with the nitrous oxide-hydrogen flame, *Analyst*, 97, 860, 1972.
29. Hunt, C.D., Halas, E.S., and Eberhardt, M.J., Long-term effects of lactational zinc deficiency on bone mineral composition in rats fed a commercially modified Luecke diet, *Biol. Trace Elem. Res.*, 16, 97, 1988.
30. Varo, P., Nurtamo, M., Saari, E., and Koivistoinen, Z., Mineral element composition of Finnish foods. III. Annual variations in the mineral element composition of cereal grains, *Acta Agric. Scand.*, Suppl. 22, 27, 1980.
31. Meacham, S. and Hunt, C.D., Boron content of commercial baby foods and beverages typically consumed by infants, *FASEB J.*, 8, A430, 1994.
32. Spivack, A.J. and Edmond, J.M., Boron isotope exchange between seawater and the oceanic crust, *Geochim. Cosmochim. Acta*, 51, 1033, 1987.
33. Zittle, C.A., Reaction of borate with substances of biological interest, in *Advances in Enzymology*, Ford, F., Ed., Interscience Publishers, New York, 1951, 493.
34. Ishikawa, T. and Nakamura, E., Suppression of boron volatilization from a hydrofluoric acid solution using a boron-mannitol complex, *Anal. Chem.*, 62, 2612, 1990.
35. Subcommittee on Laboratory Animal Nutrition, *Nutrient Requirements of Laboratory Animals*, 4th ed., National Academic Press, Washington, D.C., 1995, 173.

vitamin mix, are added together in an appropriately sized "V-shaped" plastic mixer equipped with an internal rotating "intensifier" bar. After 15 min of mixing time, the corn oil with choline chloride and *DL*-alpha-tocopherol are added and mixed for another 8 min to coat all mineral surfaces. Finally, the vitamin premix is added and *mixed* for an additional 8 min. Aliquots for subsequent mineral analysis are taken of the finished diet which is stored at –30°C in tightly capped plastic containers. To avoid staleness, the finished diet described in Table 1 should be used within 14 to 21 days after preparation. The mixing device should be washed according to the protocol outlined above immediately after final use and before mixing a subsequent batch that contains lower amounts of a dietary treatment than the previous batch.

V. ANIMAL ENVIRONMENT

In nutritional studies of boron, rodents are typically housed in all-plastic cages placed inside a laminar flow rack.[13] However, stainless steel caging apparently does not contribute appreciable amounts of boron to the local environment. Modern vivaria typically have low dust loads such that boron contamination from the air is insignificant. However, dusts created when removing droppings should be minimized. Absorbent paper under false bottom cages (designed to minimize coprophagy) to catch droppings should be changed daily. Open plastic food and water containers within the cage should be changed at least weekly. Water can be provided in plastic bottles fitted with stainless steel sipper tubes. Fresh food and deionized water (~18 MΩ · cm) should be provided daily. Wood chip bedding not analyzed for boron content should be avoided.

VI. *IN VITRO* EXPERIMENTATION

A. Media and Salt Solutions

Some form of inorganic boron is typically listed as one of the constituents of a given plant tissue culture medium. On the other hand, there are no published findings from *in vitro* studies designed to determine the physiological effects of "normal" concentrations of boron on isolated animal cells or human lines. Therefore, recommended concentrations of supplemental boron for animal cell media are unavailable. At the present time, boron is present in animal tissue culture media as a contaminant.

The only known boron analyses of media, salt solutions, or sera for use in animal cell culture[26] are presented in Table 4. The reported values do not represent certified analyses and are presented for comparative purposes only. It is not surprising that the analyzed boron content of identical media differs considerably between suppliers; the concentration of boron in animal cell culture media and salt solutions is not monitored. However, because the concentration of boron in several of those preparations seems to be remarkably low, it is possible that some of those preparations are appropriate for *in vitro* studies designed to determine the effects of boron deficiency on cell function. As described below, the addition of certain sera to low-boron media must be examined closely.

Boron analyses (Table 4) of various tissue culture media prepared according to manufacturers' instructions indicate that media shipped in liquid form in glass containers (Supplier A) may contain more boron than media shipped in dry form (Suppliers B and C). Assuming no differences in manufacture, it seems reasonable that liquid media stored in glass containers should not be used for *in vitro* cell culture studies of boron on function.

As discussed above, boron leaches readily from most glass products such that animal cells intended for *in vitro* boron research or related liquid tissue culture preparations should not come in contact with any glass surface. Demineralized, deionized water should be further purified in a quartz sub-boiling point still as described above.

B. Sera

As indicated by boron analysis of single lots of various animal sera obtained from a single supplier (Table 4), the concentration of boron apparently is considerably lower in chick or bovine calf serum than in fetal bovine, horse, lamb, pig, or, especially, rabbit serum. The sera boron values may reflect some degree of contamination from environmental sources; all sera were shipped in glass containers and stored frozen in the original containers until analysis. Based on the limited data available, it is reasonable to assume that bovine calf or fetal bovine sera are appropriate for use in boron deficiency *in vitro* studies.

VII. SUMMARY

The discovery that physiological amounts of dietary boron affect vitamin D, mineral, and bone metabolism as well as energy substrate utilization in chicks, rats, and humans was based on findings from studies that used low boron diets ($\leq$ 0.10 μg [0.009 μmol] B/g animal diet [dry weight basis] or about 0.24 mg [0.022 mmol] B/2000 kcal [8.4 MJ] of human diet) and drinking water. Boron from natural or anthropomorphic sources dissolved in potable water supplies sometimes accounts for a significant portion of total dietary boron intake. Ingested boron is considered highly available. Boron must be removed from drinking water as necessary to attain $\leq$ 0.015 μg [$\leq$ 0.001 μmol] B/mL. Formulation of boron-

TABLE 4.

Concentration of Boron in Selected Commercial Sources of Sera, Cell Culture Media, and Salt Solutions

Cell Culture Product	Boron, analyzed value (μg/mL)[a]		
	Supplier A	**Supplier B**	**Supplier C**
Sera[b]			
Bovine, calf	0.052 ±0.045[d]	—[c]	—
Bovine, fetal	0.236 ±0.018	—	—
Chicken	0.047 ±0.028	—	—
Horse, donor	0.227 ±0.033	—	—
Lamb	0.163 ±0.005	—	—
Porcine	0.126 ±0.020	—	—
Rabbit	0.653 ±0.013	—	—
Media[e]			
Basal Medium Eagle (BME), with Earle's salts without sodium bicarbonate	(L)[f] 0.022 ±0.020	(P) 0.018 ±0.005	(P) BDL[g]
Basal Medium Eagle (BME), with Hanks' salts with L-glutamine without sodium bicarbonate	—	(P) 0.023 ±0.007	(P) BDL
Dulbecco's Modified Eagle Medium (D-MEM), with 4,500 mg/L D-glucose with L-glutamine without sodium pyruvate without sodium bicarbonate	—	—	(P) 0.024 ±0.009

TABLE 4. (Continued)

Concentration of Boron in Selected Commercial Sources of Sera, Cell Culture Media, and Salt Solutions

Cell Culture Product	Boron, analyzed value (μg/mL)[a]					
	Supplier A		Supplier B		Supplier C	
F-10 Nutrient Mixture (Ham), with L-glutamine without sodium bicarbonate	(L)	0.042 ±0.007	(P)	BDL	(P)	BDL
F-12 Nutrient Mixture (Ham), with L-glutamine without sodium bicarbonate	(L)	0.059 ±0.002	(P)	BDL	(P)	BDL
Fischer's Medium, with L-glutamine without sodium bicarbonate	(L)	0.074 ±0.008		—	(P)	0.022 ±0.013
Glasgow Minimum Essential Medium (G-MEM, BHK-21), with L-glutamine without tryptose phosphate broth without sodium bicarbonate		—		—	(P)	0.029 ±0.035
Iscove's Modified Dulbecco's Medium (IMDM), with L-glutamine with 25 mmol HEPES buffer without sodium bicarbonate without ∂-thioglycerol without ß-mercapto-ethanol		—		—	(P)	BDL
Leibovitz's L-15 Medium, with L-glutamine	(L)	0.147 ±0.013		—	(P)	0.028 ±0.015
McCoy's 5A, modified, with L-glutamine without sodium bicarbonate	(L)	BDL	(P)	BDL	(P)	BDL

TABLE 4. (Continued)

Concentration of Boron in Selected Commercial Sources of Sera, Cell Culture Media, and Salt Solutions

Cell Culture Product	Boron, analyzed value (μg/mL)[a]					
	Supplier A		Supplier B		Supplier C	
Medium 199, with Earle's salts with L-glutamine without sodium bicarbonate	(L)	0.030 ±0.015		—	(P)	BDL
Medium 199, with Hanks' salts with L-glutamine without sodium bicarbonate	(L)	0.042 ±0.018		—	(P)	BDL
Medium NCTC-135, with L-glutamine without sodium bicarbonate		—		—	(P)	0.084 ±0.018
Minimum Essential Medium (MEM), with Earle's salts with L-glutamine without sodium bicarbonate	(L)	BDL	(P)	BDL	(P)	BDL
Minimum Essential Medium (MEM), with Hanks' salts with L-glutamine without sodium bicarbonate	(L)	0.079 ±0.003	(P)	0.063 ±0.011	(P)	BDL
Minimum Essential Medium (MEM) Alpha Medium, with L-glutamine with ribonucleosides and deoxyribonucleosides without sodium bicarbonate	(L)	—		—	(P)	0.061 ±0.002

TABLE 4. (Continued)

Concentration of Boron in Selected Commercial Sources of Sera, Cell Culture Media, and Salt Solutions

Cell Culture Product	Boron, analyzed value (μg/mL)[a]					
		Supplier A		Supplier B		Supplier C
Minimum Essential Medium (S-MEM), with L-glutamine without sodium bicarbonate	(L)	—	(P)	BDL	(P)	BDL
Minimum Essential Medium (S-MEM), (Joklik-modified) with L-glutamine with 10X phosphate without calcium chloride without penicillin G without streptomycin without sodium bicarbonate	(L)	—	(P)	BDL	(P)	0.068 ±0.004
RPMI Medium 1640, with L-glutamine without sodium bicarbonate	(L)	BDL	(P)	0.074 ±0.012	(P)	BDL
Waymouth's MB 752/1 Medium, with L-glutamine	(L)	0.071 ±0.024		—	(P)	0.056 ±0.019
Salt Solutions						
Dulbecco's Phosphate-buffered Saline (D-PBS), with 1,000 mg/l of D-glucose with 36 mg/L sodium pyruvate with 5 mg/L phenol red	(L)	—			(P)	0.030 ±0.015
Earle's Balanced Salt Solution (EBSS), without sodium bicarbonate	(L)	0.031 ±0.015	(P)	BDL	(P)	0.017 ±0.004

TABLE 4. (Continued)

Concentration of Boron in Selected Commercial Sources of Sera, Cell Culture Media, and Salt Solutions

Cell Culture Product	Boron, analyzed value (μg/mL)[a]					
		Supplier A		Supplier B		Supplier C
Hanks' Balanced Salt Solution (HBSS), without sodium bicarbonate with phenol red	(L)	—	(P)	0.037 ±0.009		—
Hanks' Balanced Salt Solution (HBSS), without sodium bicarbonate without phenol red	(L)	0.035 ±0.008	(P)	BDL	(P)	BDL

[a] All samples were analyzed in triplicate by inductively-coupled argon plasma emission spectroscopy.[52]
[b] All sera received in, and stored frozen in, glass containers.
[c] Dash (—) indicates that a given product was not analyzed for boron or there was insufficient statistical confidence to report the analyzed value.
[d] Mean ± standard deviation. Reported values[26] do not represent certified analyses and are presented for comparative purposes only.
[e] All media filtered (10 mL) 1.5 h after preparation through a 0.2 μm "acrodisc" filter assembly.
[f] All media and salt solutions were prepared according to manufacturer's instructions from original liquids (L) or powders (P) prior to boron analysis. All media and salt solutions received in liquid form were shipped and subsequently stored frozen in glass containers.
[g] Below detection limit (<0.015 μg/mL).

low diets exploits the unequal boron requirement, and typically, boron tissue concentration within Angiospermae, the class of plants most often utilized in rodent diets. Monocotyledon-based (for example, coconuts, corn, rice, and wheat), compared to dicotyledon-based (for example, beans, peanuts, potatoes, and soybeans) products, are lower in boron content and therefore more appropriate sources of carbohydrate, protein, or fat. Animal-based products typically contain low amounts of boron. Vitamin supplements contribute inconsequential amounts of boron although coordination of certain vitamins or their carriers with boron may be of concern. Variability among lots or different forms of a dietary ingredient, especially individual mineral element supplements, necessitates boron analysis of each batch of diet by sensitive instrumentation.

Because boron isotopic fractionation occurs in plants and varies according to the plant tissue, special care must be taken to ensure that test doses of intrinsically-labeled plant materials (for use in boron kinetics studies) is taken from similar tissues and homogenized thoroughly. The preparation of diets low in boron content requires special avoidance of seemingly negligible sources of environmental contamination. Most notably, boron is a constituent or contaminate of most hard glasses. Because boron leaches readily from glass, drinking water and wet or hydrated dietary ingredients must not come in contact with laboratory glassware even for short periods of time. Inorganic boron, a common constituent of plant tissue culture media, is present in animal media only as a contaminant at the present time. The boron content of some commercially prepared media is apparently low. Such media and certain animal sera may be appropriate for studies of the consequences of boron deficiency conducted *in vitro*.

REFERENCES

1. Loomis, W.D. and Durst, R.W., Chemistry and biology of boron, *BioFactors*, 3, 229, 1992.
2. Lovatt, C.J., Evolution of xylem resulted in a requirement for boron in the apical meristems of vascular plants, *New Phytol.*, 99, 509, 1985.
3. Lovatt, C.J. and Dugger, W.M., Boron, in *Biochemistry of the Essential Ultratrace Elements*, Frieden, E., Ed., Plenum Press, New York, 1984, 389.
4. Hunt, C.D., Dietary boron modified the effects of magnesium and molybdenum on mineral metabolism in the cholecalciferol-deficient chick, *Biol. Trace Elem. Res.*, 22, 201, 1989.
5. King, N., Odom, T.W., Sampson, H.W., and Yersin, A.G., The effect of *in ovo* boron supplementation on bone mineralization of the vitamin D-deficient chicken embryo, *Biol. Trace Elem. Res.*, 31, 223, 1991.
6. Nielsen, F.H., Mullen, L.M., and Gallagher, S.K., Effect of boron depletion and repletion on blood indicators of calcium status in humans fed a magnesium-low diet, *J. Trace Elements Exp. Med.*, 3, 45, 1990.
7. Nielsen, F.H., Gallagher, S.K., Johnson, L.K., and Nielsen, E.J., Boron enhances and mimics some effects of estrogen therapy in postmenopausal women, *J. Trace Elem. Exp. Med.*, 5, 237, 1992.
8. Hegsted, M., Keenan, M.J., Siver, F., and Wozniak, P., Effect of boron on vitamin D deficient rats, *Biol. Trace Elem. Res.*, 28, 243, 1991.
9. Hunt, C.D., Herbel, J.L., and Idso, J.P., Dietary boron modifies the effects of vitamin D_3 nutriture on indices of energy substrate utilization and mineral metabolism in the chick, *J. Min. Bone Res.*, 9, 171, 1994.
10. Hunt, C.D., Herbel, J.L., and Nielsen, F.H., Physiological amounts of dietary boron influence magnesium and calcium metabolism in the postmenopausal woman, *FASEB J.*, 8, A430, 1994.
11. Meacham, S.L., Taper, L.J., and Volpe, S.L, Effect of boron supplementation on blood and urinary calcium, magnesium, and phosphorus, and urinary boron in athletic and sedentary women, *Am. J. Clin. Nutr.*, 61, 341, 1995.
12. Nielsen, F.H., Dietary boron affects variables associated with copper metabolism in humans, in *6th International Trace Element Symposium 1989*, Anke, M., Baumann, W., Bräunlich, H., Brückner, C., Groppel, B., and Grün, M., Eds., Karl-Marx-Universitat, Leipzig and Friedrich-Schiller-Universitat, Jena, DDR, 1989, 1106.
13. Hunt, C.D. and Herbel, J.H., Boron affects energy metabolism in the streptozotocin-injected, vitamin D_3-deprived rat, *Magn. Trace Elem.*, 10, 374, 1991–1992.

36. Deitch, E.A., Intestinal permeability is increased in burn patients shortly after injury, *Surgery*, 107, 411, 1990.
37. Scudi, J.V., Bastedo, W.A., and Webb, T.J., The formation of a vitamin B_6-borate complex, *J. Biol. Chem.*, 136, 399, 1940.
38. Landauer, W., Complex formation and chemical specificity of boric acid in production of chicken embryo malformations, *Proc. Soc. Exp. Biol. Med.*, 82, 633, 1953.
39. Frost, D.V., The water-soluble riboflavin-boron complex, *J. Biol. Chem.*, 145, 693, 1942.
40. Landauer, W. and Clark, E., On the role of riboflavin in the teratogenic activity of boric acid, *J. Exp. Zool.*, 156, 307, 1964.
41. Roe, D.A., McCormick, D.B., and Ren-Tsong, L., Effects of riboflavin on boric acid toxicity, *J. Pharm. Sci.*, 61, 1081, 1972.
42. Subcommittee on Laboratory Animal Nutrition, *Nutrient Requirements of Laboratory Animals*, 3rd ed., National Academic Press, Washington, D.C., 1978, 7.
43. Greenwood, N.N. and Earnshaw, A., *Chemistry of the Elements*, Pergamon Press, 1984, 155.
44. Green, G.H. and Weeth, H.J., Responses of heifers ingesting boron in water, *J. Anim. Sci.*, 46, 812, 1977.
45. Brown, T.F., McCormick, M.E., Morris, D.R., and Zeringue, L.K., Effects of dietary boron on mineral balance in sheep, *Nutr. Res.*, 9, 503, 1989.
46. Magour, S., Schramel, P., Ovcar, J., and Mäser, H., Uptake and distribution of boron in rats: interaction with ethanol and hexobarbital in the brain, *Arch. Environm. Contam. Toxicol.*, 11, 521, 1982.
47. Pfeiffer, C.H. and Jenney, E.H., The pharmacology of boric acid and boron compounds, *Bull. Nat. Formulary Comm.*, 18, 57, 1950.
48. Vanderpool, R.A., Hoff, D., and Johnson, P.E., Use of inductively coupled plasma-mass spectrometry in boron-10 stable isotope experiments with plants, rats, and humans, *Environ. Health Perspect.*, 102 (suppl 7), 13, 1994.
49. Nomura, M., Kanzaki, T., Ozawa, T., Okamoto, M., and Kakihana, H., Boron isotopic composition of fumarolic condensates from some volcanoes in Japanese island arcs, *Geochim. Cosmochim. Acta*, 46, 2403, 1982.
50. Meacham, S.L. and Hunt, C.D., Boron content of common American foods, *FASEB J.*, 9, A576, 1995.
51. Varo, P., Lehelma, O., Nuurtamo, M., Saari, E., and Koivistoinen, P., Mineral element composition of Finnish foods. VII. Potato, vegetable, fruits, berries, nuts and mushrooms, *Acta Agriculturae Scand.*, 22(suppl), 89, 1980.
52. Hunt, C.D. and Shuler, T.R, Open-vessel, wet-ash, low-temperature digestion of biological materials for inductively coupled argon plasma spectroscopy (ICAP) analysis of boron and other elements, *J. Micronutrient Anal.*, 6, 161, 1989.

Chapter 15

Measurements of Boron in Rodent Diets and Tissues

Curtiss D. Hunt
U.S. Department of Agriculture/ARS
Grand Forks Human Nutrition Research Center
Grand Forks, North Dakota

CONTENTS

0-8493-9611-5/97/$0.00+$.50

I. OVERVIEW OF BORON ESSENTIALITY AND FUNCTION

Boron is a light, nonmetallic element known to be essential for vascular plants, diatoms, and some species of marine algal flagellates.[1,2] Advances in analytical capabilities have prompted renewed interest in determining the essentiality of boron for humans. To help identify specific boron-dependent metabolic processes in animals, considerable effort has been expended to assess boron metabolism (for example, uptake, distribution, turnover, and excretion) and the effects of dietary boron on mineral metabolism. The absence of a usable boron radioisotope has spurred development of isotope dilution methodology that is based on manipulations of the distribution of the stable isotopes ^{10}B and ^{11}B over time in body tissues.

It is now known that most (90 to 100%) ingested boron is absorbed and excreted in the urine[3–6] with increased concentrations in milk during supplementation.[7] Boron supplemented to a boron-low diet (~0.2 μg [μmol]/g) modulates mineral metabolism and does so in a species-specific manner. For example, dietary boron improves the apparent absorption and retention of calcium and phosphorus and increases femur magnesium concentrations in the vitamin D-deficient rat.[8] Dietary boron increases femoral calcium and phosphorus concentrations in vitamin D adequate, but not inadequate chicks.[9] Dietary boron also influences the concentrations of several minerals in the hearts of vitamin D-deprived rats.[9] In humans, dietary boron influences serum ionized calcium concentrations[11] as well as the urinary excretion of calcium.[5,10]

Current methods of boron analysis of biological substances reflect technological advances as well as greater appreciation for characteristics of boron geochemistry, boron speciation, and boron anthropomorphic sources. The analysis of biological substances low in boron concentration (< 1.0 μg [0.092 μmol]/g) depends upon the correct manipulation of those characteristics as described below.

II. BORON CONTAMINATION

A. Labware

Boron is either a constituent or contaminate of most laboratory glassware[11] because the element, in the form of borax, is used in the preparation of hard (borosilicate) glasses. Boron leaches readily from most glassware, especially under conditions of alkalinity.[12] Even the brief time taken to quantify volumes of demineralized, deionized water* with a standard glass volumetric pipette is sufficient exposure to glass to increase the boron content of a water sample from background concentrations[13] to half that found in rat plasma (0.050 μg [0.004 μmol]/mL).[14] The use of borosilicate glass digestion vessels for low-level detection of boron is unacceptable for obvious reasons, but, surprisingly, is sometimes overlooked. Nickel and vycor crucibles also have a high boron content (3% B_2O_3) and contaminate the sample.[15] The cost of boron-free digestion vessels, such as pure metal (i.e., zirconium or platinum) or quartz crucibles is usually prohibitive. Teflon** tubes are ideal for wet-digestion of biological materials as described below.

Glass nebulizers, which convert a liquid sample into an aerosol in, for example, an inductively coupled plasma spectrophotometer should be replaced with nebulizers made of specialized plastics, Teflon, polypropylene and platinum, or quartz. Some contact with glass cannot be avoided; only glass repetitive dispensers can withstand direct contact with concentrated nitric acid over time. Before use with acids, glass repetitive dispensers should be flushed adequately with the acid to be used to remove leached boron. A plastic repetitive dispenser with a glass plunger coated with Teflon is an appropriate device to dispense certain analytical reagents (i.e., hydrogen peroxide).

Before use, all labware [for example, Teflon and polypropylene tubes] except new plastic transfer pipettes should be cleaned by using the following method. All labware should be rinsed with tap water, soaked overnight in a soap solution ("Micro," International Products Corp., Trenton, NJ; 19.5 mL/L warm water), and subsequently sonicated for ten minutes in another EDTA (ethylenediaminetetraacetic acid)-soap solution ("Radiacwash," Atomic Products Corp., Shirley, NY, 5:40 dilution; "Liquinox," Alconox, Inc., New York, NY, 1:30 dilution). The tubes are then rinsed with hot tap water, demineralized water, and placed in a "Radiacwash," solution (5:40 dilution) for 1 h. They are again rinsed a minimum of three times each with demineralized, then deionized water. Because boron is also used in the manufacture of many detergents, soaps, cleaners,[16] special care should be taken to rinse plastic ware of all cleaning agents as described above.

To further reduce environmental boron after the washing procedure, all Teflon and polypropylene tubes are soaked overnight in 2 *M* HCl ("Baker Analyzed," J.T. Baker Chemical Co., Phillipsburg, NJ), then rinsed

* The term "demineralized" is used instead of "distilled" to denote the removal of minerals from aqueous solutions by all methods including distillation by volatilization and subsequent condensation. The term "deionized" is used to denote the further removal of small trace amounts of both cations and anions from aqueous solutions.

** Mention of a trademark or proprietary product does not constitute a guarantee or warranty of the product by the U.S. Department of Agriculture, and does not imply its approval to the exclusion of other products that may also be suitable.

and submerged in water (demineralized, deionized, ~18.0 MΩ · cm; Millipore System, "Super-Q," Millipore Corp., Bedford, MA) overnight, then finally rinsed eight times each with demineralized and deionized waters, respectively, before being dried in a glassware-drying oven (50° C).

B. Water and Analytical Reagents

The concentration of boron in surface water[16] may vary substantially between geographic locations and water sources, from negligible (≤ 0.015 μg [0.001 μmol]/mL) to relatively very high (120 μg [11.2 μmol]/mL) concentrations. Therefore, boron usually must be removed from water used in boron analysis, a process accomplished by reverse osmosis* which provides water of relatively high purity (~15.0 MΩ · cm). In my laboratory, the process of reverse osmosis is followed by further purification and deionization steps to achieve nearly "pure" water** as measured by resistivity (~18.0 MΩ · cm). These steps, in sequence, are *mixed* bed ion [cation and anion] exchange, ultraviolet light, activated carbon, second mixed bed ion exchange, and 0.2 μ filter. Boron-specific resins also effectively remove boron from solution,[18] although one such resin that contains a hydrophobic styrene backbone and a tertiary amine group will only absorb boron in the alkaline form [$B(OH)_4^-$].[19] Obviously, water should be processed to remove other elements undergoing concurrent analyses. Mannitol, a polyhydroxy compound, reacts with boric acid to form stable mannitol-boron complexes even in the presence of 6 mol/L hydrochloric acid.[20] Thus, the boron content of even medium-grade reagents (HNO_3, HCl) can be reduced to acceptable amounts by distillation in a quartz sub-boiling point still that contains mannitol (6 g [0.033 mol]/L water).[21]

III. SAMPLE PREPARATION

A. Standards

Accurate determinations of low concentrations of boron in biological substances have proven exceptionally difficult. The analytical difficulty is

* Feed water for the reverse osmosis (RO) process that contains chlorine or water-hardening cations (calcium, magnesium, and iron ions) should be processed through activated charcoal and an ion-exchange system (typical water-softening) to prolong the life of the RO filter.

** Pure water has a theoretical specific resistance of 18.2 MΩ · cm at 25°C. Analysis of water purity by the measure of resistivity is based on ionization of materials in solution and does not take into account uncharged molecules, for example, many organic molecules. In potable water, inorganic boron is essentially present only as the mononuclear species $B(OH)_3$ and $B(OH)_4^-$. The uncharged $B(OH)_3$ is the dominant species in most water (pH 1, ~100%; pH 7.4, >98%; pH 9.25, 50%; pH 11, ~0%).[17] Therefore, determination of water purity by typical resistivity measurements may not provide an accurate indication of boron concentrations in water.

exemplified by the fact that until recently there was no biological standard reference material (SRM) certified by the U.S. National Institute of Standards and Technology (NIST) for boron. Improvements in boron analysis technology has prompted the development of several NIST-certified biological standard reference materials appropriate for use during analysis of biomaterials low in boron content (0.6 to 2.8 μg/g) (bovine muscle powder [SRM 8414], whole egg powder [SRM 8415], and corn bran [SRM 8433]) or high in boron content (27.0-37.6 μg/g) (apple leaves [SRM 1515], spinach leaves [SRM 1570a], tomato leaves [SRM 1573a], peach leaves [SRM 1547]). The standard for boron isotope analysis is NIST boric acid (SRM 951).

B. Sample Dissolution

All methods of boron analysis, except neutron activation analysis (NAA), require dissolution of the sample. Most wet-ash methods employ the use of either perchloric or hydrofluoric acid, both of which have extremely toxic and/or explosive qualities.[22] Many boron compounds volatize at temperatures far below those that are required for most dry-ash procedures (≥ 450°C). Microwave bomb digestion technology contains volatized boron although special care must be taken to avoid reducing the boron concentration below detection by dilution.[23] Ashing agents (i.e., barium hydroxide or calcium oxide) may be added to facilitate sequestration of boron during dry-ashing.[24] However, these agents usually contain concentrations of boron high enough to interfere with boron determination or leave unacceptable amounts of chemical residue in the final digestate. Additional ashing agents are not required during the digestion process if the natural calcium concentration in the sample is high enough to sequester the boron (as determined by recovery of added boron). Alkali fusion methods for boron sequestration may cause damage to metal crucibles or loss of sample from splattering[25] and may require subsequent extraction of the boron by distillation.[19]

The boron halides, formed during various sample dissolution procedures, are extremely volatile boron compounds (boiling point [° C]: BF_3, –99.9; BCl_3, 12.5; BBr_3, 91.3; BI_3, 210).[26] However, except for chlorine and fluorine, the concentration of halides in substances of nutritional importance is negligible. Fluorine concentrations are typically negligible except in substances, for example, various dentrifrices, to which some form of fluoride is purposely added. Boron-chloride compounds formed by the following reaction are extremely volatile:[27]

$$B_2O_3(\text{or } [B_4O_7]^{2\pm}) + 3Cl_2 + 2C = 2BCl_3 + CO + CO_2$$

Fortunately, those volatile compounds readily decompose in water.[28] Furthermore, that reaction requires fairly high temperatures, such as those attained when a digestate is left to dry and char.[26] In the wet-ashing procedure described below, digestates remain in liquid form and boron is probably present in those digestates as $(B_4O_7)^{2-}$ or H_3BO_3, chemical species that do not form BCl_3.[27] In situations where volume reduction after digestion is required, mannitol may be added at a 10/1 ratio (mannitol to boron) to reduce boron volatilization.[29]

C. Practical Dissolution of Biomaterials Low in Boron Content

A recently developed digestion technique[30] for subsequent analysis of low-boron biomaterials by inductively coupled argon plasma emission spectroscopy (ICP) or ICP-mass spectrometry (ICP-MS) circumvents many of the previously described problems associated with boron analysis. Significant improvements of the technique since its initial publication warrants detailed description of the procedure.

Reference materials (with matrices, mass, and boron concentrations similar to that of samples) and samples (1 to 5 g based on expected boron content; total number not to exceed that which can be analyzed in one day) are weighed directly into Teflon tubes (28.5 × 104 mm; 50 mL; Nalge Co., Rochester, NY). Five to ten mL of high quality (HQ) 15.9 M HNO_3 ("Baker Analyzed" grade, J.T. Baker Chemical Co., Phillipsburg, NJ; purified in sub-boiling point quartz still [described above] and stored in Teflon bottles) are added to each tube and the tubes are capped with a Teflon screw cap (drilled and fitted with 1 to 2 mm diameter Teflon tubing (20 mm length) for release of generated gas) and allowed to sit for 48 h at room temperature. Because certain sample matrixes (for example, samples containing chocolate) induce rapid foaming reactions, it is sometimes necessary to place some of the tubes in an ice water bath to limit foaming of the sample after introduction of the acid. Tubes are then stationed in holes (tapered at bottom; 25 mm deep, 30 mm diameter) and drilled in an aluminum block that is placed over a hotplate. Sample temperatures are ramped (50° C, 3 h; 90 ° C, 3 h; 110°C, 2 to 3 h) to a temperature less than that of the constant boiling point of nitric acid [120.5° C]) each time they are placed in the heating block.

After refluxing for 48 to 72 h, samples are brought to near dryness (0.25 to 0.50 mL) on the heating block by removing the vented screwcap, and then removed from heat to avoid complete evaporation of nitric acid and recapped. Complete drying of the sample allows temperatures to rise far above that needed to induce significant boron volatilization. After cooling, 30% H_2O_2 (hydrogen peroxide, "Baker Analyzed" grade, obtained in plastic containers, J.T. Baker Chemical Co., Phillipsburg, NJ) and HQ 15.9 M HNO_3 are added sequentially in a ratio of 3:10 for a total of 12 to 13 mL and allowed to sit overnight at room temperature before heated refluxing (CAUTION: HNO_3 and H_2O_2 must not be mixed together and stored, as strong exothermic reactions may create sufficient heat over time to induce a violent explosion). After refluxing for 48 to 72 hr, samples are brought again to near dryness.

The additions and subsequent evaporations are repeated until each digestate is visibly clear of particulate matter. Fat digestion is enhanced by application of 30% H_2O_2 in a dropwise fashion directly onto heated (≤ 110°C) samples (CAUTION: This procedure is dangerous). All heating is carried out in a standard laboratory hood. Vented tubes are stored in vented cabinets when removed from the heating block. Triplicate blanks in Teflon tubes are prepared by adding and evaporating to near dryness all reagents, in exactly the same quantities and sequences as those used for the samples.

After digestion, the tubes sit overnight with 0.5 mL of 12.1 *N* HCl ("Instra-Analyzed" grade, J.T. Baker Chemical Co., Phillipsburg, NJ; purified in sub-boiling point quartz still [described above] and stored in Teflon bottles). All samples and blanks are then quantitatively transferred (with plastic 3.0 mL transfer pipettes [BA Supply Co., San Antonio, Texas]; minimum of three washes) to pre-weighed polypropylene tubes (Becton Dickinson Labware, Lincoln Park, NJ). Samples are diluted (typically ~1:5 to 10) by weight with demineralized, deionized water (~18.0 MΩ · cm; Millipore System, Super-Q, Millipore Corp., Bedford, MA). For analysis by ICP, a 1:5 to 10 dilution is appropriate for concurrent analysis of copper, iron, manganese, vanadium, and nickel (described below). To bring other selected elements (for example, calcium, magnesium, sodium, and phosphorus) within analytical range, those diluates are diluted further (1:10) by weight to obtain a total dilution of 1:100.

Final sample dilution (by weight, not volume, to avoid contact with glass volumetrics and precision errors associated with plastic labware) is achieved by drop-wise addition of demineralized, deionized water. Original sample weight, empty dilution tube weight, and the combined weights of the dilution tube and final diluate are required to calculate the final dilution. Preparation of a typical batch of 50 samples for analysis by ICP usually requires ten working days including weighing and final dilution with strict observance of appropriate safety procedures maintained throughout the entire procedure.

For boron analysis by ICP-mass spectrometry (ICP-MS), beryllium is used as an internal standard because boron and beryllium have similar masses and ionization energies.[29] A working dilution of beryllium standard is prepared by adding 1.000 mL of the original beryllium standard solution [ICP standard solution, Specpure, 1000 μg Be/mL; $Be_4O(C_2H_3O_2)_6$ in 5% HNO_3; Johnson Matthey distributed by Alfa AESAR, Ward Hill, MA] to 1000 mL of a 5% nitric acid solution (HQ 15.9 *M* HNO_3 in demineralized, deionized water [purified further in sub-boiling point quartz still]). Exactly 0.250 mL of the working beryllium standard and exactly 0.150 mL of 15.9 M HNO_3 are added by pipette (P1000 Rainin Pipetman; Rainin, Woburn, MA) to each pre-weighed polypropylene tube (specific for each model of mass spectrometer) and brought up to final weight (exactly 5.000 g) with demineralized, deionized (~18.0 MΩ · cm) water.

IV. SAMPLE EXTRACTION, ISOLATION, AND CLEAN-UP

Boron analysis by colorimetric methods requires extraction of the boron-ligand complex prior to spectrophotometric[18,25] or fluorimetric[31] analysis. Various interferents (i.e., Cu^{2+}, Al^{3+}, Fe^{3+}, K^+, F^-, or V^{5+}), when present, react with boron or with the free molecules of the ligand employed in the colorimetric method to give erroneous absorbance readings;[32] additional clean-up procedures are required in those situations.

Determination of boron isotope ratios by thermal ionization mass spectrometry (TI-MS) requires the removal of impurities in ashed samples to eliminate isobaric interferences, avoid suppression of the ionization of the boron species analyzed, maintain a good vacuum in the instrument, and avoid fractionation of the ion beams during analysis.[19] Boron is therefore extracted by ion exchange chromatography or distillation prior to analysis by TI-MS.[21] Further purification after ion exchange by converting boron to the BF^- anion, an extremely volatile compound, represents a serious disadvantage of the extraction technique.[19] The distillation technique is time consuming and suitable only to aqueous samples with relatively high boron concentrations.[19] As discussed above, boron extraction by exposure to boron-specific resins requires that all boron in the sample be present in the alkaline form $[B(OH)_4^-]$.[19]

V. BORON ANALYSIS

Several methods for the determination of boron in biological materials are briefly discussed in descending order of preference according to overall sensitivity, availability, throughput, safety, and cost.

A. Inductively Coupled Argon Plasma Emission Spectroscopy (ICP-AES)

ICP-AES technology is relatively inexpensive and provides acceptable sensitivity (≤ 0.015 μg [0.001 μmol] B/g) for the measurement of boron in most biomaterials.[30] The technology circumvents most boron speciation problems during the preparatory or analytical phases of the procedure. Sample preparation (described above) is straightforward. Because ICP-AES generates many emission lines, spectral interference may mar the analytical results. ICP-AES generates boron emission lines at 249.618 and 249.733 with the latter of greater height and therefore better suited for boron detection (not appropriate when significant amounts of iron are present in the sample).

In regards to the two forms of ICP-AES technology, sequential, compared to simultaneous, analysis of mineral elements provides better flexibility in maximizing sensitivity and minimizing interference within a given matrix. Both forms of ICP-AES technology allow for concurrent determination of several mineral elements including aluminum, calcium, copper, iron, magnesium, manganese, molybdenum, nickel, phosphorus, sodium, and zinc (with working detection limits varying considerably among these elements).

B. Inductively Coupled Plasma-Mass Spectrometry (ICP-MS)

ICP-MS combines ICP technology for ionization of minerals in solution with that of MS technology for mass determinations. The mass spectrometer is used as a powerful detector for the ICP spectrometer.[33] The mass spectrometric detection of elements extends the sensitivity of the ICP to at least 0.11 ng [10 fmol]/g for ^{10}B, and 0.40 ng [37 fmol]/g for ^{11}B.[29] Memory, the retention of a signal from the previous sample, increases with increasing boron concentration and is best accommodated with a 5 min rinse with 1% HQ HNO_3, a procedure that significantly reduces sample throughput.[29]

The radioisotopes ^{8}B, ^{12}B, and ^{13}B all have half-lives of less than one second. Thus, they have very limited value in biological studies of boron. Fortunately, there are two stable boron isotopes, ^{10}B (18.9 atom%) and ^{11}B (81.02 atom%); their naturally-unequal distribution allows ^{10}B enrichment without undue increases in background boron concentrations. Exploitation of this phenomenon permits use of isotopic dilution in the study of boron kinetics and the determination of body boron pool size. The disadvantages associated with TI-MS technology (described below) in the determination of boron body pool size and distribution are minimized with ICP-MS technology. Boron extraction is not required before analysis unless the solution contains high concentrations of dissolved salts.[19] However, the lack of a fixed natural abundance value for boron is also a problem for this technology and requires that a natural abundance ratio be determined for each sample or related data set.[29] Furthermore, the natural abundance variability prevents quantitation and calculation of isotope dilution by instrument-supplied software.[29] Finally, the matrix of the solution significantly influences the apparent isotope ratio in the sample.[19,29]

C. Thermal Ionization Mass Spectrometry (TI-MS)

Use of TI-MS provides a sensitive method for the determination of boron isotope ratios ($^{11}B/^{10}B$), and is based on the measurement of the isotopic composition of specific boron complexes, as either positive ions (PITI-MS) or as negative ions (NITI-MS).[19] Significant isotope fractionation may occur during sample analysis, a phenomenon that can be reduced by increasing the molecular mass of the species analyzed. The $CsBO_2^+$ species at masses of 308 and 309 is not as prone to instrumental mass fractionation.[19] Relatively long sample analysis time, high instrumentation costs, frequent equipment failures, and limited availability puts TI-MS technology at a serious disadvantage compared to other MS-related technologies.

D. Neutron Activation Methods

Neutron activation of ^{10}B in various biological matrices, followed by mass spectrometry (NA-MS) or in-beam thermal neutron capture prompt gamma-ray activation analysis (NA-PGAA), provides analytical sensitivity in the range of <0.010 μg (0.0009 μmol) B/g[34] or 0.3 (0.028) to 0.8 μg (0.074 μmol) B/g,[35] respectively. NA-MS technology determines the ^4He produced by the thermal-neutron reaction ^{10}B (n, alpha) ^7Li.[34] NA-PGAA technology quantifies gamma-ray emission from ^{10}B atoms in the sample during sample irradiation with neutrons.[35] Both NA techniques require access to a working nuclear reactor and the assumption that the $^{10}B/^{11}B$ ratio in each sample is "usual" and invariable.[34] These assumptions limit use of NAA because it is known that boron fractionation occurs in plants such that natural abundance boron isotope ratios in plant material are variable.[21] Furthermore, NA-MS must take into account the contributions from neutron reactions with other target nuclides (for example, in blood, nitrogen, carbon, and oxygen) to measured ^3He and ^4He.[34] NA-PGAA must also take into account the consequences of neutron scattering in hydrogenous matrices.[35] Hydrogen concentrations range from ~6% for grain products up to ~12% for fats and whiskey.[36]

E. Other Methods

Several common analytical methods for boron analysis and/or associated clean-up procedures (described above) are dependent upon, or hampered by, the exact speciation of boron in a given sample, usually either $B(OH)_3$ or $B(OH)_4^-$. For example, boron colorimetric methods measure the visible absorbance peak of various boron complexes, e.g., methylene blue-tetrafluoroborate. Unfortunately, those colorimetric methods require that boron be present as undissociated $B(OH)_3$ because only that species complexes with a given assay ligand (for example, curcumin, carminic acid, azomethine H). Determination of boron content by analysis of the anionic complex formed between boric acid and chromotropic acid by high-performance liquid chromatography (HPLC) of boron depends upon all boron being present as the $B(OH)_3$ species.[37] Furthermore, molybdenum (VI) and vanadium (V) ions cause positive errors in the determination of boron by HPLC at concentrations near 10 μmol/L (Mo, 960 μg [10.01 μmol]/L; V, 510 μg [10.01 μmol]/L).[37] Conductivity detection methods of boron-polyol compounds after ion-exclusion chromatographic separation are also dependent upon correct boron speciation.[38] Atomic absorption spectrometry (AAS) requires that all boron be converted to the elemental form. Unfortunately, certain boron compounds, (i.e., the boron carbides) are stable to 3000°C, a temperature above that produced by a graphite furnace.

VI. SUMMARY

Discovery of physiological effects of dietary boron restriction in chicks, rats, and humans has prompted efforts to assess boron metabolism (for example, uptake, distribution, turnover, and excretion) and the effects of dietary boron on mineral metabolism. Samples prepared for boron analysis must avoid contact with glassware because boron leaches readily from hard glass surfaces. Distillation of medium-grade reagents to remove boron in a quartz sub-boiling point still that contains mannitol helps control material costs associated with boron analysis. Many boron compounds volatize at temperatures far below those required for most dry-ash procedures. Attempts to sequester boron in the digestate may result in unacceptable cost, boron contamination, chemical residue, or hazardous working conditions. Wet-ash, low-temperature digestion circumvents many of these problems.

ICP-AES technology is relatively inexpensive, provides acceptable sensitivity in the analysis of most biomaterials for boron content, and circumvents most boron speciation problems. ICP-MS provides exceptional sensitivity but must adjust for the lack of a fixed natural abundance value for boron and differences in matrix structure. Boron extraction is not necessarily required before analysis by ICP-MS. ICP-MS is very useful in determining changes in the ^{10}B:^{11}B ratio and subsequently in the study of boron kinetics and the determination of body boron pool size. TI-MS is a sensitive method for the determination of boron isotope ratios ($^{11}B/^{10}B$) although significant isotope fractionation of low-molecular mass species occurs during sample analysis and relatively long sample analysis time hinders productivity. NA-MS or NA-PGAA provide excellent to mediocre analytical sensitivity, respectively. Both NA techniques require access to a working nuclear reactor, assume "normal" and invariable boron isotopic ratios in each sample, and are affected by contributions from neutron reactions with other target nuclides. Most colorimetric methods and TI-MS methodology requires the prior removal of impurities and/or specific speciation of the boron in the sample.

REFERENCES

1. Loomis, W. and Durst, R., Chemistry and biology of boron, *BioFactors*, 3, 229, 1992.
2. Lovatt, C., Evolution of xylem resulted in a requirement for boron in the apical meristems of vascular plants, *New Phytol.*, 99, 509, 1985.
3. Kent, N. and McCance, R., The absorption and excretion of 'minor' elements by man. I. Silver, gold, lithium, boron and vanadium, *Biochem. J.*, 35, 837, 1941.
4. Tipton, I., Stewart, P., and Martin, P., Trace elements in diets and excreta, *Health Phys.*, 12, 1683, 1966.
5. Hunt, C., Herbel, J., and Nielsen, F., Physiological amounts of dietary boron influence magnesium and calcium metabolism in the postmenopausal woman, *FASEB J.*, 8, A430, 1994.
6. Jansen, J., Andersen, J., and Schou, J., Boric acid single dose pharmacokinetics after intravenous administration to man, *Arch. Toxicol.*, 55, 64, 1984.
7. Owen, E., The excretion of borate by the dairy cow, *J. Dairy Res.*, 13, 243, 1944.
8. Hegsted, M., Keenan, M., Siver, F., and Wozniak, P., Effect of boron on vitamin D deficient rats, *Biol. Trace Elem. Res.*, 28, 243, 1991.
9. Hunt, C. and Herbel, J., Effects of dietary boron on calcium and mineral metabolism in the streptozotocin-injected, vitamin D_3-deprived rat, *Magnes. Trace Elem.*, 10, 387, 1991-1992.
10. Meacham, S., Taper, L., and Volpe, S., Effect of boron supplementation on blood and urinary calcium, magnesium, and phosphorus, and urinary boron in athletic and sedentary women, *Am. J. Clin. Nutr.*, 61, 341, 1995.
11. Burdo, R. and Snyder, M., Determination of boron in glass by direct current plasma emission spectrometry, *Anal. Chem.*, 51, 1502, 1979.
12. Green, G., Blincoe, C., and Weeth, H., Boron contamination from borosilicate glass, *J. Agric. Food Chem.*, 24, 1245, 1976.
13. Hunt, C., unpublished data, 1995.
14. Nielsen, F., Shuler, T., and Uthus, E., Dietary arginine and methionine effects, and their modification by dietary boron and potassium, on the mineral element composition of plasma and bone in the rat, *J. Trace Elem. Exp. Med.*, 5, 247, 1992.
15. Lohse, G., Microanalytical azomethine-H method for boron determination in plant tissue, *Commun. in Soil Sci. Plant Anal.*, 13, 127, 1982.
16. Butterwick, L., de Oude, N., and Raymond, K., Safety assessment of boron in aquatic and terrestrial environments, *Ecotox. Environ. Safety*, 17, 339, 1989.
17. Spivack, A. and Edmond, J., Boron isotope exchange between seawater and the oceanic crust, *Geochim. Cosmochim. Acta*, 51, 1033, 1987.
18. Mair, J. and Day, H., Curcumin method for spectrophotometric determination of boron extracted from radiofrequency ashed animal tissues using 2-ethyl-1,3-hexanediol, *Anal. Chem.*, 44, 2015, 1972.
19. Aggarwal, J. and Palmer, M., Boron isotope analysis. A review, *Analyst*, 120, 1301, 1995.
20. Ishikawa, T. and Nakamura, E., Suppression of boron volatilization from a hydrofluoric acid solution using a boron-mannitol complex, *Anal. Chem.*, 62, 2612, 1990.
21. Vanderpool, R. and Johnson, P., Boron isotope ratios in commercial produce and boron-10 foliar and hydroponic enriched plants, *Agric. Food Chem.*, 40, 462, 1992.
22. Gorsuch, T., Wet oxidation, in *The Destruction of Organic Matter, International Series of Monographs in Analytical Chemistry*, Belcher, R. and Frieser, H., Eds., Pergamon Press, New York, 1970g, 19.
23. Ferrando, A., Green, N., Barnes, K., and Woodward, B., Microwave digestion preparation and ICP determination of boron in human plasma, *Biol. Trace Elem. Res.*, 37, 17, 1993.
24. Szydlowski, F., Boron in natural waters by atomic absorption spectrometry with electrothermal atomization, *Anal. Chim. Acta*, 106, 121, 1979.

25. Yoshino, K., Okamoto, M., and Kakihana, H., Spectrophotometric determination of trace boron in biological materials after alkali fusion decomposition, *Anal. Chem.*, 56, 839, 1984.
26. Greenwood, N., Chapter 11. Boron, in *Comprehensive Inorganic Chemistry,* First, Bailar, J. J., Emeléus, H., Nyholm, R., and Trotman-Dickenson, A., Eds., Pergamon Press Ltd., Oxford, 1973, 665.
27. Thaler, L., Khodschaian, S., Kubach, I., Mirtsching, A., Müller, W., Pietsch-Wilcke, G., Wendt, H., Bersin, T., and von Harlem, J., *Bor,* 8th, Verlag Chemie, Gmbh., Weinheim/Bergstrasse, 1954, 1.
28. Cotton, F. and Wilkinson, G., *Advanced Inorganic Chemistry,* Interscience Publishers, Division of John Wiley and Sons, New York, 1972, 223.
29. Vanderpool, R., Hoff, D., and Johnson, P., Use of inductively coupled plasma-mass spectrometry in boron-10 stable isotope experiments with plants, rats, and humans, *Environ. Health Perspect.*, 102 (suppl 7), 13, 1994.
30. Hunt, C. and Shuler, T., Open-vessel, wet-ash, low-temperature digestion of biological materials for inductively coupled argon plasma spectroscopy (ICAP) analysis of boron and other elements, *J. Micronutrient Anal.*, 6, 161, 1989.
31. Ogner, G., Automatic determination of boron in plants, *Analyst*, 105, 916, 1980.
32. Monzó, J. and Pomares, F., Spectrophotometric determination of boron in water with prior distillation and hydrolysis of the methyl borate, *Analyst*, 113, 1069, 1988.
33. Johnson, P., Methodology for stable isotope analysis in biological materials (a review), *J. Micronutrient Anal.*, 6, 59, 1989.
34. Clarke, W., Koekebakker, M., Barr, R., Downing, R., and Fleming, R., Analysis of ultratrace lithium and boron by neutron activation and mass-spectrometric measurement of ^{3}He and ^{4}He, *Int. J. Rad. Appl. Instrum.*, 38, 735, 1987.
35. Anderson, D., Cunningham, W., and Mackey, E., Determination of boron in food and biological reference materials by neutron capture prompt-gamma activation, *Fres. J. Anal. Chem.*, 338, 554, 1990.
36. Anderson, D., Cunningham, W., and Lindstrom, T., Concentrations and intakes of H, B, S, K, Na, Cl, and NaCl in foods, *J. Food Comp. Anal.*, 7, 59, 1994.
37. Jun, Z., Oshima, M., and Motomizu, S., Determination of boron with chromotropic acid by high-performance liquid chromatography, *Analyst*, 113, 1631, 1988.
38. Jones, W., Heckenberg, A., and Jandik, P., Integrated separation schemes in chromatography. Simultaneous employment of ion exclusion and ion exchange for the separation of borate, chloride, nitrate and iodide, *J. Chromatog.*, 366, 225, 1986.

Chapter **16**

Methods for Studying Dietary Lead and Its Toxicity in Rodents

John D. Bogden
Department of Preventive Medicine and Community Health
UMDNJ — New Jersey Medical School
Newark, New Jersey

CONTENTS

I. INTRODUCTION

Rodents have been widely used to investigate lead metabolism and toxicity. Although various rodent species have been utilized, mice and especially rats have been most widely studied. The value of these species for providing insight into human lead metabolism and toxicity is well established.[1,2]

0-8493-9611-5/97/$0.00+$.50

Occupational and environmental lead exposure to people has been and continues to be widespread in numerous countries of the world. Experimental studies can be valuable for understanding lead toxicity in order to better address the substantial public health problem posed by environmental lead contamination. However, considerable care must be exercised in the design and conduct of laboratory studies of lead toxicity, since numerous factors can influence the results obtained.

II. STUDY DESIGN

A detailed written protocol is essential. In my laboratory, draft protocols are reviewed by all study participants, including senior investigators, postdoctoral fellows, technicians, and students. We usually find that we make a considerable number of changes in several draft protocols before agreeing on a final protocol. Each protocol is accompanied by a calendar that includes the dates on which specific procedures will be done and who will do them and includes a list of all reagents and equipment required. Of course, flexibility is also important, since initial study results may suggest that modifications be made while the study is in progress. Issues of particular concern in the design of studies of lead toxicity in rodents are the number of animals to be studied, random assignment to treatment groups, animal care and dosing, monitoring for expected and unwanted outcomes, diet, caging and other environmental concerns, and plans for data evaluation. In fact, identification of the inferential statistical methods to be used in data evaluation can be very useful in deciding upon the number of animals to be studied. The latter needs to be sufficient to achieve the study objectives; however, use of an excessive number of animals is more costly and unethical. If there is much uncertainty about details of the experimental design, for example the dose of lead to be used or the duration of exposure, then a small pilot study is appropriate. It is unwise to initiate a large study with doses of lead or exposure details that may not be correct for achievement of the study objectives.

Rodents should be randomly assigned to the various treatment groups. This can be done by using a table of random numbers. Since initial body weight may influence study results, it can be helpful to have comparable body weights in each treatment group. We recently studied the effect of weight loss on lead metabolism in rats.[3] Since there were three treatment groups, the weighed rats were grouped in sets of three in order of increasing body weight. Each of the three rats in a set was then randomly assigned to one of the three treatment groups, guaranteeing very similar initial body weights in each group.

Past experience of the investigators and the published literature are very helpful in deciding on the number of rodents per treatment group and the total number of animals to be studied at one time. The latter, of course, is limited by housing and staff available to the investigator. For example, if prior studies have shown that 8 to 10 animals in each of 4 treatment groups is sufficient to assess the major outcome variable of interest (e.g., systolic blood pressure or fetal growth), then this number of animals is a reasonable starting point for designing a new study. However, power analyses should always be done to provide an estimate of the statistical power available to achieve the study objectives. Of course, sometimes there are several outcome variables of interest. In this case, a decision must be made to use the number of animals required to adequately assess all outcome variables or only the major outcome(s). The latter might require fewer animals. In performing a power analysis, prior choice of the inferential statistical methods of analysis to be used is essential. If necessary, an experienced statistician should be consulted.

III. ANIMAL CARE AND DOSING

Making a wise choice of the specific procedures for dosing with lead requires some planning. In general, acute lead toxicity is of little scientific or medical significance, and chronic exposures of rodents are most relevant to human lead toxicity. Therefore, single or relatively few exposures to lead are generally of little interest. The most common methods for dosing of rodents are by the incorporation of lead salts into the diet or especially the drinking water. Due to its solubility and relative palatability, lead acetate has been the most widely used salt. Drinking water concentrations that are palatable to rats and mice are from less than 1 mg Pb/L to more than 1000 mg/L. We have given rats 100 mg/L of Pb as the acetate for up to one year and up to 250 mg/L for a month with little or no effect on drinking water consumption.[4,5,6] We start with high quality distilled-deionized water and prepare fresh lead-containing water in large volume (10 to 20 L) weekly. A small quantity of glacial acetic acid (12.5 uL/L) is added to each batch to prevent precipitation of lead carbonate by reaction of Pb^{+2} with dissolved CO_2. Development of cloudiness of the lead acetate solution indicates that the pH is not low enough to prevent lead carbonate formation. To estimate dose, drinking water consumption must be monitored at least several times per week, depending on the size of the drinking water bottles used. With bottles that hold 250 ml, measuring the volume of water consumed twice per week is usually sufficient. An advantage of the use of drinking water instead of food as the source of lead exposure is that it is more difficult to incorporate lead into food at a uniform concentration. It will also be more expensive if done by a commercial supplier ot rodent diets.

Age, gender, and duration of exposure as well as dose are obviously important considerations. For example, rats exposed to a specific dose of lead at 5 weeks of age will develop higher blood and organ lead concentrations than those exposed when 10 or 15 weeks old,[7] probably because of greater gastrointestinal absorption at younger ages.[8] In addition, the design of some studies may require *in utero* lead exposure or exposure that ends before the outcome variables of interest are assessed. For example, we recently finished a study to assess the effect of bone lead mobilization on maternal blood pressure and fetal growth during pregnancy.[9] Lead exposure ended 4 weeks prior to mating and 7 weeks before the birth of the pups, yet the lead exposure was sufficient to provide relatively high pup blood and organ lead concentrations after birth, due primarily to the long half life of lead stored in the skeleton and its equilibrium with blood.[1,10–11]

Caging of rodents during studies of lead toxicity is an important consideration. Among the decisions to be made is the choice of cages with solid or mesh bottoms. Since exposed rodents may ingest feces that will contain lead, this can confound the assessment of lead dose. Thus, cages with wire-mesh bottoms or other designs to prevent coprophagy are preferred. In our studies of pregnant lead-exposed rats, we transfer the dams to plastic cages with solid bottoms prior to delivery, since cages without solid bottoms are not suitable for newborn pups. Otherwise, we house lead-exposed rodents in cages with wire mesh bottoms.

Another concern is the environment of the animal care facility. Constant temperature and humidity, controlled light/dark cycles, and filtered air are preferred. Frequent washing of drinking water bottles and cleaning of cages are also important. It is best to purchase rodents from a supplier with a history of providing infection-free animals and house them in separate rooms from other rodents.

Lead is a ubiquitous environmental contaminant and unwanted exposures must be rigidly excluded. I know of at least one study that was hopelessly compromised because the rats were housed near a window that was frequently open and were thereby exposed to lead from auto exhaust.

It has long been known that the diet of the experimental animals is an important factor influencing lead toxicology. In particular, dietary vitamin D, calcium, iron, and magnesium can have very substantial effects on lead absorption and toxicity.[4–6,12–14] For example, we have found that increasing dietary calcium as the carbonate from 0.1% (moderately deficient) to 0.5% (normal) can markedly reduce blood and organ lead concentrations.[5,6] For male rats exposed to 100 mg/L lead in the drinking water for one year, those fed diets of 0.1% calcium had blood, brain, femur, and kidney lead concentrations that were about 3-, 5-, 8-, and 50-fold higher than those fed 0.5% calcium. Rats fed 2.5% calcium had the lowest blood and organ lead concentrations. Increases in dietary calcium, magnesium, and iron will decrease lead absorption but an increase in vitamin D will enhance it.[4–6,12–14]

Rodent diets should always be analyzed for their lead concentrations, since food will invariably contain some lead. If the experimental groups will be given very low doses of lead, then the lead present in some batches of food may be significant. Since some low level of exposure via the diet is almost unavoidable, even rodents in non-lead exposed treatment groups will have detectable blood and/or organ lead concentrations. This is especially true for organs with relatively high lead concentrations, such as bone and kidney.

Careful monitoring of rodents exposed to lead is essential. Evaluation and treatment for infections should be done. In long term experiments, monitoring for kidney disease may also be useful. In addition, regular measurements of body weight, at least weekly, can provide clues about the presence of disease or other problems that cause inadequate food ingestion. For young animals, comparison of growth curves of the various treatment groups can be useful.

IV. BIOMARKERS AND LABORATORY ANALYSES

Blood and organ lead concentrations have been the most frequently used parameters for assessment of lead exposure, dosing, and toxicity. They are also useful for relating rodent studies to human toxicity.

Prevention of lead contamination of blood and organs is essential during their collection, storage, and analysis, and techniques for blood withdrawal and organ harvesting should minimize the potential for contamination. For example, excision of organs at necropsy should be done with precleaned stainless steel instruments. A particular concern is contamination from rodent body hair, which can be reduced by wetting the hair with distilled water prior to dissection. We also quickly rinse each organ with distilled/deionized water to remove surface extracellular fluid contaminants and air dry on glassine weighing paper. Organs are stored at reduced temperatures (–20 to –80°C) in precleaned polyethylene containers. Monitoring of rodents during or subsequent to lead exposure can be achieved by periodic withdrawal of small volumes of blood (0.25 to 0.50 ml) from a tail vein. The tail is immersed in warm water (40°C) for 3 to 5 minutes and the site of blood withdrawal cleaned prior to insertion of a needle. The right and left lateral tail veins should be used alternately. For collection of larger volumes of blood (4.0 to 10.0 ml) prior to euthanasia, cardiac puncture is very effective but requires some experience to consistently do it well. The anticoagulant used should not introduce any detectable lead; we have found that heparin is usually satisfactory. In fact, heparinized and EDTA-treated vacutainers for collection of blood for analysis of lead and other trace metals are commercially available (Becton Dickinson, Rutherford, NJ). Separation of serum or plasma is not necessary, since the assessment of lead toxicity is based on whole blood due to the high lead concentrations of erythrocytes.

Flame or electrothermal atomic absorption spectrophotometry are the methods of choice for blood and organ lead analysis. Prior ashing is not required for blood lead determinations, for which procedures for flame[15] or electrothermal analysis[16] are available. The long-in-use flame procedure requires at least 1.5 ml of blood and involves chelation of the lead with ammonium pyrollidine dithiocarbamate (APDC) and extraction into methyl isobutyl ketone (MIBK). After centrifugation to separate the MIBK from the lower aqueous phase, the lead-containing MIBK is aspirated directly into the flame. Appropriate standards and quality control samples are processed in a similar manner. This procedure, however, lacks the sensitivity of the electrothermal method and should not be used if blood lead concentrations below 5 μg/dl are anticipated.

The electrothermal atomic absorption procedures require about 100 μl of whole blood, which is diluted prior to analysis with a matrix modifier containing 0.5% Triton X-100, 0.2% nitric acid, and 0.2% dibasic ammonium phosphate. Quality control can be assessed by analysis of commercially available samples. We use Bio-Rad Lyphochek Whole Blood Controls with certified lead concentrations (Bio-Rad, Anaheim, CA). Analysis of samples in duplicate or triplicate is recommended.

Organs require prior ashing before analysis by atomic absorption spectrophotometry. We prefer wet ashing and have analyzed thousands of samples using it. Briefly, weighed organs are ashed with a 3:1 mixture of concentrated nitric/perchloric acids in a perchloric acid fume hood. The use of an ordinary hood may result in the deposition of explosive perchlorate salts in the hood ventilation system. The acids must be of high purity to prevent introduction of excessive lead. We use redistilled acids (GFS, Columbus, OH), which are expensive but very low in metal contaminants.

An excess of acid must be used to avoid the possibility of an explosion, since dry perchlorate salts in the presence of unashed organic matter may ignite. Typically, 7 to 10 ml of acid mixture are sufficient to ash a 1.0-g*M* tissue sample. Ashing is conducted in the hood in open 100 ml borosilicate beakers on hot plates and usually takes less than 2 hours when the hot plate surface temperature is gradually increased to 150 to 175°C. A progressive increase in temperature helps control the reaction rate. Heating is continued until most of the excess acid has evaporated. After cooling, the residue is quantitatively transferred to a 5.0 to 25.0 ml volumetric flask using at least four portions of distilled water to ensure a quantitative transfer. One technician can easily complete the analysis of about 30 samples in a day. At least two blanks and two quality control samples are processed in the same manner. The quality control samples we use are National Institute of Standards and Technology (NIST) bovine liver (SRM 1577b) and orchard leaves (SRM 1571). The former has a certified lead concentration of 129 ng/gm, and the latter contains 45 μg/gm. Blank concentrations must be suitably low (< 10% of sample value) and subtracted from the sample concentrations in order to achieve reliable results.

We report our results on the basis of wet tissue weight, which is preferred by most investigators. If concentrations based on dry weights are needed, then the organs will have to be dried at about 85°C to constant weight. Concentrations based on ashed tissue weight are less reliable due to the often very low weights of ashed samples and/or incomplete ashing. If appropriate blank and quality control samples have not been properly used, analyses of blood and organs for lead should be considered suspect.

Other biomarkers of lead toxicity that are very useful are blood concentrations of free or zinc protoporphyrin and delta-aminolevulinic acid dehydratase. The latter is decreased by lead exposure while the former are increased.[17–19] We have found erythrocyte protoporphyrin concentrations to be a good marker of rat lead exposure and toxicity.[6]

V. CONCLUSION

Rodents, in particular rats, are a remarkably good model for studying lead metabolism and toxicity, since lead pharmacokinetics in humans and rats are similar.[1,2,20,21] However, a number of pitfalls must be avoided in the design, conduct, and interpretation of studies in rodents, as described in this review. Particular concerns are random assignment to treatment groups, appropriate dosing, caging and other environmental considerations, choice of diet, monitoring for unwanted outcomes, prevention of contamination of blood and organs during collection and analysis, and quality control of the analytical procedures.

REFERENCES

1. O'Flaherty, E.J., Physiologically based models for bone-seeking elements. II. Kinetics of lead deposition in rats. *Toxicol. Appl. Pharmacol.*, 111, 313, 1991.
2. O'Flaherty, E.J., Physiologically based models for bone-seeking elements. I. Rat skeletal and bone growth. *Toxicol. Appl. Pharmacol.*, 111, 299, 1991.
3. Han, S., Qiao, X., Simpson, S., Ameri, P., Kemp, F.W., and Bogden, J.D., Weight loss alters organ concentrations and contents of lead and some essential divalent metals in rats. *J. Nutr.*, 126, 317, 1996.
4. Bogden, J.D., Gertner, S.B., Kemp, F.W., McLeod, R., Bruening, K., and Chung, H.R., Dietary lead and calcium: effects on blood pressure and renal neoplasia in Wistar rats. *J. Nutr.*, 121, 718, 1991.
5. Bogden, J.D., Gertner, S.B., Christakos, S., Kemp, F.W., Yang, Z., Katz., S. R., and Chu, C., Dietary calcium modifies concentrations of lead and other metals and renal calbindin in rats. *J. Nutr.*, 122, 1351, 1992.
6. Bogden, J.D., Kemp, F.W., Han, S., Murphy, M., Fraiman, M., Czerniach, D., Flynn, C.J., Banua, M.L., Scimone, A., Castrovilly, L., and Gertner, S.B., Dietary calcium and lead interact to modify maternal blood pressure, erythropoiesis, and fetal and neonatal growth in rats during pregnancy and lactation. *J. Nutr.*, 125, 990, 1995.
7. Han, S., Qiao, X., Kemp, F.W., and Bogden, J.D., Age at lead exposure influences lead retention in bone. In: *Therapeutic Uses of Trace Elements*, Neve, J., Capes, P., and Lamand, M., Eds., Plenum, New York, 1996, in press.
8. Rader, J.I., Celesk, E.M., Peeler, J.T., and Mahaffey, K.R., Retention of lead acetate in weanling and adult rats. *Toxicol. Appl. Pharmacol.*, 67, 100, 1983.

9. Han, S., Eguez, M.L., Ling, M., Qiao, X., Kemp, F.W., and Bogden, J.D., Effects of prior lead exposure and diet calcium on fetal development and blood pressure during pregnancy. In *Trace Elements in Man and Animals*, Vol. 9, MRC Research Press, 1996.
10. Barry, P.S.I., A comparison of lead in human tissues. *Brit. J. Industr. Med.*, 32, 119, 1975.
11. Pounds, J.G., Long, G.J., and Rosen, J.F., Cellular and molecular toxicity of lead in bone. *Environ. Health Perspect.*, 91, 17, 1991.
12. Sobel, A.E., Yuska, H., Peter, D.D., and Kramer, B., The biochemical behavior of lead — Influence of calcium, phosphorus, and vitamin D on lead in blood and bone. *J. Biol. Chem.*, 132, 239, 1940.
13. Six, K.M. and Goyer, R.A., Experimental enhancement of lead toxicity by low dietary calcium. *J. Lab. Clin. Med.*, 76, 933, 1970.
14. Mahaffey, K.R., Environmental lead toxicity: nutrition as a component of intervention. *Environ. Health Perspect.*, 89, 75, 1990.
15. Hessel, D.W., A simple and rapid quantitative determination of lead in blood. Atomic Absorption Newsletter, 7, 55, 1968.
16. Miller, D.T., Paschal, D.C., Gunter, E.W., Strood, P.E., and D'Angelo, J., Determination of lead in blood using electrothermal atomisation atomic absorption spectrometry with a L'vov platform and matrix modifier. *Analyst*, 112, 1701, 1987.
17. Simmonds, P.L., Luckhurst, C.L., and Woods, J.S., Quantitative evaluation of heme biosynthetic pathway parameters as biomarkers of low-level lead exposure in rats. *J. Toxicol. Environ. Health*, 44, 351, 1995.
18. Davidow, B., Slavin, G., and Piomelli, S., Measurement of free erythrocyte protoporphyrin in blood collected on filter paper as a screening test to detect lead poisoning in children. *Ann. Clin. Lab. Med.*, 6, 209, 1976.
19. Burch, H.B. and Siegel, A.L., Improved method for measurement of delta-aminolevulinic acid dehydratase activity of human erythrocytes. *Clin. Chem.*, 17, 1038, 1971.
20. O'Flaherty, E.J., Physiologically based models for bone-seeking elements. IV. Kinetics of lead deposition in humans. *Toxicol. Appl. Pharmacol.*, 118, 16, 1993.
21. Leggett, R.W., An age-specific kinetic model of lead metabolism in humans. *Environ. Health Perspect.*, 101, 598, 1993.

Chapter 17

Methods in Chromium Dietary Supplementation and Deficiency

G. Stephen Morris
Maren Hegsted
Deborah L. Hasten
Departments of Kinesiology and Human Ecology
Louisiana State University Agricultural Center
Baton Rouge, Louisiana

Chromium (Cr) is a trace mineral which is acquired through the diet and is generally accepted as being nutritionally beneficial for mammals. A biological function for chromium was first suggested in 1954 when Curran[1] reported that chromium enhanced the synthesis of cholesterol and acetate in the rat liver. Three years later, Schwarz and Mertz[2] identified an extract of *tortula* yeast which improved glucose tolerance in rats suffering liver necrosis. This extract was named glucose tolerance factor (GTF), and analysis identified chromium III as its active component.[3] Subsequent work, using laboratory animals fed Cr deficient and Cr supplemented diets, supported a role for Cr and the GTF in the maintenance of normal glucose homeostasis.[4,5] Other studies reported that the growth rates of animals were lower when they were fed a Cr deficient diet and reared in an essentially Cr free environment.[6,7] Collectively, these data were interpreted to suggest that dietary Cr potentiates the action of insulin thereby contributing to the maintenance of normal glucose and lipid

0-8493-9611-5/97/$0.00+$.50

homeostasis. These data have lead the National Research Council to define Cr as a recommended component of the diet of laboratory animals and to suggest that the diet of laboratory rats, mice, guinea pigs, and hamsters contain chromium at a level of 300 ug/kg.[8]

While recognized as being of potential nutritional value, neither the essentiality of nor a daily dietary requirement for Cr has been firmly established for laboratory animals.[8] Dietary essentiality requires that a particular nutritional deficiency consistently produces a suboptimal response in an essential physiological function, and that this suboptimal response be preventable or reversible when a physiological amount of the nutrient is provided in the diet.[8] Dietary chromium fails to meet this definition of essentiality on several counts. First, a nutritional Cr deficiency has not been clearly documented in laboratory animals. Secondly, reduced dietary Cr intake has failed to consistently alter glucose homeostasis and lipid metabolism, the physiological functions thought to be regulated by Cr.[9,10] Thirdly, the addition of a physiological amount of Cr to the diet does not consistently prevent or reverse defects in glucose homeostasis thought to result from a dietary Cr deficiency.[9,11] Lastly, its specificity has been questioned because other heavy metals appear to elicit similar physiological effects.[8]

Although Cr is not currently considered to be essential, there are substantial data available which suggest a nutritional and physiological role for this micronutrient. Therefore, it seems reasonable to continue formulating and testing hypotheses regarding the essentiality of dietary Cr, particularly within the context of glucose homeostasis and insulin action. The remainder of this chapter is intended to provide information which will help researchers correctly design experiments to further our understanding of chromium as a micronutrient.

Because the diet is the principal source of chromium, adding or withholding Cr from a purified diet, one composed primarily of refined ingredients, is the most common method of regulating dietary Cr intake. Prior to 1993, the recommended basal diet for laboratory rodents was the AIN-76A semi-purified diet.[12] Since then, the AIN-93M diet has been the recommended diet for adult maintenance and the AIN-93-G has been recommended for growth, pregnancy, and lactation.[13] Relative to the AIN-76 diets, the formulations of the AIN-93 diets are substantially different in terms of the amounts and sources of carbohydrates and fats. Seaborn and Stoecker[14] have demonstrated that the source of carbohydrate can alter the uptake of Cr from the gut and its subsequent distribution. Similarly, Donaldson et al.[15] have suggested that the sources of fat used in a purified diet may actually be responsible for changes attributed to dietary Cr. It remains to be determined if and/or how the compositional changes recommended in the AIN-93 diets impact Cr absorption, retention, and physiological action. Until these issues are resolved, experimental results must be interpreted carefully and within the context of the composition of the diet.

The use of a purified diet requires that the diet be prepared from individual ingredients which are available from commercial suppliers and/or local sources. Those ingredients purchased from commercial suppliers of scientific products would be expected to have higher levels of purity. However, it is important to note that the components used to make the AIN-76 and AIN-93 diets can not be assumed to free of Cr contamination. Donaldson et al.[15] report that a high sucrose, AIN-76A diet, excluding the mineral mix, contains 40 μg of Cr/kg. Using a Cr free mineral mix (ICN Nutritional Biochemicals, Cleveland, OH) in an AIN-76A semi-purified diet results in approximately 180 μg of Cr/kg of diet (personal observation). Standard rat chow contains approximately 100 μg/kg Cr.[16] Roginski and Mertz[17] were able to keep Cr content of low chromium diets below 100 μg/kg by carefully selecting diet ingredients. Hence, it appears that some Cr contamination cannot be avoided in preparing diets, but the contamination can be kept to a minimum by utilizing appropriate, low Cr dietary components. By preparing all of the diet needed for a study at the same time, one can minimize differences in the dietary Cr content that occur when different lots of ingredients are used. As discussed below, environmental Cr is a serious issue in carrying out dietary Cr studies. For this reason, equipment used to mix these individual ingredients (mixing bowls, spatulas, whisks, and utensils) should be made of nonmetal materials or siliconized. It is the current inability to make a diet completely devoid of Cr that has rendered it impossible to produce a true Cr deficient animal model.[18]

This dietary contamination clearly points to the need to analyze the prepared diet for Cr content so that the "background" chromium content of the diet is known. Random samples of the diet should be taken and analyzed for Cr content. Graphite furnace atomic absorption spectroscopy is frequently used for Cr analysis since few reagents are used which lowers the chance for contamination, and this method is sensitive at sub ng/g or nmol/L concentrations.[19] Measurement of such low Cr concentrations requires that very strict precautions be taken to prevent sample contamination. For this reason, accurate Cr determinations may be best obtained from laboratories expressly designed to carry out such sensitive analysis.[11] Seaborn and Stoecker[14] provide methods for preparing dietary samples for Cr analysis.

Several different forms of Cr, including both naturally occurring and synthesized Cr containing compounds, have been used to supply or supplement the diets of laboratory animals. Because barley contains high levels of Cr, researchers have added this grain to the diet of laboratory animals and then measured its ability to moderate glucose homeostasis.[20] Positive results have been attributed to the Cr found in barley, but it is difficult to unequivocally link any observed physiological effect to Cr since this dietary ingredients add other unknown ingredients to the experimental diet.[21] Other studies have employed a basal diet supplemented with brewer's yeast[22] or *tortula* yeast.[23] As with barley, the addition of these Cr sources to the diet also adds additional vitamins, minerals, and other contaminates which confound rather than clarify the biological actions of Cr. Early, acute studies investigating the effects of isolated GTF on glucose metabolism produced beneficial effects on glucose metabolism.[24] Subsequent studies have produced conflicting results,[22,23,25] perhaps for several reasons. It is important to note that the structure of GTF has not yet been described, and GTF has even been questioned as the biologically active form of Cr.[18,26] It has also been argued that dietary GTF is better absorbed than other forms of Cr and is the bioactive form of Cr,[27] but these contentions remain in dispute.[18] Attempts have been made to synthesize GTF based on the assumption that it consists of Cr, nicotinic acid, and specific amino acids; however, these synthetic GTFs have not consistently demonstrated the ability to regulate glucose homeostasis.[11,18] Until these issues are clearly resolved, the use of GTF isolated from natural products does not provide the best approach to answering fundamental questions dealing with the nutritional importance of dietary Cr.

Trivalent forms of Cr have received the most study since this valance state appears to be the bioactive form of this micronutrient. Several Cr^{3+} containing compounds, including chromium chloride ($CrCl_3$), chromium nicotinate, and chromium picolinate have been used as a dietary source of Cr. Inorganic chromium ($CrCl_3$) is water soluble, and is not well absorbed, but offers an unadulterated form of trivalent Cr to the animal. Because $CrCl_3$ is water soluble, it can be delivered to animals in their drinking water as well as their feed. In either form, $CrCl_3$ has been reported to be effective in positively modulating glucose homeostasis.[6,24] Two synthetic Cr containing compounds, chromium nicotinate and chromium picolinate, have recently been used to provide a dietary Cr source.[11] Chromium nicotinate is thought to structurally resemble GTF,[28] contains nicotinic acid which is a component of the GTF,[29] and has been shown to modify glucose homeostasis.[30,31] Chromium picolinate is a tripicolinate complex which also positively impacts glucose homeostasis.[11,32] It is highly soluble in chloroform,[30] which suggests that it can more easily enter or cross the plasma membrane of cells. Previously, chromium picolinate was thought to have the greatest absorption by the gut of these three compounds, but more recent data suggest that chromium nicotinate has the greatest absorption and retention.[33] Few studies have simultaneously compared the physiological impact of these synthetic chromium containing compounds; therefore, it is currently impossible to argue that one of these forms of chromium is better able to serve as a dietary source of Cr than the other.

Several matters regarding the mechanics of diets and feeding warrant mentioning. In order to monitor Cr intake, measurements of daily food intake should be considered. These data provide information about compliance as well as weight gain in relationship to food intake. Addition of these Cr containing compounds to the diet, regardless of the concentration, does not change the palatability of the diet. Thus pair-fed and pair weighted groups are not typically required in these experiments. Cr has been added to the diet of laboratory rats in concentrations ranging up to mg/kg concentrations. Feeding programs lasting from 1 week to 18 months are well tolerated by research animals.[15,24,29] Trivalent Cr does not pose a toxic threat to the animals, even at extremely high concentrations.[8] However, large concentrations of Cr in the diet should be considered pharmacologic rather than physiologic quantities. As such, changes induced by large doses of Cr containing compounds may be the result of nonspecific pharmacologic effects.

Once dietary concerns have been adequately addressed, other factors in the experimental design of chromium experiments need to be addressed. The chromium researcher should be aware that Cr is found virtually everywhere in the environment of the laboratory animal and that such Cr can be taken up and used by the animal. Once ingested, environmental Cr is currently not distinguishable from dietary Cr,[34] therefore efforts to minimize environmental Cr exposure are necessary to ensure adequate control over Cr intake by the animal. Cages should be completely constructed of acrylic plastic, utilizing plastic water bottles equipped with glass sipper tubes and ceramic feeding containers. These actions minimize Cr contamination from housing and feeding equipment.[35] Environmental air can be contaminated with chromium; therefore, experimental animals should be housed in a "clean" room, if possible, which provides top-to-bottom flow of class 100 filtered air.[15] Animals on a low chromium diet should be housed in the upper cages of the racks. Wood shavings have been reported to be Cr free.[14] Feeding equipment should be washed, rinsed, soaked in 10% nitric acid overnight, and then rinsed again.[15] Ultrapure water, stored in plastic containers, should be provided to the animals and used for making all solutions and rinsing equipment.

Concern for environmental Cr contamination should also exist during tissue collection if these samples are to be measured for Cr content. Blood should always be drawn through siliconized needles into plastic, Cr free syringes. Solid tissue should be removed from animals using siliconized, plastic, or tantalum instruments and handled so as to avoid Cr contamination, i.e., weighed and stored in glass vials and dissected with glass knives.

Defining Cr status remains problematic in studying this micronutrient. Typically, a change in some parameter directly or indirectly linked to a nutrient is used to assess or define an animal's status with regards to that nutrient. For example, the amount of 1,25-dihydroxyvitamin D3 in the blood is used to define Vitamin D status of an animal. Chromium has not yet been identified as a component of an enzyme or as a metabolic cofactor; therefore, no direct assessment of Cr status is currently available. The advent of sufficiently sensitive methods for determining nanogram quantities of Cr in biological samples has made it possible to reliably measure Cr concentrations in a variety of tissues.[15,18] For example, measurements of urinary Cr concentrations are useful for studying Cr absorption, retention, and excretion in Cr balance studies. Urinary Cr measures have, however, failed to correlate with measures of glucose homeostasis and therefore are probably not useful in assessing Cr status.[36] The use of plasma or serum Cr to indicate status has been dismissed because serum/plasma Cr does not appear to equilibrate with other body compartments and serum levels do not correlate with other measures of glucose homeostasis.[18,36] Hair Cr concentration may provide a useful measure of Cr status, but apparently only under very rigorously controlled experimental conditions.[18] Interestingly, Donaldson et al.[15] and Jewell et al.[37] have reported that chronic consumption of a high Cr diet significantly increases kidney Cr concentration. While the physiological significance of this difference remains unknown, such differences in kidney Cr concentration might be exploited to document a relative change in chromium status brought about by a particular feeding protocol.

Data published by Verch et al.[38] suggest that determining even a relative measure of chromium status may be of significant importance at the initiation of a study. These workers reported that the chromium levels in different organs varied across different shipments of rats. Differences in initial Cr status might explain the differences we have observed in the growth of animals fed the same chromium supplemented diet but obtained from different breeding facilities of the same company. It is not unreasonable to assume that differences in chromium status at the onset of a study may significantly impact the responsiveness of laboratory animals to Cr based experimental interventions.[18] Measurement of kidney Cr levels of select animals may help to insure reasonably similar initial Cr status across experimental groups.

Because Cr is thought to modulate insulin activity and hence glucose metabolism, one might suspect that a glucose tolerance test (GTT) would provide insight into Cr status. However, an impaired GTT test can result from a variety of causes, and therefore is not useful as a diagnostic criterion for Cr status.[11] Results from a GTT are frequently used, however, to define the physiological response or adaptation to dietary Cr content.[11,24] Presumably, Cr will modulate the activity of available insulin which, in turn, will modify glucose metabolism and the rate at which glucose is removed from the blood. Simply performing a GTT does not provide definitive information on insulin status or insulin sensitivity of the animal. Therefore serum insulin levels should also be determined in conjunction with the glucose levels. The resulting data are most easily handled if expressed as total areas under the insulin and glucose curves, respectively.[16] Expressing these data as a ratio (glucose area/insulin area) will provide insight into insulin sensitivity. That is, if Cr treatment increases insulin sensitivity, then less insulin will be required to clear similar amounts of glucose from the blood and this ratio should increase. Other methods are available for defining the effects of Cr on insulin action, including the hyperinsulinemic-euglycemic clamp,[17] but these procedures are more complex and may be beyond the capability of many laboratories. Improvements in glucose transport by dietary Cr supplementation have been suggested, but direct measurements of this component of glucose homeostasis have yet to be made. It must be remembered that these measures do not provide a direct measure of Cr status, but rather an indirect measure of possible Cr actions.

Despite the fact that chromium studies continue to generate conflicting and controversial results, this micronutrient continues to be viewed as being nutritionally and physiologically beneficial. Clearly, several steps need to be taken to advance our fundamental understanding of Cr in both a normal and a pathological setting. First, a quantifiable definition of Cr status needs to be developed. Such a definition will allow different investigators to interpret and compare their findings within a common framework. Secondly, efforts must be directed toward the generation of a Cr deficient animal model. Only when such a model is available will investigators have the ability to explore Cr biology in a controlled and unambiguous setting. Thirdly, new techniques and new experimental approaches must be employed in the effort to determine the biological role of Cr. For example, the use of radioactively labeled Cr containing compounds will allow for monitoring the absorption, movement, and deposition of Cr through the body.[32] Cell culture techniques afford the investigator with the ability to manipulate the amount of Cr in an otherwise constant environment. Lastly, a complex array of extracellular and intracellular processes contributes to the maintenance of glucose homeostasis and are therefore potential targets for regulation by Cr. Experiments designed to test the ability of Cr to regulate these individual processes are clearly warranted and needed.[39]

REFERENCES

1. Curran, G.L., Effect of certain transition group elements on hepatic synthesis of cholesterol in the rat. *J. Biol. Chem.*, 210, 765, 1954.
2. Schwarz, K. and Mertz, W., A glucose tolerance factor and its differentiation from factor 3. *Arch. Biochem. Biophys.*, 72, 515, 1957.
3. Schwarz, K. and Mertz, W., Chromium (III) and the glucose tolerance factor. *Arch. Biochem. Biophys.*, 85, 292, 1959.
4. Schroeder, H.A., Diabetic like serum glucose levels in chromium deficient rats. *Life Sci.*, 4, 2057, 1965.
5. Schroeder, H.A., Chromium deficiency in rats: A syndrome simulating diabetes mellitus with retarded growth. *J. Nutr.*, 88, 439, 1966.
6. Mertz, W., Roginski, E.E., and Schroeder, H.A., Some aspects of glucose metabolism of the chromium deficient rats raised in a strictly controlled environment. *J. Nutr.*, 86, 107, 1965.
7. Schroeder, H.A., Vinton, Jr., W.H., and Balassa, J.J., Effects of chromium, cadmium and lead on the growth and survival of rats. *J. Nutr.*, 80, 48, 1963.

8. National Research Council, Nutrient requirements of the laboratory rat. In: *Nutrient Requirements of Laboratory Animals*. 4th revised edition. Washington, D.C.: National Academy of Sciences, 1995, 30.
9. Offenbacher, E.G., Rinko, C., Pi-Sunyer, F.X., The effects of inorganic chromium and brewers yeast on glucose tolerance, plasma lipids, and plasma chromium in elderly subjects. *Am. J. Clin. Nutr.*, 42, 454, 1985.
10. Flatt, P.R., Juntti-Berggren, L., Gould, B.J., and Swantston-Flatt, S.K., Effects of dietary inorganic trivalent chromium (Cr3+) on the development of glucose homeostasis in rats. *Diab. Metab.*, 15, 93, 1989.
11. Mertz, W., Chromium in human nutrition: A review. *J. Nutr.*, 124, 117, 1994.
12. American Institute of Nutrition. Report of the American Institute of Nutrition ad hoc committee on standards for nutritional studies. *J. Nutr.*, 107, 1340, 1971.
13. Reeves, P.G., Nielsen, F.H., and Fahey, Jr., G.C., AIN-93 purified diets for laboratory rodents: Final report of the American Institute of Nutrition ad hoc committee on the reformation of the AIN-76A rodent diet. *J. Nutr.*, 123, 1939, 1993.
14. Seaborn, S.D. and Stoecker, B.J., Effects of starch, sucrose, fructose and glucose on chromium absorption and tissue concentrations in obese and lean mice. *J. Nutr.*, 119, 1444, 1989.
15. Donaldson, D.D., Lee, D.M., Smith, C.C., and Rennert, O.M., Glucose tolerance and plasma lipid distributions in rats fed a high sucrose, high cholesterol, low-chromium diet. *Metabolism*, 34, 1086, 1985.
16. Yoshimoto, S., Sakamoto, K., Wakabayashi, I., and Masui, H., Effect of chromium administration on glucose tolerance in stroke-prone spontaneously hypertensive rats with streptozotocin-induced diabetes. *Metabolism*, 41, 636, 1992.
17. Roginski, E.E. and Mertz, W., Effects of chromium (III) supplementation on glucose and amino acid metabolism in rats fed a low protein diet. *J. Nutr.*, 97, 525, 1969.
18. Offenbacher, E.G. and Pi-Sunyer, F.X., Chromium in human nutrition. *Ann. Rev. Nutr.*, 8, 543, 1988.
19. Miller-Ihli, N.J. and Greene, F.E., Graphite furnace atomic absorption method for the determination of chromium in urine and serum. *J. AOAC Int.*, 75, 354, 1992.
20. Mahdi, G.S., Naismith, D.J., Price, R.G., Taylor, S.A., and Risteli, L., Modulating influence of barley on the altered metabolism of glucose and of basement membranes in the diabetic rat. *Ann. Nutr. Metab.*, 38, 61, 1994.
21. Mannering, G.J., Shoeman, J.A., and Shoeman, D.W., Effects of colupulone, a component of hops and brewers yeast, and chromium on glucose tolerance and hepatic cytochrome P450 in nondiabetic and spontaneously diabetic mice. *Biochem. Biophys. Res. Comm.*, 200, 1455, 1994.
22. Flatt, P.R., Juntti-Berggren, L., Berggren, P.O., Gould, B.J., and Swantston-Flatt, S.K., Failure of glucose tolerance factor containing Brewer's yeast to ameliorate spontaneous diabetes in C57BL/KsJ DB/DB mice. *Diab. Res.*, 10, 147, 1989.
23. Shepherd, P.R., Elwood, C., Buckley, P.D., and Blackwell, L.F., Glucose tolerance factor potentiation of insulin action in adipocytes from rats raised on a tortula yeast diet cannot be attributed to a deficiency of chromium or glucose tolerance factor activity in the rat. *Biol. Trace Elem. Res.*, 32, 109, 1992.
24. Mertz, W., Chromium occurrence and function in biological systems. *Physiol. Rev.*, 49, 163, 1969.
25. Simonoff, M., Shapcott, D., Alameddine, S., Sutter-Dub, M.T., and Simonoff, G., The isolation of glucose tolerance factors from brewer's yeast and their relation to chromium. *Biol. Trace Elem. Res.*, 32, 25, 1992.
26. Vincent, J.B., Relationship between glucose tolerance factor and low molecular weight chromium binding substance. *J. Nutr.*, 124, 117, 1994.
27. Wallach, S., Clinical and biochemical aspects of chromium deficiency. *J. Am. Coll. Nutr.*, 4, 107, 1985.

28. Cooper, J.A., Anderson, B.F., Buckley, P.D., and Blackwell, L.F., Structure and biological activity of nitrogen and oxygen coordinated nicotinic acid complexes of chromium. *Inorgan. Chim. Acta*, 91, 1, 1984.
29. Mertz, W., Toepfer, E.W., Roginski, E.E., and Polansky, E.E., Present knowledge of the role of chromium. *Fed. Proc.*, 33, 2275, 1974.
30. Evans, G.W. and Pouchnik, D.J., Composition and biological activity of chromium-pyridine carboxylate complexes. *J. Inorgan. Biochem.*, 49, 177, 1993.
31. Preuss, H.G., Gondal, J.A., Bustos, E., Bushehri, N., Lieberman, S., Bryden, N.A., Polansky, M.M., and Anderson, R.A., Effects of chromium and guar on sugar-induced hypertension in rats. *Clin. Neph.*, 44, 170, 1995.
32. Lee, N.A. and Reasner, C.A., Beneficial effect of chromium supplementation on serum triglyceride levels in NIDDM. *Diab. Care*, 17, 1449, 1994.
33. Olin, S.K., Stearns, D.M., Armstrong, W.H., and Keen, C.L., Comparative retention absorption of chromium-51 (Cr-51) from Cr-51 chloride, nicotinate Cr-51 and Cr-51 picolinate in a rat model. *Trace Elem. Electrolytes*, 11, 182, 1994.
34. Gargas, M.L., Norton, R.L., Paustenbach, D.J., and Finley, B.L., Urinary excretion of chromium by humans following ingestion of chromium picolinate — Implications for biomonitoring. *Drug Metab. Dispos.*, 22, 522, 1994.
35. Polansky, M.M. and Anderson, R.A., Metal-free housing units for trace element studies in rats. *Lab. Anim. Sci.*, 29, 357, 1979.
36. Offenbacher, E.G., Rinko, C., and Pi-Sunyer, F.X., The effects of inorganic chromium and brewers yeast on glucose tolerance, plasma lipids, and plasma chromium in elderly subjects. *Am. J. Clin. Nutr.*, 42, 454, 1985.
37. Jewell, D., Azain, M., and Wedekin, K., Dietary chromium picolinate changes kidney chromium without changing growth in rats. *FASEB J*, 9, A449, 1995.
38. Verch, R.L., Chu, R., Wallach, S., Peabody, R.A., Jain, R., and Hannan, E., Tissue chromium in the rat. *Nutr. Rep. Int.*, 27, 531, 1983.
39. Yurkow, E.J. and Kim, G., Effects of chromium on basal and insulin-induced tyrosine phosphorylation in H4 hepatoma cells: Comparison with phorbol-12-myristate-13-acetate and sodium orthovanadate. *Mol. Pharm.*, 47, 686, 1995.

Chapter **18**

Isolation and *In Vitro* Analysis of Biologically Active Chromium

Diane M. Stearns
Department of Chemistry
Dartmouth College
Hanover, New Hampshire

CONTENTS

0-8493-9611-5/97/$0.00+$.50

I. INTRODUCTION

Chromium has an interesting, albeit poorly understood, role in biology. Trivalent chromium (Cr(III)) is classified as a trace essential metal, but hexavalent chromium (Cr(VI)) is classified as a human carcinogen. Neither the process by which Cr(III) potentiates insulin function, nor the mechanism by which Cr(VI) induces tumors has been clarified. Although a metal's biological activity in terms of essential mechanisms vs. toxic mechanisms usually correlates with the metal's *in vivo* concentration, in the case of chromium its biochemical action is a function of oxidation state and ligand coordination as well as concentration.

In vitro methods have been used to study the nutritional role of chromium for essentially two reasons. *In vitro* methods can produce data that is useful toward the understanding and testing of molecular mechanisms, and many test samples can be investigated while avoiding more costly animal testing.

This review summarizes common *in vitro* techniques that have been applied to chromium nutritional research. Because one of the "holy grails" in chromium nutrition is the characterization of the ultimate form of biologically active chromium, originally labeled the glucose tolerance factor (GTF), it seemed most appropriate to describe briefly the *in vitro* methods used to date for the isolation, purification, and testing of biologically active chromium. The classic definition of GTF as a chromium compound of nicotinic acid and glutathione, which functions through the formation of a ternary complex with insulin and the insulin tissue receptor, evolved from the investigations of Walter Mertz and co-workers,[1,2] and has served as the cornerstone of the nutritional chromium paradigm. Development of this model relied in no small part on the rat adipocyte assay. Therefore, the bulk of the current chapter will describe that standard method as refined by Mertz and coworkers, as well as modifications made by other researchers, and will briefly summarize knowledge that has been gained from this assay regarding the biologically active component of GTF.

Another important *in vitro* method, the yeast fermentation bioassay, is not thoroughly described in this chapter. This is not meant to belittle the yeast assay, which has yielded results to complement and support the rat adipocyte assay, nor the researchers world-wide who have aided in its development. The decision merely reflects this volume's emphasis on rodent nutrition.

II. CHROMIUM: OCCURRENCE AND PHYSICAL PROPERTIES

Chromium is ubiquitous on the planet. Chromium content is roughly 100 ppm in rock, 0.5 to 2 ppb in water, and 0.2 to 1 ppm in plants. Air content of chromium is on the order of 0.01 to 0.03 μg/m³. Humans contain a few mg total body Cr, but tissue levels differ greatly with geographical or occupational exposures.[3] Chromium dietary intake in humans is ~ 30 to 60 μg Cr per day.[4] Chromium is used in many industrial processes, including the production of stainless steel, the manufacture of dyes and pigments, and leather tanning. In the laboratory, stainless steel can provide a source of experimental contamination of Cr.

Chromium can exist in the oxidation states of +6 to –2, but in nature Cr is found most commonly in only two stable oxidation states, trivalent, Cr(III) and hexavalent, Cr(VI). Under physiological conditions Cr(VI) exists as the tetrahedral anion chromate, CrO_4^{2-}, which crosses cell membranes readily using anion transport protein channels.[5] Intracellularly, Cr(VI) is reduced to Cr(III) by a number of species, including the small molecules glutathione and ascorbate, and many enzyme systems.[6] Under physiological conditions, Cr(III) is six-coordinate, binding ligands in an octahedral geometry. Trivalent chromium does not cross cell membranes readily, and the extent of cellular uptake is dependent on the coordinating ligands. Under normal conditions in the cell, Cr(III) is not oxidized to Cr(VI).

Relative to the other biologically relevant metals, Cr(III) is substitutionally inert. However, the hexaquo complex, $Cr(H_2O)_6^{3+}$, is known to oligomerize readily. The pKa of $Cr(H_2O)_6^{3+}$ is 4.3; thus, at physiological pH it readily loses a proton to form the hydroxide: $Cr(H_2O)_5(OH)^{2+}$. Two Cr-hydroxides can form hydroxy-bridged dimers with the loss of ligating water:

$$(H_2O)_5Cr(OH)^{2+} + (HO)Cr(H_2O)_5^{2+} \rightarrow (H_2O)_4Cr(OH)_2Cr(H_2O)_4 + 2H_2O$$

Further condensation leads to oligomers which can precipitate out of solution. This property can complicate *in vitro* experiments that use aged stock solutions of Cr(III). The commercially available salt chromic chloride hexahydrate exists as *trans*-$[Cr(H_2O)_4Cl_2]Cl \cdot 2H_2O$. In dilute aqueous solution, over time or with heating, the green species loses the chloride anions to convert to the purple $Cr(H_2O)_6^{3+}$.

III. METHODS

A. Procedures for Obtaining Biologically Active Chromium

The glucose tolerance factor (GTF) was originally reported by Schwarz and Mertz to be a unique "dietary agent required to maintain normal glucose tolerance in the rat."[7] In the original letter to the editor, a brief method was described for isolation of GTF from Brewer's yeast and pork kidney powder, and the statement was made that trivalent chromium was the active ingredient.[7] While no data were provided to demonstrate the chromium content in the isolated GTF, results were provided to show that rates of removal of glucose from the bloodstream of rats were greater in rats fed certain chromium salts compared to rats fed any of 43 other metal salts.[7] Since then, it has been taken on faith by some that GTF is a biologically active chromium compound, while others who have attempted to purify and characterize the GTF have found nothing so straightforward. Many of the *in vitro* studies have measured biological activity of isolated forms of GTF, yet its identity, and even existence, remain controversial. Therefore, it is important to briefly describe the methods used to isolate purified forms of biologically active chromium before presenting the *in vitro* methodology by which it has been evaluated.

1. *Isolation of the Glucose Tolerance Factor*

Isolation and characterization studies of the purported GTF have most commonly used Brewer's yeast, pork kidney powder, or Merck yeast extract powder as sources. The basic procedure consists of an initial extraction followed by a purification using various methods of cation- and anion-exchange chromatography.

A commonly cited study[1,8] isolated GTF from *Saccharomyces carlsbergensis*. The isolated yeast sample was extracted with 50% ethanol and filtered. Ethanol was removed *in vacuo*, and the remaining aqueous solution was stirred with activated charcoal at pH 3.5 for 1 hr and filtered. The cake was washed with 95% ethanol and water. The desired sample was eluted from the charcoal by a duplicate 3 hr treatment with 1:1 concentrated NH_4OH and diethyl ether, followed by filtration. After *in vacuo* removal of ammonia and ether, the sample was hydrolyzed by 18 hr reflux in 5 *M* HCl. The acid was removed *in vacuo*, the sample was extracted with ether, and the aqueous layer was brought to pH 3.0 with HCl. Cation chromatography on Dowex 50W-X8 was carried out with 0.1 *M* NH_4OH. Eluted fractions were monitored at 262 nm. This purification procedure was carried out with a synthetic mixture of chromium, nicotinic acid, glycine, glutamine, and cysteine. A portion of this mixture that displayed properties identical to the yeast extract was isolated by the same chromatography procedures. Both the synthetic and yeast samples were active in the rat adipocyte assay (see Section III.B.). It was therefore concluded that the biologically active component of GTF consisted of a chromium complex coordinated by 2 nicotinic acid ligands, 2 glycines, 1 glutamic acid, and 1 cysteine.

Another early report of the characterization of a chromium compound isolated from brewer's yeast (*Saccharomyces cerevisiae*) was published by Votava and co-workers.[9] Yeast cells grown in the presence of $^{51}CrCl_3{\cdot}6H_2O$

were harvested, and yeast pellets were extracted with 50% butanol. Upon separation, the aqueous phase contained 90% of the total chromium radioactivity. This portion was freeze-dried, redissolved in water, and purified on a Sephadex G-25 column with 1 m*M* Tris (tris-hydroxymethyl-aminomethane) at pH 7.3 and 50 m*M* KCl as eluant buffers. Fractions of a single band containing ^{51}Cr were combined and subjected to either cation (Dowex-50W, H^+ form) or anion exchange (Dowex-1, Cl^- form) chromatography. Cationic chromatography resulted in no portion of the sample retained on the column, with 100% of the ^{51}Cr eluting as a single band with H_2O. The anionic column was treated with a 0 to 0.5 M NaCl gradient, and produced a single band eluting at the highest salt concentration. Therefore, it was concluded that the chromium containing GTF was a single anionic species. Thin layer chromatography against standards suggested that Cr was bound by a 6 amino acid peptide containing leucine/isoleucine, proline, valine, alanine, serine, and glutamic/aspartic acid.

The above studies separated and analyzed what were concluded to be single chromium species isolated from yeast. Subsequent studies have carried out more intricate separation procedures to isolate and analyze multiple components of the yeast extract. Kupulainen and co-workers[10] incubated yeast (*Saccharomyces uvarum ex S. carlsbergensis*) with $^{51}CrCl_3$, and isolated the ^{51}Cr-containing fractions from (1) the yeast, (2) media in which yeast were incubated, and (3) fresh media treated with $^{51}CrCl_3$ in the absence of yeast. Yeast extraction was carried out with 50% ethanol, lipids were removed with petroleum ether, and aqueous solutions were purified on a Sephadex G-25 column with water elution. Chromium-containing fractions were further separated on a Dowex 50W-X8 (H^+ form) column with a 0 to 2 *M* NH_4OH gradient. A ^{51}Cr fraction that eluted with water was further purified on a Sephadex QAE A-25 (CO_3^{2-} form) column. Final purification was carried out on TLC plates with n-butanol to acetic acid to water (2:1:2). Similar cationic fractions were observed in yeast extracts, the used media, and fresh media, suggesting that those Cr complexes were not synthesized by the living yeast. One anionic fraction was observed only in the yeast and the used media samples, and thus appeared to be a result of yeast metabolism. Biological activity was not assayed. Insulin association was investigated, but none of the isolated Cr-containing fractions were able to bind ^{125}I-insulin.

A procedure using "mild conditions" for GTF extraction was reported by Mirsky and co-workers.[11] The initial extraction from Merck yeast extract powder was made with 50% butanol. The aqueous phase was dialyzed with molecular weight 3500 tubing against two volumes of water. The dialyzates were concentrated and purified on a DEAE-11 cellulose column (Cl^- form) by 0 to 2 *M* NH_4Cl gradient elution. Chromium content was reportedly measured by monitoring the absorbance at 260 or 262 nm; however, it should be noted that absorbance at these wavelengths is not unique to chromium.[8] Biological activity, as CO_2 production in yeast, was observed only for the initial fraction eluted with water. Further purification of this fraction was carried out with Dowex 50W-X8 (H^+ form) by elution with a 0 to 2 M NH_4Cl gradient, and finally with 0.25 M NH_4OH. Four major fractions showed an absorbance at 260 nm, although biological activity in yeast was only observed for the cationic band eluted with NH_4OH. It was concluded that this procedure yielded the same GTF species that had been isolated by Mertz.[1]

A procedure was reported by Haylock and co-workers[12] that used multiple chromatographic purification steps to separate 11 chromium-containing species from Merck yeast extract. The endogenous chromium content in the commercial yeast extract was found to vary tenfold between different batch numbers. Extraction with 50% butanol showed no enhanced purification relative to an aqueous extraction. Column chromatography was carried out on a series of Cellulose, Dowex 50W-X12 cation, and Dowex 1-X8 anion exchange columns using pH gradients for elution. The pH gradient techniques were reported to be more successful at separating individual species than the NH_4OH elution method (*vide supra*) and more importantly, NH_4OH treatment was reported to enhance activity of GTF fractions. Biological activity was measured by the yeast bioassay. A positive control consisted of the chromium complex of nicotinic acid, glutamic acid, and cysteine synthesized by the method of Toepfer et al.[8] None of the neutral or anionic fractions showed biological activity by the yeast fermentation assay. All cationic fractions showed biological activity; however, activity of cationic fractions was not consistent between yeast extract and the other biological sources: sage, wheat bran, peppercorns, molasses, and pork kidney powder. It was concluded that "GTF cannot be regarded as a unique species."

An exhaustingly thorough procedure was reported by Simonoff and coworkers,[13] who isolated multiple fractions of GTF from yeast (*S. cerevisiae*) cultivated in the presence of 228 ppm Cr^{3+} or in the absence of extra Cr^{3+} (0.48 ppm Cr). A 50% butanol extraction was followed by multistep chromatography on DEAE cellulose, DEAE Sephacel, Dowex 50W-X8, and Dowex 50W-X2 columns with monitoring at OD 230. Anionic fractions from the initial DEAE cellulose column were separated into 30 subfractions with DEAE Sephacel, and the initial cationic/neutral portion was separated into 20 subfractions on Dowex columns. Chromium content was measured in each fraction. Comparison of fractions isolated from the Cr-treated and untreated yeast showed that the extra Cr was largely associated with anionic components, but not with a set of specific fractions. Fractions were tested for biological activity by the rat adipocyte assay (see Section III.B).

2. *Isolation of the Low Molecular Weight Chromium-Binding Protein*

Aside from the glucose tolerance factor, a second form of endogenous biologically active chromium has been reported. Wada and co-workers[14] originally described the isolation of two forms of low molecular weight

Cr (LMWCr) from dogs treated by i.v. administration with potassium dichromate. Isolated livers were homogenized and centrifuged, and the soluble fraction was purified by ethanol fractionation, followed by multiple chromatographic separations using DEAE-cellulose (Cl^- form), Sephadex G-10 and G-15, and Dowex 50W (H^+ form) chromatography to yield purified fractions labeled LMWCr-I and LMWCr-II. The LMWCr-II fraction was ultimately purified by crystallization from 90% ethanol. It was determined that LMWCr-II was an inorganic chromium phosphate complex. The LMWCr-I fraction was determined to consist of an anionic organic Cr compound containing predominantly glutamic acid, glycine, and cysteine, which are the components of glutathione. The molecular weight of LMWCr-I was estimated at 1500. It was proposed that this protein was involved in detoxification and excretion of chromium in animals.

B. Rat Adipocyte Assay

The original rat adipocyte assay applied by Mertz and co-workers[15] measured the effect of test compounds on the insulin-induced oxidation of glucose by rat epididymal tissue. The epididymis is the convoluted ductal structure attached to the testes, in which sperm are stored. The commonly measured endpoint is the oxidation of ^{14}C-glucose to $^{14}CO_2$. Four pieces of tissue supplied by one rat allowed three tests and one control per animal. The isolation of adipose cells from rat epididymal tissue was a major improvement,[16,17] allowing many more *in vitro* tests to be run per animal. The basic procedure,[16,17] with reported modifications,[13,18] is described here. This procedure consists of four major steps: (1) the isolation of tissue from rats, (2) isolation of the adipocyte cells from the tissue, (3) incubation of the cells with test substance, insulin, and ^3H- or ^{14}C-glucose, and (4) trapping and measurement of the liberated $^{14}CO_2$.

1. Isolation of Adipocytes

The reader is referred to Rodbell[16] for the best visual description of fat cell isolation. A single rat may provide fat cells sufficient for 25 to 75 assays.[13,17] Rats are sacrificed, commonly by decapitation, and the epididymal fat pads are removed. The tissues are combined, rinsed with 0.9% NaCl, minced with scissors, and incubated with 6 to 10 mg of collagenase, or 10 mg/fat pad,[18] in 6 to 10 mL of Krebs-Ringer's phosphate (KRP) buffer, pH 7.4 (118 m*M* NaCl, 5 m*M* KCl, 1.3 mM $CaCl_2$, 1.2 m*M* $MgSO_4$, 1 m*M* KH_2PO_4, and 16.2 mM Na_2HPO_4-HCl). Other procedures[13,18] substitute KRP buffer with pH 7.4 Krebs-Ringer bicarbonate buffer (KRB), (118 m*M* NaCl, 4.7 to 5 m*M* KCl, 1.25 to 1.3 m*M* $CaCl_2$, 1.2 to 1.3 m*M* $MgSO_4$, 1.2 m*M* KH_2PO_4, 25 m*M* $NaHCO_3$, and 15 to 20 m*M* HEPES). The chosen KR buffer is millipore filtered and gassed with O_2,[17] gassed with 95% O_2/5% CO_2[13] or with 5% CO_2.[18] Collagenase digestion has also been carried out in KRB supplemented with 2% bovine serum albumin and 0.1 mg/mL dextrose.[18] Samples are incubated with collagenase at 37°C for 25 to 45 min in a water bath shaker at 150 to 200 rpm. Gentle stirring with a rod disperses fat cells from tissue. Digested tissue is filtered through silk screen or nylon mesh. Collagenase is removed by washing the adipocytes 3 to 5 times by centrifugation/resuspension with the chosen KR buffer. Some methods also include 2% albumin[13,17] and 0.1% glucose[13] in the KR buffer at this stage. Sedimented material below the floating fat cells contains capillary, endothelial, mast, macrophage, and epithelial cells.[16] Fat droplets from lysed cells float above the fat cells. Gentle aspiration is used to separate the layers of material. Adipocytes are viable for at least 4 hours.[13,17] Incubation of fat cells in glass rather than plastic results in increased cell rupture and loss of metabolic activity.[16]

2. *Glucose Oxidation in Rat Adipocytes*

None of the published studies for glucose oxidation in rat adipocytes have been carried out under identical conditions between laboratories, and readers are therefore referred to the primary literature for specific details.[13,16–20] Cells are commonly incubated in plastic vials containing hanging glass wells and capped with rubber septa. In the studies in which concentrations were specified,[18–20] fat cells were treated in samples containing 2×10^5 to 2.5×10^5 cells per 0.5 mL final volume. Solutions are in the chosen KR buffer with 2%[13,17,18] or 0.2%[20] albumin. Solutions may also contain dextrose.[17,18] Samples contain universally labeled [U-^{14}C]-glucose, insulin at a range of concentrations, and the desired test compound in a range of doses. Incubations are carried out for 1.5 to 2 hr at 37°C on a shaker, at 80 to 150 rpm under an atmosphere of 95% O_2-5% CO_2. Reactions are terminated by the addition of H_2SO_4. The liberated $^{14}CO_2$ is trapped on hyamine hydroxide-soaked filter paper during a 0.5 to 1 hr incubation, and quantitated by scintillation counting of the filter paper. Insulin potentiation is defined as the ratio of $^{14}CO_2$ released at a given insulin level with and without the test compound (*vide infra*).

3. *Lipogenesis in Rat Adipocytes*

Incorporation of glucose into lipids can be determined in the same test samples incubated in the presence of [3-^{3}H]-glucose.[19–21] Lipid is extracted from samples into a toluene-based scillicant and counted for radioactivity. An alternative method in which lipids were extracted into chloroform/methanol/ 2 *M* KCl yielded consistent results.[19] Lipogenesis is expressed as a ratio of ^{3}H-labeled lipid relative to untreated controls for a given insulin concentration.

4. *Results Obtained from the Rat Adipocyte Assay*

Several researchers have used the rat adipocyte assay to evaluate the activity of (1) "biological chromium," (2) samples of the purported glucose tolerance factor that did and did not contain chromium, and (3) synthetic chromium complexes. Rather than clarifying the biologically active form of chromium in terms of its structure and mechanism of action, results yielded from this assay have "muddied the waters" by questioning the existence of a biologically relevant chromium-containing glucose tolerance factor.

The earliest study by Mertz and coworkers[15] measured glucose uptake in epididymal tissue isolated from rats fed either a "GTF-deficient" diet of Purina laboratory chow, or the same chow supplemented with hexaurea chromium(III) chloride. Chromium content of the basal diet was not reported. Insulin-induced glucose uptake was higher in rat tissue from animals on the chromium-supplemented diet. Addition of chromic potassium sulfate ($Cr_2(SO_4)_3 \cdot K_2SO_4$) to the tissue *in vitro* also stimulated glucose uptake in the GTF-deficient rat tissue. This supported the earlier conclusion[6] that chromium was the active component in GTF.

A later study by Toepfer, Mertz, et al.[8] expanded on these results. Tests were again carried out with epididymal tissue isolated from rats fed the "GTF-deficient" diet. Extracts from Brewer's yeast were tested (see Section III.A.1), and activity was compared to that of a compound synthesized through the reaction of Cr(III) with 2 equivalents each of nicotinic acid and glycine, and 1 equivalent each of glutamic acid and cysteine. Both the yeast extract and the synthesized compound potentiated insulin's oxidation of glucose, and both samples demonstrated a sight zero-insulin effect, by enhancing glucose oxidation above control levels in the absence of insulin.

Anderson and coworkers[17] isolated adipocytes from rats fed a low Cr diet (0.04 ppm) and compared the effects of a series of chromium compounds on insulin-stimulated glucose oxidation. A crude ammonia extract of Brewer's yeast produced threefold higher glucose oxidation above control, as did chromium complexes containing glutathione and nicotinic acid; and glutathione and glycine. Both the yeast extract and the synthesized glutathione/nicotinic acid complex enhanced glucose oxidation in the absence of insulin, and the greatest dose response for both experiments was observed in the absence of insulin. Individual components of the synthesized complexes had no activity. The study concluded that only chromium with certain coordinating ligands was biologically active.

A later study by Hwang and co-workers[18] further explored the activity of chromium in different yeast sources. Extracts were isolated from three forms of chromium-containing yeast: (1) *S. cerevisiae* enriched by incubation with $^{51}CrCl_3$; (2) Chromax yeast, containing 100 μg Cr/g yeast; and (3) Difco yeast extract treated *in vitro* with $^{51}CrCl_3$. Subjection of extracts to cation exchange chromatography demonstrated nearly identical fractions in terms of chromium content for all three sources. Each extract was separated into nine fractions, and fractions from the incubated brewer's yeast and Difco yeast extract were tested in the adipocyte assay. In each case, only two fractions stimulated glucose oxidation in the absence of insulin, but no enhancement of the stimulation was observed in the presence of insulin, and activity did not correlate with chromium content. Since the Difco yeast extract had been enriched *in vitro*, it was concluded that the chromium merely bound to components of the extract and was not incorporated by a cellular process.

Yamamoto and co-workers[22] have used the adipocyte assay to evaluate the LMWCr substance (see Section III.A.2). The LMWCr isolated from bovine colostrum was observed to potentiate the effects of 5 μU/mL insulin at concentrations of 0.015 to 15 ng/mL Cr. At 1.5 ng/mL Cr the LMWCr potentiated glucose oxidation and ^{3}H-lipid incorporation for an insulin range of 0 to 50 μ*M*/mL. The activity of LMWCr was observed to be a function of Cr content.[20]

It must be noted here that it remains to be established if the LMW protein exists *in vivo* for the specific purpose of coordinating chromium, or if it is a general chelator of metals. The LMWCr species was isolated after treating animals with Cr(VI).[14,20,23] Hexavalent chromium is a mutagenic and genotoxic metal. It is not clear that the treatment of animals with this form of Cr is not merely invoking a phase II metabolism of the toxin. Other metals with no established nutritional role have been found to be chelated by "acidic proteins" in treated animals. For example, injection of lead into rats resulted in soluble and insoluble lead protein complexes in the rat kidney.[24] The soluble lead-chelating protein contained high amounts of aspartic acid, glutamic acid and glycine, as well as cysteine and tryptophan. The amino acid content is similar to the LMWCr protein isolated from rabbit liver[23] and dog liver.[14] Similar types of proteins may also be involved in the *in vivo* chelation of bismuth and gold.[25]

Simonoff and coworkers[13] isolated GTF fractions from untreated yeast and yeast incubated with Cr^{3+} (see Section III.A.1). One of the aims of their study was to determine if Cr enrichment had any effect on the biological activity of yeast GTF in the rat adipocyte assay. Adipocytes were treated with a range of 0 to 1 mg/mL of crudely purified high-Cr and low-Cr yeast extracts in the presence of 16 μU/mL insulin. Chromium content in the Cr-enriched vs. untreated yeast extracts varied by 300-fold; however, no difference was observed in insulin-potentiating activity in the adipocytes. In both cases, the crude yeast extracts exhibited a similar dose response from 0.001 to 0.1 mg/mL Cr for increased glucose oxidation, but glucose oxidation was inhibited at the highest dose of 1 mg/mL Cr. Crude extracts were further purified by chromatography, and a lack of a chromium effect was also observed with the separated cationic and anionic fractions. The anionic fraction and its further purified subfractions showed higher Cr content for both Cr-treated and untreated yeast relative to the cationic fractions, and the anionic subfractions showed higher insulin-potentiating activity than the cationic subfractions. Maximum activity consisted of ~170% potentiation for the 0.1 mg/mL treatment; however, activity of these fractions from the Cr-treated yeast were no higher than that from the untreated yeast. It was concluded that species isolated from yeast demonstrated insulin-like activity, but the results did not suggest that there was a "unique chromium-containing GTF."[13]

The above studies have resulted in two different paradigms for Cr's affect on insulin-potentiated glucose metabolism. On one side, it is argued that the active component of the purported glucose tolerance factor does not contain Cr, although Cr can enhance glucose oxidation in this assay *in vitro*. On the other side, the claim is made that only a small fraction of total Cr is active, so that biological activity does not have to correlate with Cr content, and that the Cr content of the active fraction may not be detectable.[26]

5. *Interpretation of the Rat Adipocyte Assay*

There is still some confusion as to the interpretation of results obtained from the adipocyte assay. The original definition of insulin potentiation presented by Mertz[27] was based on the effect of the test substance on a range of insulin concentrations. At a given concentration of test compound, the glucose oxidation should increase with an increasing amount of insulin. The data for a tested dose yields a curve when graphed as the amount of added insulin (X axis) vs. $^{14}CO_2$ production (Y axis) (see Figure 8 in reference 27, which graphs data from reference 15). In the absence of the test compound, an increase in $^{14}CO_2$ production would still be observed with increasing insulin. If a test compound displays an insulin-potentiating effect, then the slope of the curve in the presence of the compound should be greater than that in its absence. Of the studies summarized in Section III.B.4, results from two[15,22] fit this criteria. Two studies did not report data for a range of insulin concentrations;[13,20] one study observed no effect on $^{14}CO_2$ production with increasing insulin;[18] and one observed a *decrease* in the slope of the curves representing added test compound.[17]

Further mechanistic insight has been provided through a reinterpretation by Vincent.[28] It was noted that in previous studies,[8,17] the addition of the test compound in the absence of insulin resulted in a background increase in $^{14}CO_2$ production, i.e., the line corresponding to the presence of a test compound would not intersect the Y axis at zero for the zero insulin data point. It was observed that correcting the data of Anderson et al.[17] for the background effect of the GTF sample would actually show an *inhibition* of insulin-induced glucose oxidation. Of the studies summarized in Section III.B.4 above, only three[8,17,18] reported a zero-insulin effect. Vincent[28] noted that the zero-insulin effect may be relevant to the mechanisms of action for the different substances, reflecting a compound's influence on insulin binding to the membrane receptor vs. a post-receptor role.

6. *Complications in the Rat Adipocyte Assay*

Several experimental factors make it difficult to compare results from the adipocyte assay between laboratories. On one level, there is no general agreement as to what constitutes the active component of GTF, nor is there a standard methodology for its isolation. Different laboratories have separated different fractions by different methodologies, and have found different reactivities in the adipocytes. It has been observed that the extent of handling of the tissues during the isolation procedure, and chilling of the tissues will affect glucose metabolism.[29] The initial slicing of the epididymal tissue with stainless steel scissors may affect chromium content.[13] Therefore, it is difficult to evaluate the results between laboratories for studies that observed no effect on glucose oxidation for a test compound, without a positive control. Several studies have used synthetic chromium complexes of glutathione and nicotinic acid as positive controls,[8,17] but these compounds have not been structurally characterized, and no standard procedures exist for their synthesis.

Results may also be influenced by the use of different animal strains as sources of epididymal tissue. Also, the rat diet has been reported to have a strong effect on glucose metabolism in the isolated fat cells. Two of the above-mentioned studies reported diets of less than 0.04 ppm Cr[17] and 0.35 ppm Cr,[13] but the others cited above did not report dietary Cr content. The effect of animal diet was most clearly shown by Shepherd and co-workers,[30] who tested the insulin potentiating activity of a brewer's yeast GTF sample on adipocytes from groups of rats fed four diets that differed in protein source. All four diets contained greater than 0.3 μg Cr/g (ppm) dry weight, thus none were considered Cr deficient. The GTF sample potentiated the effects of suboptimal insulin only in fat cells from rats fed the Torula yeast diet. By assaying the diets for 23 elements, the authors concluded that the activity was due to a dietary deficiency in manganese, although manganese itself was not responsible for the insulin-potentiating activity of the GTF sample. Davies and co-workers[31] also reported that GTF activity from yeast extracts was observed in adipocytes from rats on a Torula yeast-high sucrose diet, but not from rats fed a commercial rat cube diet (Teklad, Madison, WI). Diets were found to contain from 0.47 to 0.82 ppm Cr, and thus were not defined as Cr deficient.

The above studies have laid the foundation for chromium's role in nutrition, and the rat adipocyte assay has proven to be a useful assay for the comparison of different compounds within an experiment. However, it may be difficult to further elucidate a mechanism for chromium's activity with an assay for which so many substances demonstrate insulin-like activity. Examples of other species that have been found to potentiate insulin by this assay include ornithine and ε-*N*-glutaryllysine,[31] Mn(II),[15,19] Zn(II),[19,21] Ni(II),[19] Cd(II),[19] and Cr(VI).[32] There is no nutritional role proposed for cadmium in mammals, and Cr(VI) is a mutagenic human carcinogen. Presumably, there will be a variety of molecular mechanisms through which these species induce glucose oxidation.

IV. CONCLUDING REMARKS

Many textbooks and reviews refer to a chromium-containing glucose tolerance factor. However, this summary of the literature demonstrates that even though over three decades have passed since the GTF was first proposed to be an essential cofactor for the maintenance of glucose tolerance in rats,[7] the existence of this species has yet to be experimentally verified. Indeed, the bulk of experimental evidence has weighed in against it. In order to fully assess chromium's role as a trace essential metal, and to distinguish between essential and nonessential pharmacological effects, the mechanisms for chromium's action must be elucidated. The use of *in vitro* tools such as the rat adipocyte assay and other similar systems may provide a means through which to compare the function of chromium compounds to other species with similar activity, in order to more fully explain chromium's influence on cellular signal transduction and nutrient metabolism.

REFERENCES

1. Mertz, W., Toepfer, E.W., Roginski, E.E., and Polansky, M.M., Present knowledge of the role of chromium. *Federation Proc.*, 33, 2275, 1974.
2. Mertz, W., Chromium in human nutrition: a review. *J. Nutr.*, 123, 626, 1993.
3. Versieck, J., Trace elements in human body fluids and tissues. *Critical Rev. Clin. Lab. Sci.*, 22, 97, 1985.
4. Nielsen, F.H., Chromium. In *Modern Nutrition in Health and Disease*. 8th ed., Vol. 1. Shils, M.E., Olson, J.A., and Shike, M., Eds. Lea and Febiger, 1994, 264.
5. Wetterhahn-Jennette, K., The role of metals in carcinogenesis: biochemistry and metabolism. *Environ. Health Perspect.*, 40, 233, 1981.
6. De Flora, S. and Wetterhahn, K.E., Mechanisms of chromium metabolism and genotoxicity. *Life Chem. Rep.*, 7, 169, 1989.
7. Schwarz, K. and Mertz, W., Chromium(III) and the glucose tolerance factor. *Arch. Biochem. Biophys.*, 85, 292, 1959.
8. Toepfer, E.W., Mertz, W., Polansky, M.M., Roginski, E.E., and Wolf, W.R., Preparation of chromium-containing material of glucose tolerance factor activity from Brewer's yeast extracts and by synthesis. *J. Agric. Food Chem.*, 25, 162, 1977.
9. Votava, H.J., Hahn, C.J., and Evans, G.W., Isolation and partial characterization of a ^{51}Cr complex from Brewer's yeast. *Biochem. Biophys. Res. Commun.*, 55, 312, 1973.
10. Kumpulainen, J., Koivistoinen, P., and Lahtinen, S., Isolation, purification, and partial chemical characterization of chromium (III) fractions existing in Brewer's yeast and Sabourand's liquid medium. *Bioinorg. Chem.*, 8, 419, 1978.
11. Mirsky, N., Weiss, A., and Dori, Z., Chromium in biological systems, I. Some observations on glucose tolerance factor in yeast. *J. Inorg. Biochem.*, 13, 11, 1980.
12. Haylock, S.J., Buckley, P.D., and Blackwell, L.F., Separation of biologically active chromium-containing complexes from yeast extracts and other sources of glucose tolerance factor (GTF) activity. *J. Inorg. Biochem.*, 18, 195, 1983.
13. Simonoff, M., Shapcott, D., Alameddine, S., Sutter-Dub, M.T., and Simonoff, G., The isolation of glucose tolerance factors from Brewer's yeast and their relation to chromium. *Biol. Trace Elem. Res.*, 32, 25, 1992.

14. Wada, O., Wu, G.Y., Yamamoto, A., Manabe, S., and Ono, T., Purification and chromium-excretory function of low-molecular-weight, chromium-binding substances from dog liver. *Environ. Res.*, 32, 228, 1983.
15. Mertz, W., Roginski, E.E., and Schwarz, K., Effect of trivalent chromium complexes on glucose uptake by epididymal fat tissue of rats. *J. Biol. Chem.*, 236, 318, 1961.
16. Rodbell, M., Metabolism of isolated fat cells. *J. Biol. Chem.*, 239, 375, 1964.
17. Anderson, R.A., Brantner, J.H., and Polansky, M.M., An improved assay for biologically active chromium. *J. Agric. Food Chem.*, 26, 1219, 1978.
18. Hwang, D.L., Lev-Ran, A., Papoian, T., and Beech, W.K., Insulin-like activity of chromium-binding fractions from Brewer's yeast. *J. Inorg. Biochem.*, 30, 219, 1987.
19. Yamomoto, A., Wada, O., Ono, T., Ono, H., Manabe, S., and Ishikawa, S., Cadmium-induced stimulation of lipogenesis from glucose in rat adipocytes. *Biochem. J.*, 219, 979, 1984.
20. Yamomoto, A., Wada, O., and Manabe, S., Evidence that chromium is an essential factor for the biological activity of low-molecular-weight, chromium-binding substance. *Biochem. Biophys. Res. Commun.*, 163, 189, 1989.
21. Coulston, L. and Dandona, P., Insulin-like effect of zinc on adipocytes. *Diabetes*, 29, 665, 1980.
22. Yamamoto, A., Wada, O., and Suzuki, H., Purification and properties of biologically active chromium complex from bovine colostrum. *J. Nutr.*, 118, 39, 1988.
23. Yamamoto, A., Wada, O., and Ono, T., Isolation of a biologically active low-molecular-mass chromium compound from rabbit liver. *Eur. J. Biochem.*, 165, 627, 1987.
24. Moore, J.F. and Goyer, R.A., Lead-induced inclusion bodies: composition and probable role in lead metabolism. *Environ. Health Perspect.*, 7, 121, 1974.
25. Goyer, R.A., Formation of intracellular inclusion bodies in heavy metal poisoning (lead, bismuth and gold). *Environ. Health Perspect.*, 4, 97, 1973.
26. Anderson, R.A., Recent advances in the clinical and biochemical effects of chromium deficiency. In *Essential and Toxic Trace Elements in Human Health and Disease: An Update*. Prasad, A.S., Ed., Wiley-Liss, New York, 1993, 221.
27. Mertz, W., Chromium occurrence and function in biological systems. *Physiol. Rev.*, 49, 163, 1969.
28. Vincent, J.B., Relationship between glucose tolerance factor and low-molecular-weight chromium binding substance. [Comments] *J. Nutr.*, 124, 117, 1994.
29. Winegrad, A.I. and Renold, A.E., Studies on rat adipose tissue *in vitro*. *J. Biol. Chem.*, 233, 267, 1958.
30. Shepherd, P.R., Elwood, C., Buckley, P.D., and Blackwell, L.F., Glucose tolerance factor potentiation of insulin action in adipocytes from rats raised on a Torula yeast diet cannot be attributed to a deficiency of chromium or glucose tolerance factor activity in the diet. *Biol. Trace Elem. Res.*, 32, 109, 1992.
31. Davies, D.M., Holdsworth, E.S., and Sherriff, J.L., The isolation of glucose tolerance factors from Brewer's yeast and their relationship to chromium. *Biochem. Med.*, 33, 297, 1985.
32. Goto, Y. and Kida, K., Insulin-like action of chromate on glucose transport in isolated rat adipocytes. *Jpn. J. Pharmacol.*, 67, 365, 1995.

Chapter **19**

Calcium Bioavailability Using a Rat Model

Margaret Ann Bock
Connie Weaver
New Mexico State University
Las Cruces, New Mexico

CONTENTS

0-8493-9611-5/97/$0.00+$.50

I. CALCIUM

Calcium is the most abundant mineral in the human body, constituting about 1 kg in a 70-kg man (Avioli, 1988). Bones and teeth, which contain about 99% of the calcium in the body, are primarily dependent on this mineral for their strength and structure. The remaining body calcium is distributed between the extracellular fluid and various soft tissues, where it has a variety of regulatory functions (Hunt and Groff, 1990).

Throughout life, bone is constantly being formed and resorbed. This process, called "bone remodeling," occurs more rapidly during early life and at a declining rate as one ages (Albanese et al., 1978). Therefore, an adequate intake of calcium is important not only during the years of skeletal growth and bone consolidation, but also thereafter to maintain optimal bone integrity. Calcium deposition increases and occurs in proportion to bone growth in growing children (Avioli, 1988; Tepperman and Tepperman, 1987). Of the 1 kg present in the average young 70-kg adult, only about an average of 0.55 g of this large amount is exchanged daily between the bones and body fluid. This exchange between bone and extracellular fluid is hormone regulated, with parathyroid hormone (PTH) favoring the removal of bone calcium and calcitonin favoring its deposition. The three main hormones involved in calcium homeostasis are PTH, calcitonin, and 1, 25 $(OH)_2$ D_3. PTH acts to conserve body calcium and to increase extracellular fluid calcium concentration by promoting resorption of calcium from the bone and reabsorption of calcium from the kidney tubules and increasing the rate of 1, 25 $(OH)_2$ D_3 formation which triggers active absorption of calcium from the gastrointestinal tract (GI) and plays a permissive role in the action of PTH on bone and, maybe, on the kidney as well (Tepperman and Tepperman, 1987).

Insufficient calcium intake during the period of bone mineralization is a real concern because of the high incidence of osteoporosis among elderly women and the significant correlation shown to exist between present bone density and past calcium intake. Recognizing the importance of generous calcium intake during the mineralization period of bone, the subcommittee involved with the 10th edition of the RDAs elected to extend the quantity of calcium recommended for adolescents (1200 mg/d) through age 24 (RDA, 1989). In the opinion of the subcommittee, a calcium intake of 800 mg/d is sufficient for adult women over 25 years of age through menopause and beyond. Postmenopausal osteoporosis is regarded by this group as a medical rather than a nutritional problem (Hunt and Groff, 1990).

The recommendations of the 1984 National Institutes of Health (NIH) consensus panel on osteoporosis are in contrast to the 1989 RDAs (Hunt and Groff, 1990). Consensus recommendations are a calcium intake of 1000 mg/d by estrogen-replete premenopausal women and postmenopausal women treated with estrogen. An intake of 1500 mg/d is recommended for untreated postmenopausal women (Riggs et al., 1987). Factorial estimates of calcium indicate that the levels should be closer to 2000 mg/d for 12 to 14 year old women; 1650 mg/d for 12 to 14 year old men and about 1500 mg/d for 18 to 30 year old males (Weaver, 1994).

Osteoporosis, which is a condition of reduced bone density, is the major cause of bone fractures. This condition affects 25 million individuals in the U.S. annually. Health care cost associated with these fractures exceeds $10 billion per year (Consensus Development Conference, 1993).

II. EXPERIMENTAL DESIGN

The researcher must decide what experimental design is appropriate to meet the objectives of his or her project. This is the point at which it is imperative that a statistician be involved in the design of the project so that meaningful data can be ascertained.

Some of the most common research designs used are the dose response format and the factorial design. The dose response format is used when the researcher is trying to ascertain what level of a given constituent will affect calcium absorption and status in the body. For example, the researcher may be trying to ascertain how different levels of oxalate affect the ability to absorb and utilize calcium in the diet. The factorial design is used when a researcher is attempting to ascertain the effect of multiple variables fed at multiple levels. For instance the researcher may be trying to determine the effect of feeding multiple levels of calcium (e.g., low, adequate, or high) and multiple levels of oxalate (e.g., zero, low, moderate, or high). This design will provide information related to interactions.

III. USING ANIMALS AS A MODEL

Animals are used for research for a variety of reasons. The main reason animals are used in lieu of human subjects is safety when a researcher is testing a variable with unknown effects. The second is cost. Although animal facilities are expensive to setup and maintain, they are far less expensive than human metabolic wards. Cost per subject is also lower when using animals than when using humans. In addition, animal models are needed when invasive techniques are required. Calcium status can be assessed by bone assay in animals, whereas no suitable assessment of current calcium status is available in humans.

IV. USING RATS

Rats are one of the most common animal models used. In the case of calcium research, albino rats are commonly used. When selecting a rat model, the researcher needs to carefully evaluate if the rat and the type of rat selected is the best model for the variables being tested. For instance, if one of the variables in the research being conducted is vitamin C, then a rat model of any kind is not appropriate because rats have the ability to synthesize this vitamin. Where genetics are concerned, one needs to be aware that certain types of rats and mice have genetic characteristics which may be very beneficial or very detrimental to the research being conducted. For example, some types of rats are genetically prone to be obese. Such a characteristic could have a profound impact in calcium related research because weight bearing exercise is another of the factors known to affect bone calcium status. Companies that supply animals for research should be able to give the researcher detailed information about the animals that they breed.

A. Number of Animals

If the researcher has done his or her background work and has determined that the rat is an appropriate model for the variables being tested, then the investigator must decide how many animals per group are needed to have a study which can be tested statistically. It is imperative that the researcher have a statistician involved so that the study will produce meaningful data which can be statistically analyzed. If using a cross sectional approach, usually a minimum of six animals per experimental group are essential. Generally having ten animals per group is more ideal. If using a longitudinal approach where animals are sacrificed at specified points to ascertain time related data, then the researcher generally has to insure that there are at least six animals in each group at each time point.

B. Gender

Gender of the animals being used is also critical in calcium research. Investigators doing calcium research must be aware that estrogen has a profound effect on calcium status in the bone. Because the estrogen levels fluctuate significantly in a female animal of reproductive age, calcium researchers often select male animals to control this variable. If estrogen is a critical variable in the research being done, then the researcher can utilize normal female animals compared to oophorectomized female animals. If such a comparison is made, it is important that both the normal animals and the oophorectomized animals be of the same age. It is important to recognize that using such an approach will be more costly, especially if the animals are purchased in an oophorectomized state, which is the most practical for most researchers who do not have the facilities and/or expertise to perform surgery on the animals.

C. Age

Age of the animal selected is another factor which must be considered by the researcher. Most research is done with weanling animals (21 days old). This is the age at which they are the least expensive to purchase. However, the investigator must be aware that at this age, the animal is undergoing a rapid phase of growth which will play a role in calcium uptake and utilization in the body. Some researchers elect to use older animals to remove the rapid growth variable. From a cost standpoint, the researcher may want to obtain the animals as weanlings and then feed them rodent chow until they reach the age that the scientist wishes to use. Researchers must be aware that the rat has a life expectancy of approximately 3 years which is equivalent to about 90 years in the human (Griffith and Farris, 1942). They also must be cognizant of the fact that, the older the animal is, the more difficult it is to do research without specialized germ free facilities for working with geriatric animals.

When ordering animals, the researcher should make sure that they arrive at least 3 to 5 days before the initiation of the study. This will allow the animals a period of time to recover from the stress induced by transport.

V. HOUSING OF ANIMALS

The type and sophistication of the housing needed for animals in calcium research is dependent on the tissues being collected and analyzed. To have meaningful data, animals should be housed individually so that information about food intake and waste for each animal can be ascertained. Individual housing removes confounding factors such as a dominant animal getting more food than those which are more subordinate. It also eliminates the coprophagy of another animal's fecal material.

Unless doing research involving minerals which can be derived from stainless steel, cages should be made of stainless steel, if possible. This is especially true if the researcher is using other minerals which may be derived from other types of metals as one or more of the variables in the research being conducted.

Cages should be suspended from a rack and should have mesh bottoms. There should be a solid shelf below each tier of cages. This will, for the most part, cut down on the amount of coprophagy that a given animal can do using its own feces. There should be enough room between the bottom of the cage and the shelf so that the animal cannot reach through the bottom of the cage and pick up fecal material from the shelf.

If doing analyses such as calcium balance, then the researcher will need what are typically called metabolic cages which allow for the quantitative collection of urine and feces. Fecal collection screens and urine diversion funnels can be made for many suspended cage systems. If the researcher knows that he or she is going to be collecting such data on a regular basis, then he or she should consider buying metabolic cages when setting up the animal laboratory. Both the fecal collection screens and urine diversion funnels should be stainless steel just like the cages. Urine collection bottles which are placed under the diversion funnels should be glass (unless boron is one of the variables) or nonreactive plastic.

Type and placement of feeding cups is also another issue which needs to be considered. Cages which are not specifically made for metabolic studies are typically designed so that the feeding cup is an independent entity which is set on the bottom of the cage or hooked to a side of the cage that is mesh. Such a configuration presents some special problems. The most significant of the problems is the contamination of feces which are accumulating on the shelf or a fecal collection screen below a given cage. Such an arrangement makes it almost impossible to determine wastage of diet that has been thrown out of the food cup of a rat that digs in its food container. It also makes it almost impossible to get reliable values for fecal calcium. To partially solve this problem, the researcher may use a feeding cup with a lid that has a hole that is only large enough for the animal to insert its head. To further reduce the problem of throwing food, the researcher can place a ring of metal or glass on the top of the food with a hole large enough for the animal to insert its head. An alternative is a perforated disk on top of the food. As the amount of food goes down in the feeding cup, the ring or perforated disk will go down further in the container which may help to deter digging in the food cups. Some researchers will use a combination of both the lid and the weighted ring. A more ideal arrangement is having the food cup suspended in a chute which has a catch tray under it. Indeed, this is the configuration that is noted in many cages which are designed for metabolic studies. However, even this configuration is not foolproof. The animal can still dig and throw food that does not get caught by the catch tray under the feeding chute. To solve this problem, some researchers use "diapers" made from paper towels which are suspended from paper clips under the chute to catch food which is thrown out while digging. It is important to remember that the feeding cup, lid, and weighted ring (if used) should be of materials (e.g., stainless steel or glass [unless doing boron work]) that will not add a source of variance to the study. Also the researcher must be aware that he or she cannot use a material that the animal can chew and, thereby, add an unwanted variable.

Cages should be housed in a secured facility in a room that allows for control of physical conditions. Facilities should meet standards which are established for animal research such as those which guide researchers in the U.S. (Animal and Plant Health, 1994). It is especially important that the researcher be able to control temperature and light. Temperature is an important variable to control because having the temperature too high may decrease the food intake of the animal. If the temperature is too cold, then the animals may eat more. Also under these conditions, the animal may have to shiver to keep warm, which means that the animal will be expending energy and all of the nutrients required to make and utilize the energy that it would not expend under normal conditions. Light is an issue when using rats because rats are nocturnal animals that eat under dark conditions. In addition, many factors in the body are influenced by light vs. dark conditions. Typically the temperature is controlled to about 20 to 25°C and the light cycle is often set for 12 hr light and 12 hr dark (Animal and Plant Health, 1994). If the research involves body constituents that are affected by light or dark, it is important that the researcher establish the light and dark cycles so that work (filling food cups; checking water; cleaning up waste material) with the animals is done at a time that the lights would normally be on.

At the time the researcher is ready to initiate the study, cages, food cups, and water bottles should be washed and then rinsed with distilled/deionized (DD) water. At this point, animals should be randomly assigned to the various control and experimental groups. To overcome the potential that cage position on the rack becomes a variable, cages of a given group should be arranged vertically on the rack so that part of the animals in the group are at the top of the rack, part are in the middle, and the remainder are at the bottom.

VI. CONTROL AND EXPERIMENTAL DIETS

Diets will be based on the objectives of the project. However, there are dietary factors which must be considered when formulating diets.

A. Protein

First of all, the researcher must consider the ingredient or ingredients which will supply protein in the control and experimental diets. Unless protein is one of the variables being tested, the level of protein should be consistent from diet to diet. In other words, the diets should be isonitrogenous. Nutrient Requirements of Laboratory Animals (NRC, 1995) indicate that the level of protein should be at least 5 g/100 g of diet for maintenance and 15 g/100 g for growth and reproduction. Levels between 15 and 20% are typically seen. Researchers must also consider whether the protein source will be animal, vegetable, or a combination of the two. If a combination is used, then the researcher should strive to have the ratio of animal to plant protein consistent from one diet to another. Casein is one of the typical sources of protein utilized. However, there are two factors related to using casein which must be considered. First of all, casein does not supply sufficient methionine which must be supplemented in a diet which relies on casein as the sole source of protein in the diet (NRC, 1995). Secondly, one must be aware that purified diets may be a source of artifacts in data which have nothing to do with the variables being tested.

Level of protein in a given ingredient is frequently ascertained by a Kjeldahl nitrogen determination and then multiplying the nitrogen data by a factor consistent with the food being analyzed. To insure accuracy of the assay, researchers should run a standard with a known nitrogen level such as urea. Calculating a coefficient of variation will also help the researcher ascertain procedural error.

B. Calcium

In research involving calcium, the researcher must carefully consider the source or sources of calcium in the diet. If using a calcium salt, the quantity of calcium from one calcium salt to another will vary significantly. The degree to which the calcium in a given salt is bioavailable will also vary from one calcium salt to another. In 1980, Ranhotra et al. (1980) conducted a study using nine calcium sources (Carbonate; Chloride; Hydroxide; Monophosphate; Sulfate [dihydro]; Oxide; Lactate; Acetate; Propionate and Yeast Food [$CaHPO_4$; $CaSO_4$; NH_4Cl; KBr; salt and corn flour]). All were found to be bioavailable. However, the quantity of calcium, which was analyzed using atomic absorption (AA), varied from 10.9 to 58.4% (Table 1).

TABLE 1.

Calcium Sources

Calcium Source	Percent Calcium
Carbonate	37.1
Chloride	31.0
Hydroxide	48.4
Monophosphate	15.5
Sulfate (dihydro)	21.5
Oxide	58.4
Lactate	12.4
Acetate	22.2
Propionate	19.6
Yeast Food[a]	10.9

[a] [$CaHPO_4$; $CaSO_4$; NH_4Cl; KBr; salt and corn flour]

Adapted from Ranhotra et al., 1980.

Some researchers elect to use natural carriers of calcium (e.g., nonfat dry milk) as the modality for getting calcium into the diet. However, the researcher must be aware that such ingredients may contain constituents which may enhance or inhibit (e.g., oxalates) calcium absorption. Often that is the objective of the research (e.g., factors in spinach which enhance or inhibit the absorption of calcium inherent in the spinach). However, such constituents may be sources of confounding variables that are undesirable.

Some researchers elect to use stable or radioactive isotopes of calcium. If doing so, the researcher can use either an intrinsic labeling system or an extrinsic labeling modality. Intrinsic labeling is accomplished while growing the food which will be the carrier of the calcium. This may be done in a variety of ways through addition of the stable or radioactive isotope to the hydroponic solution in which the food is growing or through stem injection of the plants while they are growing. Stable isotopes are necessary for studies in growing children or pregnant women or for use where the food must be prepared commercially for human use. For animal models, the cost of stable isotopes vs. radioisotopic Ca is usually not justified.

In addition to the source or sources of calcium in the diet, the researcher must also consider the amount of calcium that is used in the diet. The Nutrient Requirements of Laboratory Animals (NRC, 1995) indicate that the calcium needs of the rat are 0.5 g/100 g of diet. Some researchers indicate that maximum weight gain can be attained if the diet includes at least half or more of this amount (Bernhart et al., 1969).

C. Other Minerals

Because of potentials for interaction between minerals or competition for carriers, other minerals being put in the diets are often controlled. To do so, the researcher needs to ascertain the amount of a given mineral that is present in the various dietary ingredients and the amount which is present in the mineral salt that the researcher is going to use to add additional amounts of the mineral to the diet. Once such information is ascertained, then the researcher can add the mineral salt at a level that will bring the mineral up to a predetermined level so that all diets have the same amount of the mineral in question. In calcium research, the total level of phosphorus is usually controlled because of the ability of phosphorus to bind with calcium and render the calcium unavailable. Draper et al. (1972) point out that a ratio of 2:1 of calcium to phosphorus is more effective than a 1:1 ratio to prevent osteoporosis. Other minerals that are often controlled in calcium research are magnesium, zinc, and iron.

Both sodium and potassium are added directly to the diets. Unless there is a reason to control such elements because they are being used as variables, one typically adds sodium (0.05 g/100 g diet) in the form of sodium chloride at the rate of 0.13 g/100 g of diet. Potassium (0.36 g/100 g diet) is typically added as potassium chloride at the rate of 0.69 g/100 g of diet (NRC, 1995).

For those trace elements which are not being controlled, the researcher can make a mix using the following procedure.

1. Enter the names of the minerals which are going to be included in the Trace Element Mix in the first column (Column A) of Table 2 outlined below.
2. List the compound which is going to be used to provide the mineral in question in the second column (Column B) of the outlined table. Be sure to include any waters of hydration (if applicable).
3. In Column C, list the molecular weight of the element that you have listed in Column A in the top half of the square and the molecular weight of the entire compound including the waters of hydration (if applicable) in the lower half of the square.
4. Using the formula at the bottom of the table, calculate Column D.
5. Using the Nutrient Requirements of Laboratory Animals (NRC, 1995), list the amount of the nutrient outlined in Column A required per 100 g of diet.
6. Using the second formula presented at the bottom of the table, calculate Column F. This figure will be in milligrams. Data should be converted to grams.
7. Add up all of the figures from Column F. This will be the total amount of the trace element compounds that will be provided through the Trace Element Mix. Subtract this figure from 1 (one). This will give you the amount of sugar (sucrose) that will need to be added to make one gram of the Trace Element Mix.
8. The researcher needs to determine the total amount of the various diets (control and experimental) needed for the entire study. [To ascertain

the amount needed for each version of diet, the researcher needs to be aware that on average a rat will consume 15 to 20 grams (♂) or 10 to 15 grams (♀) of diet per day (NRC, 1995). To account for digging, it is recommended that the researcher calculate using at least 20 grams/day/animal. Example: If you have a control and three experimental diets with 10 animals per group and the experiment is going to last 4 weeks, you will need 22,400 grams of diet for all groups for the entire 4 week period. This represents 224 100-gram units. Therefore, all figures, including the figure for sugar, ascertained in Column F need to be multiplied by 224 so that the researcher will make sufficient Trace Element Mix for all diets for the duration of the experiment.]

9. When making the Trace Element Mix, measure all ingredients on an electronic balance. Use stainless steel measuring utensils. All ingredients should be put in a mortar which is large enough to hold the amount to be mixed. The compound should be mixed thoroughly with a pestle so that a homogeneous mixture is attained. The mixture should be stored in a freezer until diets are ready to be made. Some researchers avoid using plastic bags or glass bottles for storage because trace elements tend to stick to the sides of the container.
10. When the researcher is ready to mix diets, the Trace Element Mix should be brought to room temperature before measuring. One gram of the mixture is used for every 100 g of diet made.

Note: Trace elements which are typically included in the mix are Copper, Manganese, Iodine, and Chromium. The researcher must also remember to account for the sugar used in the Trace Element Mix when calculating the carbohydrates for the diet.

D. Vitamins

Until recently, vitamins other than vitamin D have not typically been research variables being tested. Recent research has focused some attention on vitamin K because of its role in osteocalcin formation or carboxylation. Unless either of these vitamins or another vitamin are variables related to the research, a vitamin mix which contains vitamins in amounts sufficient for the rat is used. An example of such a mix is AIN Vitamin Mixture 76 or Vitamin Diet Fortification Mixture. Unless instructions on the label indicate to the contrary, these mixes are added to the diet at the rate of 2.2 g/100 g diet.

If any vitamin is being controlled, the researcher will have to have diet ingredients which are devoid of the vitamin or ingredients must be assayed to ascertain the amount present and then add the vitamin at levels consistent with the study objectives. Such an approach will most likely necessitate that each vitamin be added individually. Amounts needed for the rat are detailed in the Nutrient Requirements of Laboratory Animals (NRC, 1995).

TABLE 2.
Trace Element Mix Calculations

Column A Mineral	Column B Compound Used	Column C MW of Element / MW of Compound	Column D Percent of Mineral[a] Listed in Column A	Column E NRC Requirement for the Mineral/100 g of Diet	Column F NRC Requirement in terms of Mineral Compound[b]
Example Mg	$MgSO_4 \cdot 7H_2O$	24.312 / 246.50	9.86	40 mg	405.56 mg (0.406 g)
					Total Amount of Various Compounds Needed
					Amount of Sugar Needed = 1- Total Amount of Various Compounds Needed

[a] $\frac{\text{MW of Element}}{\text{MW of Compound}} * 100 = \%$ of Mineral in Compound

[b] $\frac{\text{MW of Element}}{\text{MW of Compound}} = \frac{\text{MW of Element Required/100 g Diet}}{\text{X MW of Compound Required to Supply Required Amount/100 g Diet}}$

Note: MW = Molecular weight of element or compound
X = Unknown quantity of compound

E. Lipids

Fat in the diet can be contributed by many sources. Typical sources include oils, hydrogenated fats, and animals fats. Oils used include corn and soybean oil. If fat is a variable, the researcher may want to use a variety of fats to obtain specified levels of monounsaturated, polyunsaturated, and saturated fatty acids. The Nutrient Requirements for Laboratory Animals

(NRC, 1995) indicate that the diet should contain at least 5% fat. Information from the 1978 NRC (NRC, 1978) indicate that diet acceptance is better when the diet is about 30% fat. The researcher needs to remember that fat must be supplied at a level that will meet at least the essential fatty acid needs of the rat (NRC, 1995).

F. Carbohydrates

Information from the Nutrient Requirements of Laboratory Animals (NRC, 1995) indicates that glucose or gluconeogenic precursors should be added to the diet of the rat. Because of potential problems with diarrhea, dietary sugar content cannot be too high. Poor performance has been noted in animals fed lactose or galactose as the carbohydrate source (NRC, 1995). It appears that raffinose, melibiose, mannoheptulose, rhamnose, and xylose sugars are not utilized by the rat (NRC, 1995). Well utilized starches include wheat, maize, rice, and cassava (NRC, 1995). Utilization of a combination of carbohydrates such as corn starch and sucrose are noted in the literature.

Having a fiber source in the diet causes concern in calcium research. The concern stems from the fact that certain types of fiber and fiber related components have the ability to bind calcium. Fiber sources containing oxalates are a good example. However, the degree that bioavailability is affected varies from one fiber source to another. Particle size as well as the fermentability of the fiber source will also impact bioavailability. In addition, the amount of the fiber source will affect nutrient intake and caloric intake. If the amount of fiber is too high, the diets may be excessively diluted. Unless fiber is one of the variables being tested, typical levels of fiber are 5% of a nonnutritive fiber such as cellulose (NRC, 1995).

G. Mixing Diets

When decisions have been made related to all of the ingredients that will be included in the various diets, then the researcher is ready to make the diets. All of the diet needed for the duration of the study should be made for each group before initiation of the feeding part of the study. To determine how much diet that can be made in a given batch, the researcher must know the capacity of the mixing instrument which will be used. The mixing instrument should be something like a commercial size mixer which is equipped with a stainless steel bowl and paddle. Dry ingredients such as the protein source(s) and carbohydrate source(s) should be measured first and put in the bottom of the mixing bowl. Ingredients needed in large quantities should be weighed on a top loading electronic balance. Utensils for measurement of ingredients should be made of stainless steel.

Once these ingredient are in the mixing bowl, a small indentation should be made in the center of the dry ingredients. Ingredients which are needed in small amounts (e.g., vitamins and minerals) should be measured on an electronic balance using stainless steel utensils. Such ingredients should be put in the well in the middle of the dry ingredients. Lastly, the fat source should be added. If using a liquid fat source, the researcher may have to establish an amount of time that he or she lets the container with the measured amount of fat drip into the mixing bowl so that the level of fat is fairly consistent from one diet to another. Once all ingredients are in the mixing container, the researcher needs to cover the top of the bowl leaving only an opening for the paddle to attach to the mixer. The mixer should be started on a slow speed to allow micronutrients to be mixed in and not thrown out of the bowl. Once the diet is mixed enough that ingredients will not fly around in the bowl, the mixer should be turned on a moderate speed and mixed for about 45 minutes. All diets should be mixed for the same amount of time. A given diet should be placed in a labeled container, sealed, and placed in a freezer until initiation of the study. Once the study is initiated, the researcher should keep only enough diet thawed and stored under refrigeration for several days of feedings so that the diet will not become rancid or have microbial growth. A given diet should be brought to room temperature before measuring into the food cups.

Diets containing ingredients with radioactive materials must be handled based on the half life of the isotope(s) being used. Two radioisotopes of calcium (^{45}Ca and ^{47}Ca) are commonly used. ^{47}Ca has a half life of 4.53 days; ^{45}Ca has a longer half life of 164 days. Diets are usually made soon after harvest of an intrinsically labeled food or soon after extrinsic labeling. Food containing radioisotopes and diets containing such foods must be handled and stored using regulations covering radioactive materials.

VII. FEEDING REGIMENS

Several feeding regimens are utilized in calcium research. The most common are *ad libitum*, pair feeding, and meal trained feeding. When feeding *ad libitum*, animals are allowed access to the food cup 24 hours per day, except during refilling of the food cups and during a fasting period just prior to doing analytical tests (e.g., blood draws). The biggest disadvantage to *ad libitum* feeding is the potential lack of consistency in diet intake from one group of animals to another. This disadvantage is particularly important when feeding a dietary regimen that may increase or decrease dietary acceptance and/or intake.

Pair feeding is another feeding technique which is used. The technique usually involves feeding animals in groups of two. One of the two animals in a given pair is typically the control while the second is the animal receiving one of the experimental diets. Generally the animal fed the experimental diet is fed a specified amount of diet, usually an amount that the animal will consume in one feeding or the amount that the animal will consume in a 24 hour period. Then the control animal of the pair is fed an amount identical to that consumed by the experimental animal. This technique provides the researcher with an avenue for having like intakes between the control and experimental groups. However, it should be apparent that such a technique is very labor intensive and difficult to do if the number of experimental variations is large (e.g., > 1 or 2).

Meal training is a feeding technique that is used when the researcher wants the animal to consume all of what is presented at a given feeding. This may be used as a component of paired feeding described above. Meal training is particularly useful when feeding intrinsically or extrinsically labeled food components or diets. Such a technique helps to insure that the animals will consume all of the isotope. Like pair feeding, this technique tends to be labor intensive. Getting the animal to eat all of the food that is presented without digging is also a challenge with this feeding technique.

Not only must the animal have food, it must have access to a hydrations source. Typically the animals are hydrated with water. Because tap water is a potential source of calcium, feeding studies are usually done using at least distilled deionized water which in most cases is provided to the animal on an *ad libitum* basis. Water bottles should be made from glass [unless doing boron work] or nonreactive plastic. The bottles should have sipper tubes and stoppers which are made of materials that will not add a confounding variable to the research and cannot be chewed by the animals. Bottles need to be checked daily. Water should be changed frequently to prevent growth of microorganisms.

VIII. DURATION OF STUDY

A review of the literature reveals that there is no consistency in the duration of calcium bioavailability studies. Typically studies which employ *ad libitum* feeding last for four weeks. Studies involving radioisotopes are typically done using meal trained rats. Meal training requires several days for the rats to adapt to the total consumption of food in a short period of time. Regardless of the duration of the study or the feeding regimen employed, animals should be obtained far enough in advance to allow for a period of acclimation after the stress of transport.

IX. COLLECTION AND HANDLING OF SAMPLES

A. Data and Samples Typically Collected

The data and samples to be collected by the researcher during and/or at the end of the study are dependant upon the objectives of the study. Typical data which are collected include:

1. Food Fed
2. Food Wasted
3. Food Consumed (Food Fed – Food Wasted)
4. Calcium Intake (Food Consumed x mg Ca/g diet) (Ca/g diet may be obtained by calculation or by analysis [Analysis should provide more representative data])
5. Initial Body Weight
6. Weekly Body Weight
7. Body Weight Gain (Final Body Weight – Initial Body Weight)
8. Diet:Gain Ratio (Total Food Intake/Body Weight Gain)

B. Calcium Balance

If the objectives involve ascertaining Calcium Balance, the researcher will need to collect the following data, samples, and analyses:

1. Calcium Intake (See Previous Discussion)
2. Feces for a Specified Period
3. Fecal Weight
4. Urine Volume for a Specified Period
5. Fecal Calcium Excretion
6. Urinary Calcium Excretion

Once that data is available, then the researcher can calculate:

1. Absolute Calcium Balance (Ca Intake – [Fecal Ca Excretion + Urinary Ca Excretion])
2. Relative Calcium Balance (Ca Intake – [Fecal Ca Excretion + Urinary Ca Excretion])/Calcium Intake

C. Bone Calcium

1. Bone
2. Bone Weight

3. Bone Calcium
4. Total Bone Calcium

D. Collection of Anthropometric Data and Samples

1. Diet Intake

Food cups should be checked daily. The total amount of food fed and total amount of food wasted should be recorded weekly. Diet intake can be calculated by subtracting the total amount wasted for each animal from the amount fed for a given animal. Diet intakes for each animal in a group should be totaled and divided by the total number of animals in the group to get an average intake for a given week. Diets should be weighed on an electronic balance.

2. Calcium Intake

Once diet intake is determined, the researcher then can determine the amount of calcium in the diet or can use a calculated value. Using that information, the researcher can multiply the amount of calcium in 1 gram of diet times the total grams of diet consumed and determine the amount of calcium ingested by the animal. Average for the group can be ascertained as detailed under diet intake.

3. Body Weight

Animals should be weighed on a weekly basis using an electronic balance. To get an accurate weight, the animal should be confined in a perforated chamber that has been tared on the scale prior to enclosing the rat. Weekly weight gain can be determined by subtracting the weight at the beginning of the week from the weight recorded at the beginning of the following week. Total weight gain can be ascertained by subtracting initial body weight from the final body weight.

4. Diet to Gain Ratio

Diet:Gain Ratio can be calculated by dividing total weight gain into total food intake. This gives the researcher information about the amount of diet needed to support a given unit of weight gain.

5. Feces

Feces are collected to provide information on one avenue of calcium output vs. calcium intake. Feces are generally collected on paper placed on the shelf below the cage or on a fecal collection screen that is part of a metabolic cage. The biggest disadvantage to collection of feces on the paper below the cage is the potential for mixing of feces from the animals in adjacent cages. When this happens, which is highly likely, the reliability of the information gathered will be questionable.

Generally, feces are collected for a specified period of time. A 7-day period is noted. Unless doing a longitudinal study, feces are often collected the last week of the study. If a comparison is desired, feces can be collected during both the first and last 7 days of the study.

During the collection period, feces from each animal should be collected daily and placed in a container labeled for that rat. The container should be refrigerated until the fecal collection period is done. Once fecal collection is done, feces should be dried, weighed, and ground. Depending on the quantity of feces collected, grinding can be done using a Wiley mill if there is a large amount; small amounts should be done in a mortar and pestle. Once feces are dried and ground, they can be stored at room temperature. Calcium is analyzed using atomic absorption spectrophotometry (AA). The general instructions are described in Table 3.

6. Urine

Metabolic cages equipped with diversion funnels and catch bottles are needed to collect urine. If doing calcium balance, urine is collected for the same amount of time and for the same periods as fecal collection. Collection bottles should be emptied at least daily to prevent growth of microorganisms. To further inhibit growth of microbes, a couple of drops of hydrochloric acid can be added to the catch bottles. When catch bottles are emptied, urine for a given rat should be poured in a labeled storage bottle (nonreactive plastic or glass [unless doing boron work] with a tight lid). Storage bottles should be stored under frozen conditions until time for analyses.

At the time of urinary calcium analysis, the total quantity of urine from each rat for the collection period should be ascertained using a graduated cylinder. Care must be taken not to contaminate a given sample with urine from a previously measured sample.

Urinary calcium can be analyzed using AA. Samples can be prepared using the method described by Willis (1961).

7. Blood

In calcium research, blood can be analyzed to provide information about serum Ca and various hormones and/or vitamins. Hormones and/or vitamins which may be analyzed are parathyroid hormone (PTH); calcitonin; 25 (OH) cholecalciferol (25 (OH) D_3); 1, 25 $(OH)_2$ cholecalciferol (1,25 $(OH)_2$ D_3); and osteocalcin. All of these components play a role in Ca absorption from the GI tract, reabsorption from the kidney tubule, deposition in bone, and resorption from the bone.

TABLE 3.

Calcium Analysis via Atomic Absorption Spectrophotometry

Ashing

1. Dry crucibles (Remember to include a blank) for ashing samples in moisture oven;
2. Weigh crucible which has been cooled in a desiccator;
3. Add sample (bone; diet) to weighed crucible and record weight of crucible and sample - Subtract weight of crucible to get the weight of sample;
4. Char sample over a Bunsen burner - Be sure that material does not flame (Note: Some researchers dry bones overnight in a vacuum oven instead of charring);
5. Ash samples in a muffle furnace using temperatures recommended for Ca analysis;
6. Making sure that a white ash has been attained, cool crucible with ash in a desiccator and then weigh and record weight of crucible and ash - Subtract weight of crucible to get weight of ash;
7. Return weighed crucibles to the desiccator until ready to put samples into solution;

Solutions

1. Prepare solutions (Conc HCl; 2 *N* HCl; 0.5 *N* HCl and 0.5% La_2O_3 or $LaCl_3$) needed for getting samples into solution;
2. Make Standard Solutions;
3. Under a hood, add 10 ml Conc HCl to each crucible and evaporate on a hot plate until nearly dry (DO NOT BURN);
4. Cool crucibles and add 20 ml 2 *N* HCl to each crucible;
5. Bring crucibles with 2 *N* HCl to a gentle boil on a hot plate;
6. Pour contents of crucible into a funnel lined with ashless filter paper that is suspended over a 100 ml volumetric flask;
7. Rinse crucible and filter paper with D/D water;
8. Add 10 ml 0.5% La_2O_3 or $LaCl_3$ (IF 2nd DILUTION NOT REQUIRED) (Samples with large amounts of calcium (e.g., bone) will require a second dilution);
9. Bring flask up to volume with D/D water;
10. If 2nd dilution is required, transfer an aliquot from the original solution to a second volumetric flask (50 ml);
11. Add 5 ml 0.5% La_2O_3 or $LaCl_3$;
12. Bring second flask up to volume using 0.5 *N* HCl;

Atomic Absorption Analysis

1. Follow the instructions for determining calcium using the AA Spectrophotometer in the researcher's facility.
2. Researchers should check to make sure the instrument is zeroed on a regular basis during analyses. Checking between each group is advised.
3. If the instrument does not calculate the calcium, the formula is as follows:

Note: Without Second Dilution

$$\text{ug Ca/g or ml sample} = \frac{(A)(B)}{C}$$

With Second Dilution

$$\text{ug Ca/g or ml sample} = \frac{(A)(D/E)(B)}{C}$$

A = Concentration from the standard curve; B = Volume of original dilution in mls; C = Weight of original sample in g or ml; D = Volume of second dilution in mls; E = Volume of aliquot transferred from original solution.

Researchers should read all procedures which require blood for the analysis prior to the blood draw. Close attention should be paid to:

1. What fraction of the blood the analysis involves and
2. Whether the fraction must be fresh or frozen for analytical work in the future. When it is time for the blood draw, animals should be fasted overnight unless instructed to do otherwise in a given procedure.

There are several methods for drawing blood. These include heart puncture (Griffith and Farris, 1942), retroorbital bleed (American College of Laboratory Animal Medicine, 1980), and tail venepuncture (Omaye et al., 1987). Each will require experienced personnel to do the draw.

Because the amount of blood and, therefore, serum which can be obtained from a rat is limited, especially if the animal is young, researchers need to consider carefully what analytical procedures will be done using blood. They should also pay close attention to what information the data will provide and whether the information will be physiologically relevant. For example, the researcher should carefully consider the need to do serum calcium. Because of the homeostatic processes which keep serum calcium within normal limits, using what little serum that is available to do serum Ca might not be the best analysis for providing feedback about alterations related to Ca metabolism.

Once the decisions have been made related to the analyses that will be done on blood, the researcher should prepare to do the blood draw. Unless a procedure indicates to the contrary, the animal should be fasted overnight prior to the blood draw. To reduce stress in the animal, rats should be mildly anesthetized using something like CO_2 or ether, unless otherwise instructed in the methodology. Ether should only be used in a well ventilated room equipped to prevent electrical sparks that might cause the ether to ignite. The procedure to follow for a heart puncture is described by Griffith and Farris (1942). Retroorbital draws should be done using a hepranized capillary tube (American College of Laboratory Animal Medicine, 1980). The technique for drawing blood via the tail vein is described by Omaye et al. (1987).

After the blood is drawn, it should be transferred to a labeled serum separation tube (unless whole blood is required) and allowed to stand for ≈ 30 minutes. It should then be centrifuged. Following centrifugation, the serum should be transferred to labeled test tubes that have caps. Test tubes should be stored frozen unless an analytical procedure requires that the analysis be done on fresh serum.

Serum calcium is generally done on fresh serum. Researchers should check the instruction manual associated with their AA prior to the blood draw to ascertain how the blood should be prepared for the AA analysis. Procedures are also described in *Clinical Chemistry: Principles, Procedures, and Correlations* (Bishop et al., 1985).

Although seldom done in rat studies, osteocalcin, 25 (OH) D_3, and 1,25 $(OH)_2$ D_3 are done using RIA procedures on serum which has been stored at low temperatures. Researchers should contact chemical companies related to RIA kits for these analyses.

8. Bone Analyses

Because 99% of the body's calcium is stored in the bone, analysis of the calcium content of bone is one of the principle analyses done in calcium research. Although there are references to analysis of calcium in caudal vertebrae which are considered by some to be a more sensitive indicator of calcium status, the femur is the bone generally removed and analyzed.

Removal of the femur may be done at sacrifice or may be done in the future if the animal carcass is stored under frozen conditions. If the animals have not been previously marked or tagged, such marking or tagging should be done just prior to sacrifice. If just starting out in research using an animal model, the researcher should obtain a book on rat anatomy so that the correct bone will be removed. Also the researcher should consistently use the same femur from one animal to another when removing the bone. The researcher should remove the femur of each animal in a group in sequential order. Care should be taken not to break the bone and to have all of the components of the bone (e.g., epiphyseal plate) for each animal. Once the bone has been removed, it should be placed in order in an indented tray that has been labeled. The researcher should place the tray of bones in a secure place where the bones can dry for two to three days. After this period of time, all muscle and connective tissue should be removed from each bone. Care should be taken to insure that bones remain in the appropriate numerical order within a group. Once the extraneous material has been removed, the bones should be tied in sequential order on a string taking care to mark the number one position on the string with something like a double knot. Only one group should be on a given string and only one group should be put in a given Soxhlet thimble. Bones can also be defatted individually using a Soxhlet or Goldfish extraction apparatus. Fat then should be extracted from the bones using a fat extraction procedure. Because the fat is in the center of the bone, fat extraction over an extended period will be necessary. Fat extraction is necessary because the amount of fat in the bone varies from bone to bone and animal to animal. Following fat extraction, bones, which are still tied on the string, should be allowed to air out until the extraction solvent has dissipated. After making sure that all of the extraction solvent has evaporated, a given string of bones should be placed on a glass petri dish or watch glass and vacuum dried. When the bones are removed from the vacuum oven, they should be placed in a desiccator and allowed to cool to room temperature. Bones can be stored in the desiccator until the researcher is ready to continue the analysis. From this point the researcher should follow the procedure outlined in Table 3 or a suitable procedure related to ashing of bone and AA analysis.

Stable isotope analysis in diet or tissues is difficult. One of several types of mass spectrometry are used. For stable isotopes of calcium, the most commonly used methods are thermal ionization mass spectrometry (Yeargey et al., 1984) and fast atom bombardment mass spectrometry (Smith, 1983).

Two radioisotopes of calcium, ^{45}Ca and ^{47}Ca, are commonly used. ^{47}Ca is a gamma emitter with a half life of 4.53 days. Sample preparation is simple for determining ^{47}Ca activity. Diet, bones, and other tissues or excretion can be ground and added to a test tube and counted in a gamma counter. A constant geometry minimizes counting error which can be achieved by using a constant volume. Rats can be counted in a whole body counter immediately after administration or ingestion of the isotope and periodically for a week or more to determine absorption and retention of ^{47}Ca. ^{45}Ca is a beta emitter which requires more sample preparation but has the advantage of a longer half life of 164 days. For counting ^{45}Ca activity, bones or other samples are digested by ashing or wet digesting in acid. An aliquot is combined with scintillation cocktail and counted in a liquid scintillation counter. Both ^{47}Ca and ^{45}Ca can be used together as oral and intraperitoneally administered isotopes to determine absorption. The amount of oral isotope, as a percent of the dose, compared to the intraperitoneally administered isotope appearing in the femur 24 hours after administration is a measure of fractional absorption. Alternatively, two groups of rats can be used and only one isotope where one group receives the dose orally and the other group receives it by intraperitoneal injection.

REFERENCES

Albanese, A.A., *Bone Loss: Causes, Detection and Therapy.* Alan R. Liss, New York, NY, 1978.

American College of Laboratory Animal Medicine, *The Laboratory Rat. Vol. II: Research Applications.* Baker, H.J., Lindsey, J.R. and Weisbroth, S.H., Eds. Academic Press, Inc., Orlando, FL, 1980.

Animal and Plant Health Inspection Service, USDA Animal Welfare Title 9 Part 1 Office of Federal Register. National Archives and Records Administration, Washington, DC, 1994.

Avioli, L.V., Calcium and Phosphorus. Shils, M.E. and Young V. Eds. In: *Modern Nutrition in Health and Disease, 7th ed.* Lea and Febiger, Philadelphia, PA, 1988.

Bishop, M.L., Duben-Von Laufen, J.L., and Fody, E.P., *Clinical Chemistry: Principles, Procedures, Correlations.* J.B. Lippincott Company, Philadelphia, PA, 1985.

Bernhart, F.W., Savini, S., and Tomarelli, R.M., Calcium and phosphorus requirements for maximal growth and mineralization of the rat. *J. Nutr.*, 98, 443, 1969.

Consensus Development Conference, Diagnosis, prophylaxis and treatment of osteoporosis. *Am J. Med.*, 94, 646, 1993.

Draper, H.H., Sie, Ten-Lin. and Bergan, J.G., Osteoporosis in aging rats induced by high phosphorus diets. *J. Nutr.*, 102, 1133, 1972.

Greene, E.C., Anatomy of the Rat. Hafner Publishing Co., New York, NY, 1955.

Griffith, J.Q. and Farris, E.J., *The Rat in Laboratory Investigation*. J.B. Lippincott Company, Philadelphia, PA, 1942.

Hunt, S.M. and Groff, J.L., *Advanced Nutrition and Human Metabolism*. West Publishing Co., St. Paul, NM, 1990.

National Research Council (NRC), *Nutrient Requirements of Laboratory Animals. 3rd ed.* National Academy of Sciences, Washington, D.C., 1978.

National Research Council (NRC), *Nutrient Requirements of Laboratory Animals. 4th ed.* National Academy Press, Washington, D.C., 1955.

Omaye, S.T., Skala, J.H., Gretz, M.D., Schaus, E.E., and Wade, C.E., Simple method for bleeding the unanesthetized rat by tail venopuncture. *Lab Animals*, 21, 261, 1987.

Ranhotra, G.S., Lee, C., and Gelroth, J.A., Expanded cereal fortification: Bioavailability and functionality of various calcium sources. Nutrition Reports International 22(4), 469, 1980.

Recommended Dietary Allowances, 10th ed. National Academy Press, Washington, D.C., 1989.

Riggs, B.L., Wahner, H.W., Melton, L.J., Richelson, L.S., Judd, H.L., and O'Fallon, W.M., Dietary calcium intake and rates of bone loss in women. *J. Clin. Invest.*, 80, 979, 1987.

Smith, D.L., Determination of stable isotopes of calcium in biological fluids by fast atom bombardment mass spectrometry. *Anal. Chem.*, 55, 2391, 1983.

Tepperman, J. and Tepperman, H.M., *Metabolic and Endocrine Physiology.* 5th ed. Yearbook Medical, Chicago, IL, 1987.

Varma, A., *CRC Handbock of Atomic Absorption Analysis*. Vol. 1, CRC Press, Boca Raton, FL, 1985.

Weaver, C.M., Age related calcium requirements due to changes in absorption and utilization. *J. Nutr.*, 124, 1418S, 1994.

Willis, J.B., Determination of calcium and magnesium in urine by atomic absorption spectroscopy. *Anal. Chem.*, 33, 556, 1961.

Yeargey, A.L., Viera, N.E., Covell, D.G., and Hansen, J.W., Calcium metabolism studied with stable isotopic tracers. Turnlund, J.R. and Johnson, P.E., Eds. In: *Stable Isotopes in Nutrition*. American Chemical Society, Washington, D.C., 1984.

Chapter 20

Vanadium: Quantitation, Essentiality, and Pharmacologic and Toxicological Studies in Rodents

Brooke D. Martin
Robert A. Felty
Diane M. Stearns
Karen E. Wetterhahn
Department of Chemistry
Dartmouth College
Hanover, New Hampshire

CONTENTS

0-8493-9611-5/97/$0.00+$.50

I. INTRODUCTION

A. Essentiality

When the American Institute of Nutrition revised its standardized rodent diet in 1993,[1] one of the significant changes was the addition and alteration of the levels of several elements, including the addition of the so-called ultratrace (<10 ng/g)[2] element vanadium. Vanadium was included with chromium, fluoride, boron, arsenic, nickel, lithium, and tin as elements whose essentiality has not been firmly established but whose absence may have negative effects on growth and may reduce an animal's ability to deal with experimental stress, toxins, or carcinogens. Reversing the effects of vanadium deficiency in terms of human or animal nutrition requires the addition of, what Nielsen has termed, "physiological" amounts of vanadium;[3] namely, an amount normally available in the diet which restores biological indicators to those of a well-fed animal. This level has been found to be in the part per billion level when measured in human serum[2] or possibly lower.[4] Vanadium is present in a wide variety of plant tissue at 1 to 10 mg/kg dry weight, with a few notable exceptions and at lower levels in animal tissue[5] and is therefore ubiquitous within most diets. Diets comprised of processed food such as cereals and refined raw materials such as powdered milk, contain on average a higher vanadium content.[4] This is attributed to the use of vanadium as a component in alloys including the stainless steel of equipment used for food processing and preparation, although it is not known if the additional vanadium is in a form available to the animal. To completely remove vanadium from the diet in a controlled manner, therefore, requires diets made from highly purified materials. Dietary vanadium requirement is also dependent on the design of the diet; with chloride, iodide, iron, copper, chromium, ascorbic acid, sulfur-containing amino acids, and riboflavin influencing the ability to observe adverse effects due to vanadium deprivation.[6] In one instance, slow growth and the metabolism of iron in vanadium-deprived rats could not be reproduced in the same laboratory due to variations in iron content and other aspects of the diet,[2] most notably the absence and surfeit of ascorbic acid. The variability of response to dietary vanadium is such that Nielsen has suggested that the essentiality of vanadium may only become noticeable in instances where the diet or health of the animal is in some way suboptimal, but warns that, in these instances vanadium may be exerting a pharmacologic rather than nutritional effect. Based on studies with vanadium and iodine on thyroid metabolism in rats, Nielsen has suggested compromised thyroid function as a possible reason for increased vanadium demand in mammals.[7]

B. Analytical Chemistry of Vanadium

The determination of elements as essential at the ultratrace level was not possible until recent advances in analytical chemistry techniques allowed the accurate determination of trace elements such as vanadium at the parts per million and parts per billion level. As a result, the reported tissue and serum concentrations of vanadium in healthy human subjects vary three[8] to four orders of magnitude. After reviewing 28 publications covering 18 years of research, Heydorn concluded that the reproducibility of any one laboratory's results depended upon the technique used for analysis and the experience of the laboratory in trace metal analysis, although the latter is usually a reflection upon the choice of instrument for analysis. The low concentration of vanadium in biological samples is consequently sensitive to low level contamination during the collection and analysis procedures. The question of essentiality of an element relies heavily on the ability of the researcher to remove vanadium from the diet and observe a restoration of health of the animal upon the addition of vanadium at environmentally available levels back into the diet, in a controlled and consistent manner. Due to the ubiquity of vanadium in the environment, the potential for the inadvertent addition of vanadium to the diet through the use of processed foods, and the effects of other dietary components such as iron and phosphate on vanadium metabolism, there is still little evidence of reproducible deficiency effects in the literature, giving rise to uncertainty over the essentiality of this element. The determination of normal serum, tissue, and dietary levels of vanadium is dependent upon accurate analysis of vanadium in biological samples and the supplementation of vanadium into a deficient diet has to be in subpharmacologic amounts.

C. Insulin-Mimetic Properties of Vanadium

While there is still debate over the necessity and effect of vanadium at physiological levels, there is an established effect of vanadium at levels described by Nielsen[2] as pharmacologic, namely in the part per million or higher range. At these levels, vanadium behaves as a drug, alleviating an abnormality caused by something other than a deficiency of the element. There are a number of pharmacological effects reported for this element. It is a potent enzyme inhibitor, a competitor for phosphate binding sites,[9] and it affects systems used for biological processing of ionic species, such as the sodium phosphate pump and the iron transport and storage proteins.[2,9]

In view of the interest in vanadium as an essential element, one of the more interesting pharmacologic effects currently being pursued is the use of vanadium complexes as insulin mimetic agents. At the turn of the 20th century, vanadium was used to treat a variety of diseases, including diabetes.[10] After the discovery of insulin, vanadium was no longer used as a treatment for diabetes and research on the effects of the element was quiet until the mid-1970s when it was discovered that vanadium could act as a phosphate analogue in many biological phosphorylation reactions analogous to those regulated by insulin in glucose metabolism.[9,10] Vanadate (pentavalent vanadium) was subsequently shown to mimic almost all the known effects of insulin in insulin-responsive cells such as rat and human adipocytes,[10,11] including stimulation of glucose transport and oxidation and secondary effects such as increased calcium influx into cells. The exact role of vanadium in eliciting an insulin mimetic response is not known. As insulin binding to the receptor is not affected in the presence of vanadium; it may be involved in the subsequent phosphorylation reactions.[11] Rats made diabetic through the destruction of the β-cells of the pancreas by the administration of streptazotocin are used as a model for Type II diabetes, as they exhibit both hyperglycemia and hyperinsulinemia as well as insulin resistance in muscle, liver, and adipose tissue.[10] Sodium metavanadate in drinking water (0.2 mg/mL) can normalize blood glucose levels and reverse tissue damage, but daily weight gain is only 30% of that achieved with insulin.[10] The genetically diabetic BB Wistar rat is subject to juvenile onset (60 to 100 days) diabetes,[12] and characterized by many of the symptoms of human Type I diabetes including hyperglycemia, hypoinsulinemia, ketonuria, and weight loss deficiency, and eventually succumbs to many of the secondary complications of the autoimmune disease including ketoacidosis, dehydration, and death[12,13] without treatment. In this model, vanadium is able to correct hyperinsulinemia, enable the animal to respond to oral glucose loads, and decrease plasma glucose in insulin deprived rats.[11] Therefore the BB Wistar rat could also be a good *in vivo* model for the study of vanadium as an insulin substitute or supplement in Type I diabetes.

D. Toxicity of Vanadium

All of the essential elements are toxic if given in high enough concentrations. In addition, the safe and adequate levels of a trace essential element may also depend on the extent to which it is affected by the presence of other elements, thereby blurring the distinction between the toxic and essential elements.[14] In the case of vanadium, the toxic effects are usually considered to be due to industrial respiratory exposure as high concentrations of this element had previously been administered in the diet.[15] Nonetheless, chronic exposure through very high dietary concentrations of vanadium was toxic in rats, calves, and chickens.[13] Toxicity is believed to be oxidation-state dependent for vanadium, although this is affected by *in vivo* metabolism.[10,11] A green-black discoloration of the tongue is associated with acute vanadium toxicity in some cases.[16] A basic question concerning investigators of the insulin-mimetic properties of vanadium as their studies progress toward the prospect of human clinical trials[17,10] is the ability of animals to metabolize the introduced substance such that it never reaches toxic concentrations. The concern is particularly pertinent to the transition metals considered ultratrace elements, especially in the case of vanadium as barely detectable amounts appear to be capable of stimulating quite potent effects.[9] In the cases where a definitive biochemical interaction for an element is known, such as iron and zinc, it is easy to understand and monitor the body's ability to sequester for storage or dispose of excess minerals. In the case of vanadium, modes of storage, transport, and disposal are not as well known and accumulation and toxicity must be studied via a compilation of circumstantial evidence. The use of rodent models allows toxicity and accumulation of vanadium to be readily assessed both histologically and through analysis of vanadium content of tissues. Route of administration is also an important factor due to differences in absorption through different tissues.[8] Therefore, not only the element but its final chemical composition needs to be assessed toxicologically. In addition, vanadium is an element capable of strong pharmacologic effects *in vivo*. Thus, it is not possible to ignore potential toxic effects of the element, and it has been acknowledged that more work is required to understand vanadium toxicity in addition to the questions of essentiality and pharmacology.[3]

E. Aims of This Chapter

Although recently added to the American Institute of Nutrition rodent diet,[1] the nutritional requirements for vanadium have not been definitively established, nor has a specific enzymatic role for the element in mammals been determined. The World Health Organization defines an element as "essential if its deficiency reproducibly results in an impairment of a function from optimal to suboptimal."[5] Vanadium cannot, at this stage, be said to fit that criteria due to inconsistencies in experimental design and the precise analytical chemistry required to determine exact vanadium levels at the trace level. Nielsen[18] has reviewed the need for dietary consistencies in order to see a requirement for vanadium. Nonetheless, an appropriate dose of vanadium to establish normal serum vanadium levels still needs to be established to avoid giving rise to pharmacologic effects of the element. Although the level required to elicit a pharmacological effect is only 0.5 to 3.0 μg/g diet, Nielsen estimates that this may be 10 to 100 times the levels normally available.[3] Therefore, the essentiality of vanadium has not been firmly established despite over 150 years of knowledge and study of vanadium in the human diet.[5] Indeed, until a precise biological role for the element is determined, it may not be possible to establish whether the element is essential or not. Nonetheless, this element will continue to be studied in the whole animal due to its many interesting pharmacological effects, its ubiquity in the environment, and its increasing industrial applications. The purpose of this chapter is not to summarize the work done toward establishing the essentiality and biochemistry of this potentially essential element as that has been done several times recently.[3,6,14,19] Rather, this chapter is intended to provide some insight into the analytical tools available to those working at establishing vanadium requirements at low levels and animal models which may be useful for studying the role of vanadium insulin-mimetic drugs. Therefore the three areas of research covered are (1) the analytical chemistry required to quantitate vanadium for essentiality, pharmacologic, and toxicity studies; (2) the use of diabetic rat models for research into vanadium's insulin-mimetic properties; and (3) the potential toxic effects of dosing animals with artificially high levels of vanadium.

II. THE ANALYTICAL CHEMISTRY OF VANADIUM IN BIOLOGICAL SAMPLES

Vanadium is present in the earth's crust with the ubiquity of zinc and is quite evenly distributed.[16] The burning of fossil fuels concentrates vanadium in the fly ash and soot from where it is often extracted and purified. Vanadium is also used in the production of steels which are used for high speed mechanical tools, which are hard, or which are corrosion- and temperature-resistant, in inks, and as catalysts. Exposure to vanadium outside of the industries involved in the extraction and processing of vanadium is usually confined to the use of the aforementioned manufactured goods containing vanadium, through vanadium-containing airborne particulate matter, and through the diet. Dietary exposure is at the part per million to part per billion level, concentrations are at the ng/m^3 level above large cities in the U.S., and from 0.2 to 100 μg/L in North American drinking water.[16] Tissue and biological fluid levels of vanadium are correspondingly low and therefore extremely sensitive to contamination. This risk of contamination increases with handling adding to normal airborne and environmental levels of vanadium, easily giving rise to erroneous results. It therefore becomes extremely important to carry procedural blanks through all stages of analysis. The precision between replicates should also be determined and reported to ensure the reliability of the analysis.

Heydorn's review covering 18 years of analysis of vanadium levels in human serum found that the level of vanadium varied by four orders of magnitude depending on the experience of the laboratory in trace element determinations and the analytical technique used.[4] Heydorn established that, in order to accurately and reproducibly determine vanadium, an instrument or technique had to be capable of detection limits below the part per billion level. Of the methods surveyed by Heydorn, those that gave results which were reproducible from laboratory to laboratory were neutron activation analysis with pre-irradiation separation (NAA) and electrothermal (graphite furnace) atomic absorption spectroscopy (GF-AAS). At that time, inductively coupled plasma mass spectrometry (ICP-MS) was not capable of trace element detection at this level. Since then, high resolution ICP-MS machines have become available with detection limits well below the part per billion level.

Heydorn also cited the experience of the laboratory in trace element determination as a contributing factor to the interlaboratory accuracy and reproducibility of the results obtained. The experience of the laboratory correlated strongly with the choice of method of analysis. It was also evidenced by reporting of trace element clean preparation of all containers and other equipment with which the sample came into contact, and quality control techniques such as sample replicates and intermittent analysis of procedural blanks.

As Graphite Furnace-Atomic Absorption Spectroscopy is the most accessible of the techniques to be described, it is the most likely technique to be carried out in the laboratories of the researcher and, as such, will be covered in some detail. The requirements for unusual and costly equipment for Neutron Activation Analysis, along with the requirement for extensive handling of radioactive samples, means that it will normally be carried out in specialized laboratories. The technique is described here due to its use in establishing the standard levels of vanadium in several tissues and biological fluids.[20–23] High Resolution Inductively Coupled Mass Spectroscopy is a technique which has been commercially available for inorganic analysis in the last few years.[24] Because of this, it is still fairly costly and not yet widely accessible. Nonetheless, the extreme sensitivity and speed of the technique ensures its future place in trace element analysis in biological systems and the technique is described here in that expectation. The preparation of samples in a trace element clean environment is described in some detail as contamination can ruin any sample, regardless of the sensitivity of the technique.

to measurement obviates the need for such a step. It had been reported that nitrate, nitric acid, phosphate, and iron interfered with the analysis of vanadium by GF-AAS, but investigation into this phenomenon by Manning and Slavin could not reproduce the effect.[35] The shape of the vanadium peak was slightly affected by the presence of iron.[35] They also found some interference from lanthanide chloride and some interference from 500-fold excess of tungsten has also been reported.[34]

Despite the extra time for sample preparation and risk of contamination, it is necessary to digest and mineralize biological samples in order to create a homogeneous solution for injection into the graphite furnace. Vanadium is freed from protein complexes, and it is possible to mineralize most polymeric vanadium complexes with the exception of the stable V_2O_5 complex. Throughout the sample preparation, some thought should also be given to the final acid concentration being injected into the furnace as excessive injection of acid will substantially reduce the graphite tube lifetime, necessitating recalibration and increased time and expense.

Because of the effects other elements in the sample are able to have on every step of the furnace cycle and subsequent interference with the shape and size of the vanadium absorption peak, it is extremely important that the *standards* used to generate the standard curve be prepared in a solution matched as closely as possible to the sample matrix. In order to achieve this match, some researchers use the method of additions to generate a standard curve. In this method, standard aliquots of increasing concentration are doped into a sample and a standard curve generated by extrapolation of the curve, in order to account for unknown matrix effects. Slavin[37] does not recommend this method as background errors are not necessarily accounted for and the addition of too low concentrations of vanadium leads to additional error. Instead, they recommend the construction of a series of standards in a solution of the matrix modifier and, if desired, the major components of the biological sample to be analyzed, e.g., ~120 m*M* NaCl for mammalian blood samples.[38] If it is desired, it is possible to obtain standard biological samples with which to check the technique and sample handling. However, while underlining the potential importance for a particular chemical species with similar valency and chemical bonding to be used in both the reference and laboratory samples, a brief summary by Parr and Stoeppler[39] included no examples of rodent tissues nor fluids. A summary of standard reference material and sources may be found in a review by Cortes Toro et al.[20] Summaries of standard reference material available in China,[21] Poland,[22] and Europe[23] are also published, although the European database lists samples available from 18 countries.

With standard pyrolytically coated graphite tubes, it has been reported that GF-AAS has a detection limit of around 2 parts per million vanadium in plasma[16] or lower,[35,37] depending on sample size. Sensitivity

A. Trace Element Clean Techniques

The principal concern when measuring a trace element at very low levels is to minimize losses and contamination at every stage of the procedure from collection to the final analysis. Each of the three analytical methods below is sensitive to contamination at several stages but all three can be ruined by poor sample collection and preparation.

1. Sources of Contamination

Sources of contamination exist at every stage from the sample collection to analysis. The skin and sweat of the sampler contain minerals at trace levels which can contaminate the sample for collection. It is recommended to use disposable latex gloves rather than colored rubber gloves which were found to be contaminated with metals.[25] All other disposable equipment, such as colored pipette tips are also a potential source of contamination. Disposable plastics are often available guaranteed trace-element free. It is also[25] suggested that the constant use of pipettes on acidic solutions gives rise to mineral contamination through the degradation of the metal working parts. Pipette tips are available with filters and all metal parts should be regularly removed and washed.

In addition to the skin and sweat of the researcher, the skin of the animal may contain a higher level of mineral contamination so care should be taken not to introduce contamination from the skin to tissue samples collected from the animal. If fluids are to be drawn from the animal, it is recommended to wash the skin with 0.1 *M* HCl followed by ethanol before inserting the needle or catheter.[26] Surgical instruments and needles were also shown by Neutron Activation Analysis to introduce contamination into biological samples[26] and, while vanadium was not tested, another principal component of stainless steel alloys, chromium, was a major contaminant. Aitio et al.[26] found that siliconization of needles reduced contamination and that contamination was greatly reduced after the first 20 mL. Results were compared against a "blank" of a plastic catheter but the trace element clean preparation of the catheter was not described, thereby giving rise to the possibility of slight contamination from that source. The availability of quartz, glass, titanium, plastic, and tantalum instruments[26] are suggested as possibilities if contamination from these sources becomes problematic.

Air can contain ng/m^3 levels of vanadium above large cities.[2,5,16] Ideally, samples should be prepared away from stock sources of vanadium, e.g., food or chemicals for addition to drinking water, under a constant flow of filtered air. Standards for all techniques should absolutely be prepared under clean room conditions with trace element clean plasticware and implements to provide longer term reproducibility of results. Clean room conditions are those flushed with a laminar flow of filtered air and contamination from dust and other particulates minimized by the use of airlocks and dust-free outer clothing.[2]

Reagents are also a source of trace element contamination. When blood and other biological fluids are used as metal concentration indicators, it is often desirable to use preservatives and anticoagulants. These should be carefully analyzed for vanadium levels prior to use if the manufacturer has not specified the level. Nitric acid is a suitable preservative for urine specimens[26] and is available as a trace element pure reagent. It is possible to further purify acids by subboiling from trace element pure plasticware to a similar receptacle. Subboiling distillation is a purification method whereby the liquid is evaporated by infrared heating at the liquid surface to avoid cotransport of impurities through the formation of aerosols. Freshly prepared subboiled acids are extremely pure.[2] Samples may also be preserved by freezing.

All containers and receptacles for the tissue and solution samples are a potential source of metal contamination. Common borosilicate glass contains extremely high metal content and should be avoided for all trace element analytical work.[2] Pyrex glass, with a high SiO_2 content, is less contaminating[27] but it was also demonstrated to leach minerals into samples.[2] Only pure SiO_2 can be expected to be trace element free. However, in order to avoid the leaching of minerals from glasses into biological samples, it is common to use the more economical plasticware for all containment. Moody and Lindstrom[28] found polyethylene (polythene) and teflon containers to have the lowest trace element content and it is these polymers which have gained wide acceptance for trace element research. Polymer containers are expected to be less of a contaminant than glass or metallic containers, however metal-based catalysts are used in the synthesis of the polymer and remain in the matrix. Moody and Linstrom[28] found, however, that little of the matrix metal is leached into solution. It is recommended to use standard cleaning procedures, even on new containers, to remove all surface adsorbed minerals. Laxen and Harrison[29] assessed several washing procedures to determine the most effective procedure without consuming large amounts of time or tying up extraordinary numbers of containers in ongoing washing procedures. They recommended a 48 hour soaking with 10% nitric acid, followed by rinsing with nanopure water, draining, and drying, as a preliminary procedure for both new bottles and for routine cleaning.

The digestion of samples is important for the analysis of vanadium, despite the inherent risks of contamination, in order to obtain a homogenous sample for analysis. This need for homogeneity is greater than that for obtaining a solution sample as all three methods are capable of operating on solid samples. Digestion of the sample also aids the removal of interfering species such as chloride ions or the carbon of the biological matrix. Digestion of a sample can be carried out by wet or dry ashing. To replace the need for digestion, or prior to it, to increase efficiency, it may be necessary to mechanically homogenize the sample. The use of a conventional blender with stainless steel blades would invariably introduce unwanted contamination. A suggested alternative is to freeze the sample with liquid nitrogen and grind it in a teflon ball mill.[30] It is also possible to use a Stomacher laboratory blender as described by Baten et al.[31]

Dry ashing involves high temperature combustion of the material, usually under a stream of oxygen to prevent the formation of insoluble vanadium carbides.[27] The furnace can be an open muffle furnace under atmospheric pressure or an enclosed oxygen bomb under high pressure. Losses from the open furnace should be mostly bulk material losses such as the fly-ash as under oxidizing conditions with some loss of the refractory V_2O_5 complex once ashing temperatures exceed 500°C. To avoid loss of analyte and contamination from atmospheric vanadium present, enclosed combustion systems are recommended.[2] The ash can then be dissolved in mineral acids.[27]

Wet mineralization involves acid digestion of the sample with oxidizing mineral acids — usually nitric acid to avoid increasing the chloride concentration of the sample — and heating the sample. Full digestion of organic material will only occur in the presence of strongly oxidizing material such as hydrogen peroxide with evolution of CO_2.[30] The formation of excess vanadium carbide through high temperature reactions with the organic millieu is thereby prevented. In addition, in acidic solution, VO^{2+} is a stronger oxidant than nitric acid[27] so hydrogen peroxide is also required to convert vanadium to its pentavalent state if the reagent can be obtained as the ultratrace pure reagent. Digestion of the sample will be greatly enhanced if pressurized vessels are used for digestion.[30]

Wet ashing in an open system is inherently susceptible to losses; however, these may be minimized by setting up a reflux system and optimizing conditions with respect to temperature and duration of reflux. Closed systems are preferred to minimize losses and to reduce the quantity of reagents used by increased efficiency of decomposition.[2] High purity polytetrafluoroethylene (PTFA), Teflon PFA (Perfluoroalkoxy vinyl ether), or quartz "bombs" are commercially available, although Hoffman[25] found problems cleaning older bombs of contaminating elements. This was resolved through the use of quartz inserts. Freezing of the entire container and contents upon completion of digestion was also suggested in order to prevent bulk material losses occurring by opening a container still under pressure. Using a PTFA decomposition bomb, it was determined that heating the unit above 280°C for 2 hours was sufficient to completely digest organic material[2] for 0.1 to 5 g of organic sample. As conventional PTFA bombs are not capable of withstanding the pressures brought about by heating above 200°C, longer times are recommended. High pressure bombs are also commercially available.[2] Heating can also be achieved through microwave radiation. In specimens with low organic content, UV light and hydrogen peroxide[27] can also be used. Wet mineralization is the more commonly used method for creating the homogeneous solution required for injection into the graphite furnace. For the acid digestion, it is not recommended to use HCl nor $HClO_4$ unless the chloride ions are to be removed later due to potential problems of increased NaCl in the furnace, and increased chloride ion concentration for neutron activation analysis and high resolution ICP-Mass Spectroscopy. Nitric acid is usually used for digestion of biological material and for mineralization for trace elements such as vanadium, and has the advantage of being an oxidizing acid. Nitric acid is also the recommended acid for digestion of samples for analysis by ICP-Mass Spectroscopy due to its ionization properties.[2]

Digestion method will depend on the tissue or fluid being analyzed. Bone can be cryogenically fractured and ashed after washing in organic solvents such as acetone to remove fat and tissue.[27] Soft tissue can be directly ashed, or lyophilized and ashed, and biological fluids can be lyophilized prior to digestion to concentrate the lower vanadium levels.[27]

Commercial tissue solublizers are also available but have long (48 hour) digestion times and may interfere with some analytical methods. Sansoni and Panday suggest their utility for GF-AAS spectroscopic use.[27] As with all reagents, they should be assessed for trace element purity prior to use. They should also be useful for neutron activation analysis and high resolution ICP-Mass Spectrometry provided they are free of vanadium to the part per trillion level and are comparatively free of chloride.

One final contamination is from one part of the animal to the other. It is well known that blood and tissue concentrations of trace elements can differ greatly.[32] Vanadium accumulates in the kidney, liver, and bone and contamination from these organs may affect results of other tissues or fluids. Even within the blood, elements can be quite differently distributed between the erythrocytes and the plasma.[26] It is not possible to completely separate tissue samples from the surrounding connective tissue, blood vessels, nerves, and extracellular fluids. Sansoni and Panday[30] suggest thawing material to remove blood or a quick wash in deionized water. When assessing the distribution of an element throughout the animal, it is important, as much as is possible, not to contaminate the tissue of interest by contact with the fur of the rodent or its biological fluids and vice versa.

2. *Losses of Analyte and Water*

Erroneous results can arise through the loss of sample and through the concentration of trace elements by uncounted water losses from the sample. Evaporation of sample is the most obvious source of water loss leading to an apparent increase of the mineral content of the sample. Proper sealing and storage should reduce loss from this source. Water losses may also occur by expiration through thin or permeable container walls[26] or through poor integrity of the container seal. The use of a humidity chamber to reduce these losses has been found effective, provided water is not then absorbed; however, transpiration losses should not be a problem for short term (one to two weeks) storage if care is used.

Samples may also be lyophilized and frozen after accounting for wet weight in order to preserve them and to account for water losses.

There is some concern that over-aciduous cleaning may activate adsorption sites on the polymer and trace elements can actually be lost from solution. This is not considered as large a problem for biological samples but Laxen and Harrison[29] also mention that they did not notice adsorption losses with polyethylene. Vandecasteele and Block confirm this observation for polyethylene and note that quartz also has a low tendency to adsorb elements.[2] When analyzing for trace metals leached from plastics, Moody and Lindstrom discovered that the use of HCl acid to clean the plasticware increased the chloride concentration in the plasticware.[28] Because of this inward diffusion of chloride ions, they used a second washing with nitric acid. Laxen and Harrison recommend a washing procedure which does not require the use of HCl, thereby obviating any concern arising from increased chloride ion concentration which may interfere in all three analytical procedures described below.[29]

Losses may also occur through precipitation when vanadium is trapped within the flocculent or precipitate. This is particularly pertinent if sulfuric acid is used to digest the sample, as the sulfate can form insoluble matrices with the alkaline earth metals which may coprecipitate the vanadyl cation.[30] Biological material contains comparatively high concentrations of calcium which precipitates out as $CaSO_4$ in the presence of sulfuric acid. The VO^{2+} moiety is a calcium mimic and is coprecipitated by adsorption onto the sparingly soluble $CaSO_4$. Sulfuric acid should not be used to digest samples to be used for the analysis of vanadium.

3. *Pre-Concentration of Vanadium*

There are several chelating agents available which are used to concentrate vanadium by chelation in the digested homogenate prior to extraction into organic solvent. Vanadium can be analyzed from the organic solvent or re-extracted into an aqueous phase at an appropriate pH. Vanadium chelators include dialkyldithiocarbamates (pH 3), cupferron (the ammonium salt of *N*-nitrosophenylhydroxylamine) and 1-(2-pyridylazo-2-naphthol).[2] As these are reported as chelators of pentavalent vanadium,[2,27] digestion procedures should be carried out which will fully oxidize all vanadium to this state. Oxine (8-hydroxyquinoline) and its derivatives are reported as useful vanadium chelators without the valence-specificity.[2] These chelators are not vanadium specific and will copurify other elements along with vanadium. They do, however, separate vanadium from major interfering ions such as chloride and sodium ions. Seiler[27] describes the extraction procedures using chlorinated organic solvents which may cause unnecessary contamination due to chloride ion.

Ion exchange techniques can also be used to isolate vanadium from many matrix elements. Vanadium is retained by Chelex-100 (iminodiacetate) in acetate buffer at pH 5 to 6.[2] Attention in this regard is drawn to a recent publication warning that Chelex-100 is occasionally shipped containing an unidentified impurity which can be simply removed by washing in nanopure water for 15 minutes.[33] Extraction to concentrate vanadium increases the sample's preparation time substantially in addition to increasing the risks of contamination and loss of analyte. It should be possible to analyze most biological samples containing vanadium to a sufficient sensitivity without using such methods of concentration, particularly as biological fluids which are notoriously low in vanadium can have their volume reduced by lyophilization.

From the many sources of contamination and loss described above, it is easy to see the value of preparing full procedural blanks to identify levels of vanadium which are due to contamination from any one of a vast number of sources. Procedural blanks should be analyzed intermittently throughout the experiment to assay for losses arising with time which need to be accounted for and hopefully removed from the sample experiment.

B. Flameless (Graphite Furnace) Atomic Absorption Spectroscopy

Of the three methods presented here, graphite furnace atomic absorption spectroscopy (GF-AAS) is probably the most accessible in terms of availability of equipment and supplies, and the widespread use of the technique. The technique itself is based upon the comparatively simple concept of taking a sample, heating it uniformly such that it is all converted to an atomic vapor which is then analyzed by measuring the absorption of light from a single line of a lamp containing a pure element. Experimental conditions are chosen to ensure that the integrated absorbance at the chosen wavelength of light is linearly proportional to the number of vanadium atoms in the sample. There are a number of AAS procedures which are now considered important to obtaining accurate and reproducible results at low concentrations.

Ordinary graphite furnaces are made from graphite containing subdomains of graphite sheets and are therefore reactive at the edges of each graphite microplate which constitutes the bulk material. In addition to the chemical reactivity possible at these edges, ordinary graphite will expand upon heating and become porous to the heated atomic vapor, thereby losing both sensitivity and accuracy. Therefore, it has become almost standard to use graphite furnaces which have had a dense, continuous layer of pyrolytic graphite deposited on the ordinary graphite tubes at high temperatures. *Pyrolytically coated tubes* are not porous to the elemental vapor at the high temperatures required for AAS and are also substantially less reactive. This is particularly important for vanadium, which forms a stable carbide from which it must be dissociated prior to analysis.[34]

The theory behind graphite furnace AAS requires that the temperature and rate of vaporization be consistent between samples and standards.

There are a number of procedures which can ensure this, including the requirement to matrix match standards and samples as much as possible, as vaporization of the analyte will depend on the compounds present in the sample. For many elements, it is suggested to insert a platform, into the graphite tube furnace, known as the L'vov platform, after the original designer of GF-AAS.[35,36] The sample is deposited on the platform rather than the walls of the furnace, thereby delaying vaporization of the sample until the temperature of the walls and the gas have reached an equilibrium. This somewhat more costly procedure is not recommended for vanadium as it is one of the more refractory elements and requires *rapid heating* (2000°C/s) to high atomization temperatures. The L'vov platform is omitted from the furnace as it delays heating of the sample and broadens the peak, thereby decreasing the accuracy of the integrated peak area.[35] Instead, a high power stage is used to bring the furnace to the final temperature as rapidly as possible, followed by a power reduction once the final temperature has been reached. In addition, a cooldown stage whereby the furnace temperature is restored to room temperature between the char (1500°C) and atomization (2700°C) is also recommended.[35] In this way, the furnace is heated rapidly by radiative energy and losses due to conduction and convection, which take a short but still finite time, are minimized. This very rapid and uniform heating of the graphite furnace is important for vanadium as high temperatures must be maintained to degrade the vanadium carbide.[34] Suspending the flow of argon through the graphite furnace during the atomization step only is also recommended to avoid replacing hot atomic vapor with cold argon.[37] To drive the rapid heating sequences and to analyze the resultant absorbance peak Slavin[37] also recommends the use of a spectrometer equipped with fast digital electronics, rather than the slower analogue circuits in some older machines. All preheating (drying ~140°C), char (~1600°C), cooling (room temperature), atomization (2600^{+} °C), cleaning (2700^{+}°C), and final cooling times, as well as the ramp and holding times, should be optimized for each instrument.

In GF-AAS there is some contribution to the signal due to other elements in the atomic vapor arising unavoidably by the volatilization of the matrix and scattering effects. Modern atomic absorption equipment is equipped with a continuum correction mechanism to account for *background effects*. This can be based upon the absorption from a background lamp or a more sensitive method known as Zeeman-effect or Zeeman correction. Deuterium lamps were generally used to correct the background effect, however the deuterium arc has insufficient energy above 300 nm to adequately correct for elements such as vanadium, which is analyzed by measuring the absorption at 318.4 nm.[34,37] To counter this problem, background correction can be measured with a tungsten iodide lamp. Nonetheless, using a separate source for background correction can still lead to occasional overcorrection errors.[37] The Zeeman-effect background correction method uses a magnetic field to split either the lamp source or the signal. In this way the correction and sample signal arise from the same source and over- and under-correction is extremely rare, although theoretically possible. It has been strongly recommended that the possibility for Zeeman correction be considered when analyzing trace elements at the low concentrations found for vanadium in biological samples.[34]

The *matrix* in which the vanadium sample is analyzed consists of the *digested biological sample* and other elements which have been added to improve the vanadium signal. Despite the background correction problems that can be brought about by the matrix, the addition of these *carefully chosen matrix chemicals* can add considerably to the resolution of the vanadium peak. The matrix is modified both to stabilize the analyte to higher temperatures if required or to change the properties of the principal matrix. NaCl or the chloride anion is a major component of biological samples, often present at millimolar concentrations. As NaCl often produces a large background signal at certain wavelengths, it is desirable to convert it to material which will be driven off in the ashing cycle, below 500°C. The addition of NH_4NO_3 volatilizes NaCl as $NaNO_3$ and NH_4Cl. For the refractory elements such as vanadium, it is suggested to add $Mg(NO_3)_2$ as a matrix modifier, as the element is stabilized to still higher temperatures by imbedding the vanadium in the MgO, avoiding loss of the element prior to the atomization stage.[37] Manning and Slavin[35] report that the shape of the absorption peak is sharpened, increasing the accuracy of the integration of the peak area and the background correction. Another matrix modifier which is used is palladium (Pd).[37] Commercial supplies of Pd nitrate have been reported as frequently contaminated with other elements at the trace level.[37] Slavin[37] therefore recommends using the metal powder and dissolving it in a minimum amount of concentrated nitric acid for use as a matrix modifier. Schaller[16] also recommends the use of sodium fluoride in order to optimize the form of the atomization peak, and uses an ammonia/ammonium chloride buffer.

All of these matrix modifications can lead to a greater sensitivity and accuracy of GF-AAS for the analysis of trace concentrations of vanadium. Nonetheless, each modification adds considerable time to the procedure and increases the risk of contamination. Therefore, before fully modifying the digested biological sample, the sensitivity of the spectrometer for measuring vanadium in unmodified sample and the shape of the absorption peak should be determined. Matrix modification can be introduced as required. The most frequently recommended modification is the addition of 50 to 100 ppm magnesium nitrate.[34,35]

As biological samples are likely to contain a lot of organic material, it has been suggested that the samples be charred in the presence of O_2 to avoid building up carbonaceous residue which is capable of forming excess vanadium carbide.[34] The oxygen stream should, however, be flushed away with argon prior to the furnace reaching 1000°C to extend the life of the graphite furnace. Oxidative digestion of the samples prior

may be increased by concentration through the use of chelating agents, provided contaminants are not introduced during the procedure.

C. Neutron Activation Analysis Techniques

Until the recent advances in inductively coupled plasma mass spectrometry, neutron activation analysis was considered the definitive method for determining exact amounts of vanadium in biological samples with great sensitivity and accuracy. In addition, after the initial sample preparation, irradiation of the vanadium to be analyzed in the sample renders the sample immune to further contamination. Neutron activation analysis (NAA) involves irradiating the sample with high energy neutrons to form an unstable isotope with a specific half life and measuring the energy of the resultant radiation. As vanadium can be considered to be essentially 100% ^{51}V, irradiation will result in all of the vanadium present being converted to ^{52}V, which produces a single readable photopeak, thereby achieving, in theory, the maximum possible sensitivity for an element. Unfortunately, the biological matrix from which vanadium is to be analyzed usually contains large amounts of sodium, chlorine, and potassium which are also irradiated to the excited state by neutron activation and which have photopeaks whose energies swamp that of vanadium. In addition, the excited states are much longer lived than that of vanadium with $t_{1/2}$ of 0.5 to 15 hours compared with 3.75 min for ^{52}V.[40] The sodium excited state ^{24}Na with a half-life of 14.929 hours poses the largest problem in biological matrices.[41] Therefore, extraction procedures must be performed in order to be able to see the vanadium signal. Indeed, Heydorn[4] found that NAA without radiochemical separation gave results which were too high.

The short lifetime of the ^{52}V radioactive isotope reduces the analysis time as the irradiation time can be as short as 3 to 10 min.[4,27] One of the principal advantages of this technique is therefore mitigated by the short lifetime of the excited state. For longer lived isotopes, the biological sample can be irradiated and all extraction procedures from wet- or dry-ash digestion of the sample and radiochemical separation can be subsequently conducted. For detection of vanadium, a compromise must be reached between irradiating untreated material to avoid contamination and isolation of the vanadium while the signal is still readable.[42] This has been achieved by dry-ashing prior to irradiation, followed by post-irradiation wet-ashing and cupferron extraction[42] and post-irradiation wet-ashing of a lyophilized sample followed by organic and aqueous 8-hydroxyquinoline extraction[4] among others.[27] If all post-irradiation extraction procedures are completed within two half-lives, i.e., 7 min, there is still sufficient radiation to achieve the sensitivity required to measure the low levels of vanadium found in biological fluids.[4] In the absence of interference, this technique is capable of detecting as little as 74 pg of vanadium.[41]

Despite the inherent difficulties in using NAA for the analysis of vanadium in biological samples, the sensitivity of this technique has ensured its place in the establishment of many of the vanadium levels in the international biological standards.[41,21] The requirement of a particle accelerator or nuclear reactor and the need to chemically modify radioactive samples has meant that this technique is only performed in specialized laboratories equipped to carry out such tasks in relative safety.

D. High Resolution-Inductively Coupled Plasma Mass Spectrometry (HR-ICP-MS)

The principal idea behind all the mass spectrometry techniques is to completely ionize a sample, separate the resulting ions by mass, and analyze the relative peak areas for each mass. Mass spectrometry, which uses an argon plasma for ionization, is known as inductively coupled plasma mass spectroscopy (ICP-MS). The advantage of the inductively coupled plasma technique for analysis of vanadium is that the argon plasma, at an ionization temperature of 7500 to 8000 K,[2] is able to completely ionize the sample introduced into it with minimal preparation. As for neutron activation analysis, it is matrix interference which causes problems for the analysis of this element. In biological samples, chloride anion concentration can be of the order of 0.1 *M*.[38] Either when the sample is introduced into the argon plasma or in the relatively high pressure, moderate temperature interface region, the chloride and oxygen, e.g., water, matrix elements form a $^{35}Cl^{16}O^+$ peak of 50.96377 amu. This peak needs to be resolved from the isobaric peak $^{51}V^+$ peak at 50.94396 amu.

When discussing ICP-MS, resolution (R) is calculated as $M/\Delta M$; where M is the average mass of the ion producing the peak, ΔM is the mass difference between the two ions one wishes to distinguish. It can be seen that instrumentation would require a resolution of 2572 to resolve vanadium from this common matrix interference. Alternatively, resolution is discussed with regard to peak width of the element of interest as this must be sufficiently narrow so as not to interfere with adjacent signals. In this way, resolution is absolute and independent of the presence or absence of other peaks in close proximity. By this definition, ΔM in the equation above refers to the peak width. In order to establish a practical perspective for resolution, peak width is usually measured at 10% or, more stringently, 5% of the peak height and the intrumental resolution required is correspondingly slightly lower. When this is the resolution between two peaks, it is referred to as the 10 or 5%, "valley definition."

After ionizing the sample, mass spectrometers then separate the resultant ions by m/z, or mass-to-charge ratio. Quadrupole mass analyzers are commonly used to select ions of a particular mass-to-charge ratio to reach the detector. Such ICP-mass spectrometers can resolve ions which differ by 0.3 to 0.8 amu,[2] which is insufficient sensitivity to resolve $^{51}V^+$ from the isobaric ClO^+ interference. Therefore, in biological samples part per billion to part per trillion vanadium signals were swamped by the ClO^+ matrix interference peak and conventional quadrupole inductively coupled mass spectrometers were not able to detect vanadium without prior separation of the element from the interfering matrix elements. As the abundance of the ^{51}V isotope is almost 100%, the entire sample contributes, effectively, to a single peak and it is not possible to simply analyze a different isotope as it is for elements such as iron.

Instead of using a quadrupole to analyze the ions generated, it is now possible to use machines with a double focusing magnetic sector mass spectrometer. Double focusing analyzers use a magnetic field and electrostatic analyzer in tandem following a high accelerating voltage to select ions with respect to their mass-to-charge ratio, m/z, without the energy spread due to the different kinetic energies of the ions.[2] By using the double focusing analyzer and other instrumental design changes[43] required to operate the machine at high potentials to increase ion transmission, it has become possible to obtain high resolution ICP-mass spectrometers capable of resolutions of up to 7500[44] to 15,000,[2] although a resolution of 3000 is sufficient for the detection of vanadium in biological samples.[45] Those wishing to read a more complete description of the instrument and its capabilities are referred to the article by Feldmann et al.[24] Double focusing analyzing equipment has an additional advantage that very few ultraviolet photons reach the detector as this requires multiple deflections through the instrument.[2] Therefore, the background reading is also extremely low.

Distinguishing the vanadium peak from the ClO^+ peak inherent in the ionization of biological samples is quite trivial by HR-ICP-MS, as can be seen in Figure 1 for a solution of 1 ppb vanadium in a 0.10 M NaCl concentration which mimics the serum concentrations of both elements.[38] Some idea of the low background count can also be seen in Figure 1, although it may not be possible to distinguish the 1 to 2 counts per second level. The extremely high sensitivity coupled with almost negligible background means that the HR-ICP mass spectrometry equipment is capable of detection limits in the parts per trillion to parts per quadrillion level. The technique is also useful over this entire range with a high dynamic working range covering 10 orders of magnitude. As can be expected, the extreme sensitivity of this technique makes it imperative to analyze procedural blanks which have been processed identically to the sample but contain no sample in order to detect background vanadium levels as contamination will easily be detected. It is also possible to analyze all reagents to confirm or to test their ultratrace purity. An advantage apart from the extreme sensitivity is that very little sample is required — as little as 10 to 50 μL using a high efficiency nebulizer, and the technique is very rapid so that many samples can be analyzed in a short period of time.

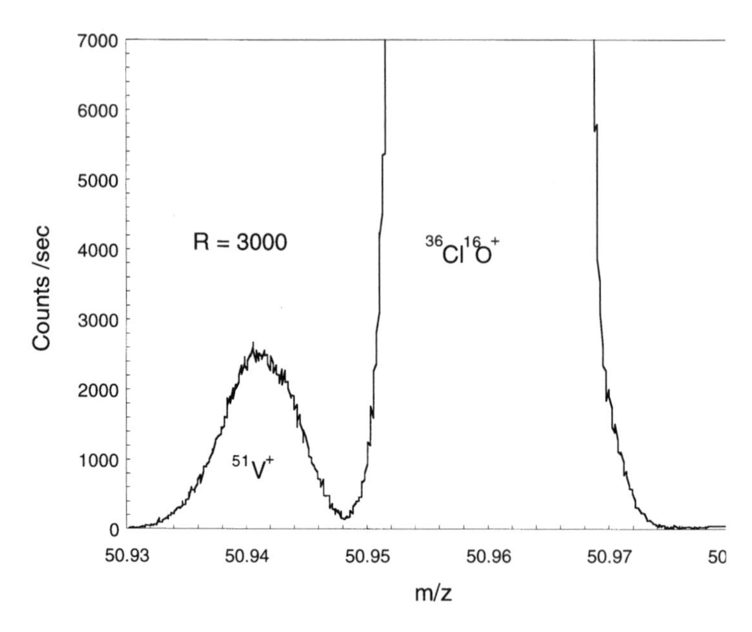

Figure 1
HR-ICP-MS spectrum of 1.0 ng/g vanadium in a solution of 0.1 *M* NaCl on an ELEMENT HR-ICP Mass Spectrometer (Finnigan MAT, Bremen, Germany) with a Micro Concentric Nebulizer (CETAC, Omaha, NE) at R = 3000, sample flow rate 20 μL/min, dwell time 100 msec (Courtesy of Dr. Bjorn Klaue).

Despite the clear advantages of this technique, there remain some difficulties. The resolution is excellent but is still not able to resolve true isobaric interference, i.e., ions of same mass. Should interferences of this nature arise, standard extraction techniques would have to be employed. As there are no common isobaric interferences to three decimal places for $^{51}V^+$,[46] this is not a concern for the analysis of vanadium. Using high resolution mass spectrometry, it is possible to resolve the vanadium peak from matrix interference due to serum levels of chloride. If the chloride ion concentration were artificially increased, e.g., if sample were digested in HCl or $HClO_4$, the ClO^+ peak width will naturally become greater and start to impinge upon the resolution. Nonspectral interferences may also be a concern in HR-ICP-MS. These arise when the constitution of the sample differs greatly from that of the standards used to generate the standard curve, e.g., sea-water contains extremely high sodium concentrations, changing the instrumental conditions such that the behavior of the ion beam is no longer consistent.[47] This is not a significant problem for biological samples, but standards should be matched as closely as possible to the samples to be analyzed to ensure that the behavior of both samples and standards in the ion beam is as similar as possible.

III. VANADIUM AS AN INSULIN-MIMETIC: ANIMAL MODELS

Since the rediscovery over 20 years ago that vanadium compounds have insulin-mimetic properties with regard to glucose, cholesterol, and triglyceride metabolism *in vitro*,[48] numerous *in vivo* studies using rat models have been reported. The bulk of these studies involve the use of Wistar rats made diabetic by injection with streptozotocin (STZ) followed by oral intake of aqueous vanadyl [V(IV)], vanadate [V(V)], or ligand-containing vanadium compounds.[32,49,50–55] Streptozotocin-induced diabetic rats are believed to be a suitable animal model because this system reflects the symptoms of both insulin-dependent diabetes mellitus (IDDM or type I) and non-insulin-dependent diabetes mellitus (NIDDM or type II). Recently, studies have appeared in the literature in which vanadium compounds have been introduced to the spontaneously diabetic BB Wistar rat.[56,57] The BB rat was discovered in 1974 in a colony of outbred Wistar rats at the BioBreeding Laboratories in Ottawa, Canada.[12] A major advantage of using such a model is that BB rats can provide data relevant to human IDDM. There are a number of analogies that exist between human IDDM and clinically, morphologically, and immunologically similar disorders in BB rats. They can also be bred for the purpose of studying and manipulating inheritance and their genome can also be altered.

Results obtained by manipulating a number of experimental variables have contributed to determining a range of conditions best suited to the study the role of vanadium as an insulin mimetic. Diabetic control is highly dependent on vanadium dosage,[32,50,51,53–55] as well as the type of vanadium compound administered[32,50,54] since different synthetic vanadium-based drugs have different thresholds of toxicity. The degree of toxicity also depends on the amount of time that rats are given to adjust to a high vanadium diet. A treatment protocol must allow for a period of adjustment over the course of several weeks.[32,52–54] Section IV of this chapter describes vanadium toxicity studies in rat models in more detail.

Based on the published reports of investigations into the insulin-mimetic effects of orally administered vanadium,[32,49–55] a generalized protocol for studying the effects of vanadium on rats made diabetic with STZ is described here. Male Wistar rats (150 to 300 g) are injected with STZ (45 to 75 mg/kg, intravenously in 0.9% NaCl or in citrate buffer). A second group is injected with isotonic saline or citrate buffer to serve as an age-matched control group. Three days after STZ administration, plasma glucose levels are measured. Animals exhibiting glucose levels greater than 13 to 14 m*M* are considered diabetic. Rats are separated into 4 groups: untreated control, vanadium-treated control, untreated STZ-diabetic, and vanadium-treated STZ-diabetic rats. Animals are maintained under alternating 12-hr light and dark cycles with free access to food and drink for the duration of the study. Untreated control and STZ-diabetic rats are given distilled water in place of water solutions containing vanadium. Studies utilizing vanadyl compounds administered orally in distilled drinking water found concentrations between 0.25 to 1.00 mg/mL sufficient to produce good diabetic control. Under these conditions, many rats are still euglycemic even 13 weeks after vanadyl treatment removal.[52] In studies that used vanadate compounds, lower concentrations are generally administered due to problems with toxicity.[53,54] VO_3^- at a concentration of 0.2 mg/mL in the drinking water lowered the amount of glucose to nearly normal levels and was optimal for achieving stable normoglycemia in STZ-rats over a period of several weeks. Recent attention has been focused on vanadyl and vanadate bound to organic ligands such as cysteine and 3-hydroxy-2-methyl-4-pyrone (maltol) which have also been found to reduce blood glucose levels.[58–60] Effective subtoxic dosage has been shown to be modulated by the complexing ligand.[58,59]

The small number of studies involving oral administration of vanadium to BB rats involves initially controlling diabetes with insulin followed by supplementation with vanadyl (0.25 mg/mL) in the drinking water due to the extreme nature of diabetes in these animals.[56,57] Insulin dosages were gradually lowered and glucose levels monitored each day. Studies then proceeded by either maintaining the animals at the lowest insulin dosage that resulted in a normal glycosourea test[56] or increasing the vanadyl concentration of the stabilized rats to 0.5, and finally, 0.75 mg/mL.[57] Data derived from rat models of IDDM should be interpreted with caution since heterogeneity exists not only among species but also among the various rodent colonies around the world. Different colonies house sublines that vary substantially with respect to the frequency and severity of disease as well as the immunological characteristics of the animals. The BB rat is known to exists in at least 24 inbred and 2 outbred lines.[12] Any future studies involving insulin mimetic properties in the BB rat should take these factors into account.

IV. TOXICITY STUDIES OF VANADIUM IN RODENTS

The toxicity of vanadium, like all other metals, is dependent on the chemical speciation and route of exposure. In general, V(V) is more toxic than V(IV). This is believed to be due to the ability of V(V) to cross cell membranes more easily than V(IV), although intracellular vanadium undergoes reduction and exists predominantly as V(IV).[61] Oral ingestion is less toxic than inhalation because of limited gastrointestinal absorbance.[3] As expected, injection of vanadium results in the highest levels of toxicity. Human exposure to vandium through the diet is low, and inhalation risk is limited to occupational settings and air pollution. The medicinal use of vanadium complexes for their insulin-potentiating effects may result in an uncharacteristically high exposure. Therefore, several toxicity studies have been carried out in order to assess toxic effects of the insulin-potentiating vanadium compounds, and the majority of these studies have been carried out in rats.

Because the methods used to explore the toxicity of vanadium are not unique to the metal, less attention will be paid to the methodological details. Rather, this section briefly summarizes the types of assays that have been useful in the assessment of vanadium toxicity when carried out in parallel with nutritional studies, and emphasis is placed on oral ingestion. This is not intended to be an exhaustive literature summary of vanadium toxicology in rodents, but rather an indication of the kinds of toxicological indicators which are being used to assess the toxic side effects of oral vanadium administration. The observations of toxic endpoints vary with the form of vanadium tested, the dose, and route of exposure. Studies have generally been aimed at evaluation of liver and kidney damage, hypertension, and hematological effects. Mechanisms of vanadium-induced tissue damage[62] and lipid peroxidation[63] have been reviewed.

The relative effectiveness and toxicity of three forms of vanadium: sodium metavanadate ($NaVO_3$), sodium orthovanadate (Na_3VO_4), and vanadyl sulfate ($VOSO_4{\cdot}5H_2O$), were tested in male Sprague-Dawley streptozotocin-diabetic and nondiabetic rats.[64] Rats were treated with vanadium in drinking water for 2 weeks, at doses of 0.20 mg/mL $NaVO_3$, 0.50 mg/mL Na_3VO_4, and 1.1 mg/mL $VOSO_4{\cdot}5H_2O$. Vanadyl sulfate was the most effective in decreasing blood glucose levels. A related study found similar results at doses of 0.15 mg/mL $NaVO_3$, 0.23 mg/mL Na_3VO_4, and 0.31 mg/mL $VOSO_4{\cdot}5H_2O$ in rats treated for 28 days.[65] In this study, $NaVO_3$ was the most effective against diabetic symptoms. Another study treated virgin female Sprague-Dawley streptozotocin-diabetic and nondiabetic rats with 1.2 mM (or 0.15 mg/mL) sodium metavanadate ($NaVO_3$) in drinking water for 16 days[66] and found a similar decrease in plasma glucose in diabetic rats and decreased plasma insulin concentrations in treated nondiabetic rats vs. nontreated rats. Reported negative side effects for all studies included death, weight loss, and vanadium accumulation in tissues. The second study[65] also observed increased serum concentrations of urea and creatine. In the third study, vanadium treatment correlated with increased lipid oxidative damage as measured by thiobarbituric acid reactive substances (TBARS) in both diabetic and control animals.[66]

In order to mitigate some of the toxic side effects of oral vanadium administration while taking advantage of the improved glucose homeostasis, the effect of the vanadium chelator Tiron (sodium 4,5-dihydroxybenzene-1,3-disulfonate) on the tissue accumulation of vanadium was studied.[67] Male Sprague-Dawley streptozotocin-treated rats treated with 0.20 mg/mL sodium metavanadate ($NaVO_3$) in drinking water for 2 weeks with 0, 125.6, 314, or 628 mg Tiron/kg/day by gavage. The chelator did not reduce vanadium's ameliorative effects on hyperglycemia, hyperphagia, and polydipsia, but did decrease vanadium tissue accumulation. Toxicity of the chelating agent was not assessed.

Longer term studies have also been carried out. Administration of ammonium metavanadate (NH_4VO_3), vanadyl sulfate ($VOSO_4$), and bis(maltolato)oxovanadium(IV) at 0.19, 0.15, and 0.18 mmol V/kg/day, respectively, to normal male Wistar rats in drinking water for 12 weeks did not produce detectable hematological toxicity by measurement of hematocrit, hemoglobin, reticulocyte percentage; erythrocyte, leukocyte, and platelet counts, and erythrocyte osmotic fragility.[68] Nondiabetic rats exposed to 1, 10, or 40 μg/mL sodium metavanadate ($NaVO_3$) in drinking water for 6 to 7 months presented evidence of blood hypertension, kidney damage, and defective potassium and calcium metabolism.[69] A still longer term study treated male Wistar streptozotocin-diabetic and nondiabetic rats with vanadyl sulfate ($VOSO_4$) at 0.5 to 1.5 mg/mL in drinking water for 1 year.[32] Plasma concentrations of aspartate aminotransferase, alanine aminotransferase, and urea were measured at 3 month intervals, and at 16 weeks after cessation of treatment. Enzyme and urea levels did not differ between vanadium-treated rats and their respective controls (diabetic or nondiabetic). No significant differences were observed in organ/body weight ratios after 1 year of treatment in nondiabetic, vanadium treated rats and their nondiabetic controls. Lung, heart, liver, kidney, and adrenal weight/body weight ratios were significantly decreased in vanadium treated diabetic rats with respect to diabetic controls. No morphological abnormalities were observed by autopsy. Vanadium treatment did reduce the incidence of urinary stones in the nondiabetic group. Vanadium was found to accumulate in tissues of normal and diabetic rats, and was still detectable at 16 weeks post-treatment.

The mutagenicity, carcinogenicity,[70] and teratogenicity[70,71] of vanadium compounds have been reviewed. Vanadium is weakly mutagenic, but not clastogenic. The carcinogenicity of the element has neither been established nor ruled out. A variety of teratogenic endpoints have been observed in animals treated with vanadium compounds at high doses. Immunotoxicity of ammonium metavanadate has been observed *in vitro*[72] and in animals.[73]

V. CONCLUDING REMARKS

The principal advantage of the rodent model is the ability to carry out investigations of pharmacology and toxicity such as those described above while being able to correlate whole animal results such as weight gain or loss, biochemical indicators, and death with firm analytical data such as tissue and serum concentrations of vanadium. Constant improvements in the analytical techniques for vanadium allow this data to be acquired with ever-increasing speed and accuracy. There are some in the field who feel that the question of the essentiality of vanadium can be determined by developing and correlating the analysis of worldwide human vanadium serum levels.[42] It may then be possible to determine whether the current four order of magnitude variation in the reported human serum vanadium concentration is an analytical artifact hiding a true serum level of an essential element or a true fluctuation, indicating environmental exposure rather than a physiologically modulated demand. Newer and more sensitive analytical techniques such as high resolution mass spectrometry may also be able to distinguish true vanadium deprivation from apparent deficiency through the analysis of animal tissue and environmental factors such as feed and drinking water.

Human serum is one of the few biological materials which has been assayed for vanadium by GF-AAS, NAA, and by high resolution ICP-MS. Table 1 summarizes many results for human serum vanadium levels of non-industrially-exposed subjects by these techniques. There are many more results than those summarized in the table as Heydorn's criteria[4] were followed to eliminate results which varied too greatly within the experiment or varied too greatly from the average result. Cornelius and Versieck[74] also point out that factors such as sampling whole blood instead of blood serum and the sampling of control, nonexposed subjects in a contaminated workplace have also lead to errors and therefore results based on these data should not be considered as accurate measures of endogenous human serum vanadium levels. The summarized results of Table 1 begin to show some level of agreement between the different techniques. However, as pointed out by Cornelius and Versieck for graphite furnace AAS[74] and by Heydorn for neutron activation analysis,[4] the reported results are at the limit of detection for these techniques, even in the absence of matrix interference. These limits of detection were also established at comparatively high sample volumes. By contrast, the levels reported in Table 1 are well within the detection limits of high resolution ICP mass spectrometry techniques with much lower sample volumes. In addition, HR-ICP-MS is able to analyze biological samples for vanadium in the presence of the biological matrix, obviating the need for lengthy extraction procedures. All three techniques are capable of measuring vanadium in biological samples at levels as low as those of human serum vanadium levels. There is still, however, some considerable disagreement over the actual levels as evidenced by the approximate order of magnitude discrepancy between the results in Table 1, which themselves have been drawn from strictly edited literature values. Due to its ubiquity, graphite furnace-AAS will retain its predominance in trace element tissue analysis. However, when highly accurate data is required, more accurate techniques such as NAA with radiochemical separation are used and, perhaps in the future, HR-ICP-MS will be used.

The insulin-mimetic and toxicological studies are also by no means fully representative of all the work done in the field. Much excellent work has been carried out using intravenous, intraperitoneal, and intratracheal injection of vanadium. However, this work has focused on only the oral administration of vanadium in order that analogies may be drawn to nutritional studies. The oral administration studies described above demonstrate a side effect in all vanadium pharmacological studies of tissue accumulation of the element. Other observed toxic side effects of the

TABLE 1.

Detection Limits and Vanadium Human Serum Results for Graphite Furnace AAS, NAA with Radiochemical Separation and High Resolution ICP-MS with a High Efficiency Nebulizer

Technique	Vanadium Level in Human Serum	Detection Limits
Flameless (GF)-AAS	< 0.5–0.6 ppb[4]	10[35]–30[1] pg (absolute)
NAA (with radiochemical separation)	0.10–1.00 ppb[4]	74 pg (absolute)[41]
High resolution ICP-MS	0.83 ± 0.09 ppb[45]	0.2 pg (absolute, 20 μL sample with high efficiency nebulizer)[45]

element include death, weight loss, diminished organ-to-body-weight ratios, biochemical indicators such as urea and creatine levels, and lipid peroxidation. Nonetheless, vanadium clearly displays an insulin mimetic effect in these studies. Therefore, much of the current work in the field of insulin-mimetic vanadium complexes is concentrated upon the search for more effective vanadium complexes so that the pharmacologic and toxic doses can be further separated.[11,59,75] The monitoring of background as well as pharmacologically and toxicologically elevated serum and tissue levels of vanadium, in addition to biochemical indicators of the efficacy or toxicity of the element, provides useful indicators of appropriate doses for insulin-mimetic and for nutritional studies.

The discovery of the genetically insulin-dependent rodent models has provided researchers with an invaluable model for investigations into the insulin-mimetic response of vanadium. The availability of such a rodent model allows pharmacologic studies to be carried out in conjunction with full toxicological studies.

The distinction between nutritional requirements, pharmacologic effects, and toxicity is still somewhat grey. Researchers determining the nutritional requirement for vanadium, as well as those investigating the pharmacological effects of the element, require both the appropriate animal models to study the effects of newly designed drugs and the analytical techniques with which to study them.

ACKNOWLEDGMENTS

This work was supported by Public Health Service Grants CA45735, ES07373, and ES07167, awarded to K.E.W. by the National Cancer Institute and the National Institute of Environmental Health Sciences. Thanks is given to Drs. Alan Shaver and Barry Posner of McGill University of Canada for useful discussion on the rat models for studying vanadium insulin-mimetic compounds. The authors gratefully acknowledge the insight and support of Dr. Bjorn Klaue for the analytical chemistry section and for generating the diagram of vanadium analysis by HR-ICP-MS. Thanks also to Dr. Michael Hingston, Dr. Joel Blum, and Dr. Kent Sugden for help in preparing the manuscript.

REFERENCES

1. Reeves, P.G., Nielsen, F.H., and Fahey, G.C. Jr., AIN purified diets for laboratory rodents: final report of the American Institute of Nutrition *ad hoc* writing committee on the reformulation of the AIN-76A rodent diet. *J. Nutr.*, 123(11), 1939, 1993.
2. Vandecasteele, C. and Block, C.B., *Modern Methods for Trace Element Determination*, John Wiley and Sons, Chichester, New York, 1993.
3. Nechay, B.R., Nanninga, L.B., Nechay, P.S.E., Post, R.L., Grantham, J.J., Macara, I.G., Kubena, L.F., Phillips, T.D., and Nielsen, F.H., Symposium Summary: Role of vanadium in biology. *FASEB J.*, 45, 123, 1986.
4. Heydorn, K., Factors affecting the levels reported for vanadium in human serum. *Biol. Trace Elem. Res.*, 26, 541, 1990.
5. Vanadium in *Environmental Health Criteria No. 81*, World Health Organization, Geneva, 1988.
6. French, R.J. and Jones, P.J.H., Minireview. Role of vanadium in nutrition: metabolism, essentiality and dietary considerations. *Life Sci.*, 52, 339 1992.
7. Nielsen, F.H., New essential trace elements for the life sciences. *Biol. Trace Elem. Res.*, 26, 599, 1990.
8. Nielsen, F.H., Vanadium in *Trace Elements in Human and Animal Nutrition* 5th ed. Mertz, W., Ed. Academic Press Inc., Orlando, 1987, Chap. 9.
9. Simons, T.B.J. Vanadate — a new tool for biologists. *Nature*, 281, 337, 1979.
10. Shechter, S. and Shisheva, A., Vanadium salts and the future treatment of diabetes. *Endeavour*, 17(1), 27, 1993.
11. Lönnroth, P., Eriksson, J.W., Posner, B.I., and Smith, U., Peroxovanadate but not vanadate exerts insulin-like effects in human adipocytes *Diabetologia*, 36, 113, 1993.
12. Crisá, L., Mordes, J.P., and Rossini, A.A., Autoimmune diabetes mellitus in the BB Rat. *Diabetes Metab. Rev.*, 8, 9, 1991.
13. Rossini, A.A., Handler, E.S., Mordes, J.P., and Greiner, D.L., Animal models of human disease. Human Autoimmune diabetes mellitus: Lessons from BB rats and NOD mice — *caveat emptor. Clin. Immunol. Immunopath.*, 74, 2, 1995.
14. *World Health Organisation Technical Report Series* No 532, Geneva, 1973.
15. Kendrick, M.J., May, M.T., Plishka, M.J., and Robinson, K.D., Vanadium in biological systems in *Metal Ions in Biological Systems*, Ellis Harwood Series in Inorganic Chemistry, Marcel Dekker, New York, 1992.
16. Schaller, K-H., Vanadium in *Trace Element Analysis in Biological Specimens: Techniques and Instrumentation in Analytical Chemistry*, Vol. 15, Herber, R.F.M. and Stoeppler, M., Eds. Elsevier Science BV, 1994, Chap 24.
17. Yale, J-F., Lachance, D., Bevan, A.P., Vigeant, C., Shaver, A., and Posner, B.I., Hypoglycemic effects of peroxovanadium compounds in Sprague-Dawley and diabetic BB rats. *Diabetes*, 44, 1274, 1995.
18. Nielsen, F.H., The importance of diet composition in ultratrace element research. *J. Nutr.*, 115, 1239, 1985.
19. Nielsen, F.H., Vanadium in mammalian physiology and nutrition in *Metal Ions in Biological Systems* Vol. 31, Vanadium Its Role in Life, Sigel, H. and Sigel, A., Eds., Marcel Dekker, New York, chap. 16, 1995.

20. Cortes, Toro, E., Parr, R.M., and Clements, S.A., Biological and environmental reference materials, nuclides and organic microcontaminants — a survey *Report IAEA/RL/128* (Rev. 1), Vienna, Austria, 1990.
21. Chifang, C., Present status and future trends in biological and environmental reference materials in China. *Fresenius J. Anal. Chem.*, 345, 93, 1993.
22. Dybczyński, R., Polkowska-Motrenko H., Samczyński, Z., and Szopa, Z., New Polish certified reference materials for multielement inorganic trace analysis. *Fresenius J. Anal. Chem.*, 345, 99, 1993.
23. Klich, H. and Walker, R., COMAR — The international database for certified reference materials. *Fresenius J. Anal. Chem.*, 345, 104, 1993.
24. Feldmann, I., Tittes, W., Jakubowski, N., Stuewer, D., and Giessman, U., Performance characteristics of inductively coupled plasma mass spectrometry with high mass resolution. *J. Anal. At. Spec.*, 9, 1007, 1994.
25. Hoffman, J., Experience with the sources of contamination when preparing samples for the analysis of trace metals. *Frezenius Z. Anal. Chem.*, 331, 220, 1988.
26. Aitio, A. and Järvisalo, J., Sampling and Storage in *Trace Element Analysis in Biological Specimens*, Herber, R.F.M. and Stoeppler, M., Eds. Elsevier Science BV, 1994, Chap 1.
27. Seiler, H.G., Analytical procedures for the determination of vanadium in biological materials in *Metal Ions in Biological Systems* Vol. 31 Vanadium Its Role in Life, Sigel, H. and Sigel, A., Eds., Marcel Dekker, New York, Chap. 16, 1995.
28. Moody, J.R. and Lindstrom, R.M., Selection and Cleaning of Plastic Containers for Storage of Trace Element Samples. *Anal. Chem.*, 49(14), 2264, 1977.
29. Laxen, D.P.H. and Harrison, R.M., Cleaning Methods for Polythene Containers Prior to the Determination of Trace Metals in Freshwater Samples. *Anal. Chem.*, 53, 345, 1981.
30. Sansoni, B. and Panday, V.K., Sample treatment of human biological materials in *Trace Element Analysis in Biological Specimens*, Herber, R.F.M. and Stoeppler, M., Eds. Elsevier Science BV, 1994, Chap 2.
31. Baten, A., Sakamoto, K., and Shansudden, A.M., Long-term culture of normal human colonic epithelial cells *in vitro*. *FASEB J.*, 6, 2726, 1992.
32. Dai, S., Thompson, K.H., Vera, E., and McNeill, J.H., Toxicity studies on one-year treatment of non-diabetic and streptozotocin-diabetic rats with vanadyl sulfate. *Pharmacol. Toxicol.*, 75, 265, 1994.
33. Van Reyk, D.M., Brown, A.J., Jessup, W., and Dean, R.T., Batch-to-batch variations of Chelex-100 confounds metal-catalysed oxidation. Leaching of inhibitory compounds from a batch of Chelex-100 and their removal by a pre-washing procedure. *Free Rad. Res.*, 23, 533, 1995.
34. Slavin, W., *Graphite Furnace AAS: A Source Book*, The Perkin Elmer Corp. 2nd ed., 1991.
35. Manning, D.C. and Slavin, W., Factors influencing the atomization of vanadium in graphite furnace AAS. *Spec. Acta*, 40B, 461, 1985.
36. L'vov, B.V., Electrothermal atomization — the way toward absolute methods of atomic absorption analysis. *Spec. Acta*, 33B, 153, 1978.
37. Slavin, W., Graphite furnace AAS, in *Trace Element Analysis in Biological Specimens*, Herber, R.F.M. and Stoeppler, M., Eds. Elsevier Science BV, 1994, Chap 3.
38. Lodish, H., Baltimore, D., Berk, A., Zipursky, S.L., Matsudaira, P., and Darnell, J., *Molecular Cell Biology*, W.H. Freeman and Co., New York, 3rd ed., 1995.
39. Parr, R.M. and Stoeppler, M., Reference materials for trace element analysis in *Trace Element Analysis in Biological Specimens*, Herber, R.F.M. and Stoeppler, M., Eds. Elsevier Science BV, 1994, Chap 11.
40. Versieck, J., Neutron activation analysis. *Methods Enzymol.*, 158, 267, 1988.
41. Versieck, J., Neutron activation analysis in *Trace Element Analysis in Biological Specimens*, Herber, R.F.M. and Stoeppler, M., Eds. Elsevier Science BV, 1994, Chap 7.
42. Byrne, A.R. and Kučera, J., Radiochemical neutron activation analysis of traces of vanadium in biological samples: A comparison of prior dry-ashing with post irradiation wet-ashing. *Fresenius J. Anal. Chem.*, 340, 48, 1991.

43. Gießmann U. and Greb, U., High Resolution ICP-MS — a new concept for elemental mass spectrometry. *Fresenius J. Anal. Chem.*, 350, 186, 1994.
44. Tittes, W., Jakubowski, N., Stüwer, D., Tölg, G., and Broekaert, J.A.C., Reduction of some selected spectral interferences in inductively coupled plasma mass spectrometry. *J. Anal. At. Spec.*, 9, 1015, 1994.
45. Moens, L., Verrept, P., Dams, R., Greb, U., Jung, G., and Laser, B., New high-resolution inductively coupled plasma mass spectrometry technology applied for the determination of V, Fe, Cu, Zn, and Ag in human serum. *J. Anal. At. Spec.*, 9, 1075, 1994.
46. Finnigan Mat ICP-MS Interferenz Tabule Revision 2 9th May 1995.
47. Kim, Y.S., Kawaguchi, H., Tanaka, T., and Mizuike, A., Non-Spectroscopic Matrix Interferences in Inductively Coupled Plasma Mass Spectrometry. *Anal. Chem.*, 61, 2031, 1990.
48. Tolman, E.L., Barris, E., Burns, M., Pansini, A., and Partridge, R., Effects of vanadium on glucose metabolism *in vitro*. *Life Sci.*, 25, 1159, 1979.
49. Orvig, C., Thompson, K.H., Battell, M., and McNeill, J.H., Vanadium compounds as insulin mimics, in *Metal Ions in Biological Systems*, Sigel, H., and Sigel, A., Eds., Marcel Dekker Inc., New York, 575, 1995.
50. Yuen, V.G., Orvig, C., and McNeill, J.H., Comparison of the glucose-lowering properties of vanadyl sulfate and bis(maltolato)oxovanadium(IV) following acute and chronic administration. *Can. J. Physiol. Pharmacol.*, 73, 55, 1995.
51. Cam, M.C., Faun, J., and McNeill, J.H., Concentration-dependent glucose-lowering effects of oral vanadyl are maintained following treatment withdrawal in streptozotocin-diabetic rats. *Met. Clin. Exp.*, 44, 332, 1995.
52. Blondel, O., Simon, J., Chevalier, B., and Portha, B., Impaired insulin action but normal insulin receptor activity in diabetic rat liver: effect of vanadate. *Am. J. Physiol.*, 258, E459, 1990.
53. Ramanadham, S., Brownsey, R.W., Cros, G.H., Mongold, J.J., and McNeill, J.H., Sustained prevention of myocardial and metabolic abnormalities in diabetic rats following withdrawal from oral vanadyl treatment. *Metabolism*, 38, 1022, 1989.
54. Meyerovitch, J., Farfel, Z., Sack, J., and Shechter, Y., Oral administration of vanadate normalizes blood glucose levels in streptozotocin-treated rats. *J. Biol. Chem.*, 262, 6658, 1987.
55. Heyliger, C.E., Tahiliani, A.G., and McNeill, J.H., Effect of vanadate on elevated blood glucose and cardiac performance of diabetic rats, *Science*, 227,1474, 1985.
56. Battell, M.L., Yuen, V.G., and McNeill, J.H., Treatment of BB rats with vanadyl sulphate. *Pharmacol. Commun.*, 1, 291, 1992.
57. Ramanadham, S., Cros, G.H., Mongold, J.J., Serrano, J.J., and McNeill, J.H., Enhanced *in vivo* sensitivity of vanadyl-treated diabetic rats to insulin. *Can. J. Physiol. Pharmacol.*, 68, 486, 1990.
58. Yuen, V.G., Orvig, C., Thompson, K.H., and McNeill, J.H., Improvement in cardiac dysfunction in streptozotocin-induced diabetic rats following chronic oral administration of bis(maltolato)oxovanadium (IV). *Can. J. Physiol. Pharmacol.*, 71, 260, 1993.
59. Yuen, V.G., Orvig, C., and McNeill, J.H., Glucose-lowering effects of a new organic vanadium complex, bis(maltolato)oxovanadium (IV). *Can. J. Physiol. Pharmacol.*, 71, 263, 1993.
60. McNeill, J.H., Yuen, V.G., Hoveyda, H.R., and Orvig, C., Bis(maltolato)oxo-vanadium (IV) is a potent insulin mimic. *J. Med. Chem.*, 35, 1489, 1992.
61. Zaporowska, H. and Wasilewski, W., Mini Review: Haematological effects of vanadium on living organisms. *Comp. Biochem. Physiol.*, 102C, 223, 1992.
62. Dafnis, E. and Sabatini, S., Biochemistry and pathophysiology of vanadium. *Nephron*, 67, 133, 1994.
63. Stohs, S.J. and Bagchi, D., Oxidative mechanisms in the toxicity of metal ions. *Free Rad. Biol. Med.*, 18, 321, 1995.

64. Domingo, J.L., Gómez, M., Llobet, J.M., Corvella, J., and Keen, C.L., Improvement of glucose homeostasis by oral vanadyl or vanadate treatment in diabetic rats is accompanied by negative side effects. *Pharmacol. Toxicol.*, 68, 249, 1991.
65. Domingo, J.L., Gómez, M., Llobet, J.M., Corvella, J., and Keen, C.L., Oral vanadium administration to streptozotocin-diabetic rats has marked negative side effects which are independent of the form of vanadium used. *Toxicol.*, 66, 279, 1991.
66. Oster, M.H., Llobet, J.M., Domingo, J.L., German, J.B., and Keen, C.L., Vanadium treatment of diabetic Sprague-Dawley rats results in tissue vanadium accumulation and pro-oxidant effects. *Toxicology*, 83, 115, 1993.
67. Domingo, J.L., Sanchez, D.J., Gomez, M., Llobet, J.M., and Corbella, J., Oral vanadate and Tiron in treatment of diabetes mellitus in rats: improvement of glucose homeostasis and negative side-effects. *Vet. Hum. Toxicol.*, 35, 495, 1993.
68. Dai, S., Vera, E., and McNeill, J.H., Lack of haematological effect of oral vanadium treatment in rats. *Pharm. Toxicol.*, 76, 263, 1995.
69. Boscolo, P., Carmignani, M., Volpe, A.R., Falaco, M., Del Rosso, G., Porcelli G., and Giuliano, G., Renal toxicity and arterial hypertension in rats chronically exposed to vanadate. *Occup. Env. Med.*, 51, 500, 1994.
70. Léonard, A. and Gerber, G.B., Mutagenicity, carcinogenicity and teratogenicity of vanadium compounds. *Mutat. Res.*, 317, 81, 1994.
71. Domingo J.L., Metal-induced developmental toxicity in mammals: a review. *J. Toxicol. Environ. Health*, 42, 123, 1994.
72. Cohen, M.D., Parson, E., Schlesinger, R.B., and Zelikoff, J.T., Immunotoxicity of in vitro vanadium exposures: effects on interleukin-1, tumor necrosis factor-α, and prostaglandin E_2 production by WEHI-3 macrophages. *Int. J. Immunopharmacol.*, 15, 437, 1993.
73. Mravcova, A., Jirova, D., Janci, H., and Lener, J., Effects of orally administered vanadium on the immune system and bone metabolism in experimental animals. *Sci. Total Environ.*, Suppl. 1, 663, 1993.
74. Cornelius, R. and Versieck, J., Deterination of vanadium in tissues and serum. *Clin. Chem.*, 28(7), 1708, 1982.
75. Yale, J-F., Lachance, D., Bevan, A.P., Vigeant, C., Shaver, A., and Posner, B.I., Hypoglycemic effects of peroxovanadium compounds in Sprague-Dawley and diabetic BB rats. *Diabetes*, 44, 1274, 1995.

Chapter 21

Nickel-Low Diet Formulation and Tissue Nickel Measurement

Eric O. Uthus
U.S. Department of Agriculture/ARS
Grand Forks Human Nutrition
Research Center
Grand Forks, North Dakota

CONTENTS

I. INTRODUCTION

Although nickel was first suggested to be nutritionally essential in 1936, strong evidence for its essentiality did not appear until the 1970s.[1] Many studies in the 1970s, however, gave inconsistent signs of nickel deprivation. Some of the inconsistencies can be linked to improper environmental conditions, others, to suboptimal diets.[2,3] Some of the diets were low in copper[4] or iron,[4,5] others marginal in vitamins,[6] and in some studies with ruminants, the diets were low in protein and supplemented with urea.[7–9] Because rumen bacterial urease is a nickel-dependent enzyme,[10] some of the discrepancies could have been caused by effects of dietary nickel on rumen bacteria rather than the host animal. Also, many of these earlier studies on

0-8493-9611-5/97/$0.00+$.50

nickel essentiality used nickel supplements that were probably pharmacologic; see Nielsen[1-3] for further discussion. Beginning in the mid- to late 1970s, nutritionally-adequate diets have been used to study the consequences of nickel deprivation. Also, the concentration of nickel used to supplement the control animals is considered physiologically relevant.

As suggested by Nielsen,[1] the nickel requirement for most monogastric animals is probably less than 200 ng/g diet. Thus, diets used in studies of the essentiality of nickel must be carefully constructed to avoid nickel contamination. Experience has shown that calcium salts are a major source of nickel contamination. This negates the use of diets based on AIN-76 or AIN-93 formulations. To overcome the use of calcium salts in mineral mixes, F. H. Nielsen at the USDA, ARS, Grand Forks Human Nutrition Research Center developed a dried skim milk and corn based diet that has been used to study nickel essentiality in chicks and rats. This diet has been modified over the years since its inception; the version used in the most recent nickel studies is shown in Tables 1 to 6.

TABLE 1.
Basal Diet

Ingredient	g/kg diet
Dried skim milk[a]	375.00
Ground corn, acid washed[b,c]	497.25
Corn oil[d]	100.00
dl-α-tocopherol[b]	0.06
Vitamin mix	4.19
Mineral mix	20.00
Ferric sulfate mix	1.00
L-cystine[e]	2.00
Choline chloride[b]	0.50
Total:	1000.00

[a] Carnation, purchased locally.
[b] ICN (Costa Mesa, CA).
[c] Acid washing procedure is described in text.
[d] Mazola, purchased locally.
[e] Ajinomoto (Teaneck, NJ).

II. DIET PREPARATION

The main component of the diet used in nickel deprivation studies is acid-washed ground corn. Acid washing significantly reduces metal contamination in the corn. The acid-washing procedure follows. In a chemical hood, 12 L of water (Super Q, Millipore, Bedford, MA)* are added to 5 kg

* Mention of a trademark or proprietary product does not constitute a guarantee of warranty of the product by the U.S. Department of Agriculture and does not imply its approval to the exclusion of other products that may also be suitable.

TABLE 2.
Vitamin Mix

Ingredient[a]	g/kg diet
Vitamin K_1 (phylloquinone)	0.00075
Nicotinic acid	0.0150
Biotin	0.0002
Thiamine·HCl	0.0027
Pantothenic acid, Ca salt	0.0080
Vitamin B_{12}[b]	0.0500
Pyridoxine·HCl	0.0075
Riboflavin	0.0050
Inositol	0.0050
Folic acid	0.0040
Glucose	4.09185
Total:	4.19

[a] All vitamins ICN (Costa Mesa, CA).
[b] Vitamin B_{12} (triturate, 0.1% in mannitol).

TABLE 3.
Mineral Mix

Ingredient	g/kg diet
ZnO[a]	0.0062
$MnSO_4 \cdot 4H_2O$[b]	0.0812
$CuSO_4 \cdot 5H_2O$[b]	0.0500
KI[b]	0.0005
Na_2SeO_3[b]	0.0006
$(NH_4)_6Mo_7O_{24} \cdot 4H_2O$[b]	0.0040
$Cr(C_2H_3O_2)_3 \cdot H_2O$[c]	0.0020
NaF[b]	0.0060
H_3BO_3[b]	0.0060
NH_4VO_3[b]	0.0005
$Na_2HAsO_4 \cdot 7H_2O$[a]	0.0050
$NaSiO_3 \cdot 9H_2O$[a]	0.5000
Corn, acid washed	19.3380
Total:	20.00

[a] Reagent grade, J.T. Baker, Phillipsburg, NJ.
[b] Puratronic, Alpha/JMC, Ward Hill, MA.
[c] Reagent grade, Fisher, Pittsburgh, PA.

of ground corn (ICN, Costa Mesa, CA) in a 5 gallon plastic pail and 2.5 L of concentrated HCl (reagent grade, J. T. Baker, Phillipsburg, NJ) are added carefully. The contents are mixed with a mechanical mixer (plastic stirrer) for 30 min. After letting the mixture settle for 15 min, the supernatant is decanted and discarded (as are all the following supernatants). About 6 L of Super Q are added to bring the volume up to the approximate starting volume and the contents mixed for 30 min. After the mixture has settled for 15 min, the supernatant is decanted and Super Q is added to the

TABLE 4.

Calculated Protein, Fat, and Individual Amino Acid Content of Basal Diet

Dietary Component	% in Skim milk	Added by Skim milk g/kg diet	% in Corn	Added by Corn g/kg diet	Added g/kg diet	Total g/kg diet	Requirement Growth[a] g/kg diet	Requirement Maintenance[a] g/kg diet
Protein	33.0	123.75	8.7	44.943		168.693	150	50
Fat	0.9	3.375	3.9	20.147	100	123.522	50	50
Tryptophan	0.45	1.688	0.1	0.517		2.205	2	0.5
Histidine	0.84	3.15	0.2	1.033		4.183	2.8	0.8
Lysine	2.3	8.625	0.2	1.033		9.658	9.2	1.1
Leucine	3.3	12.375	1.1	5.682		18.057	10.7	1.8
Isoleucine	2.1	7.875	0.4	2.066		9.941	6.2	3.1
Phenylalanine	1.6	6	0.5	2.583		8.583	10.2	1.9
Tyrosine	0.82	3.075				3.075		
Methionine	1.0	3.75	0.18	0.930		4.68	9.8[c]	2.3
Cystine	0.42	1.575	0.18	0.930	2	4.505[b]		
Threonine	1.7	6.375	0.4	2.066		8.441	6.2	1.8
Valine	2.4	9	0.4	2.066		11.066	7.4	2.3
Arginine	1.1	4.125	0.5	2.583		6.708	4.3	

[a] Requirements (growth and maintenance) from reference.[11]

[b] Total methionine + cystine is 9.2 g/kg diet. Although this is less than the 9.8 g/kg diet suggested by the 1995 National Research Council,[11] it is greater than the 6.0 g/kg diet suggested by the 1976 National Research Council[12] and the 8.22 g/kg diet suggested by the AIN-93G.[13] To meet the 1995 National Research Council recommendations, methionine or more cystine could be added.

[c] Requirement (growth) for methionine + cystine.

TABLE 5.

Mineral Content of Basal Diet

Mineral	Ingredient	g/kg diet	% Mineral in compound	Content in Skim milk µg/g	Added by Skim milk g/kg diet	Total g/kg diet	Requirement Growth[a] g/kg
Ca				13000	4.875	4.875	5.0
Mg				1110	0.416	0.416[b]	0.5
P				10160	3.81	3.81	3.0
K				13350	5.006	5.006	3.6
Na				5250	1.969	1.969	0.5
Cl				1130	4.2375	4.238	0.5
S				3000	1.125	1.125	
Cu	$CuSO_4 \cdot 5H_2O$	0.05	25.45			0.01273	0.005
F	NaF	0.006	45.24			0.00271	
I	KI	0.0005	76.45			0.00038	0.00015
Fe	Ferric sulfate mix	1	3.5	6	0.00225	0.03725	0.035
Mn	$MnSO_4 \cdot 4H_2O$	0.0812	24.63	2	0.00075	0.02075	0.010
Se	Na_2SeO_3	0.0006	45.65			0.00027	0.00015
Zn	ZnO	0.0062	80.35	25	0.00938	0.01436	0.012
Mo	$(NH_4)_6Mo_7O_{24} \cdot 4H_2O$	0.004	54.34			0.00217	0.00015
Cr	$Cr(C_2H_3O_2)_3 \cdot H_2O$	0.002	21.04			0.00042	
B	H_3BO_3	0.006	17.50			0.00105	
V	NH_4VO_3	0.0005	43.55			0.00022	
Si	$NaSiO_3 \cdot 9H_2O$	0.5	9.89			0.04945	
As	$Na_2HAsO_4 \cdot 7H_2O$	0.005	24.01			0.00120	

[a] Requirements (growth) from reference.[11]

[b] The only source of magnesium in the diet is from the skim milk. The amount added by the skim milk meets the 1978 National Research Council recommendation of 0.4 g/kg diet.[12] To meet the 1995 recommendation,[11] the addition of $MgSO_4 \cdot 7H_2O$ is suggested (reagent grade or better, 1.0 g/kg diet).

TABLE 6.

Vitamin Content of Basal Diet

Vitamin	Ingredient	g/kg diet	Content in Skim milk μg/g	Added by Skim milk mg/kg diet	Total mg/kg diet	Requirement Growth[a] mg/kg diet
Vitamin K	Vitamin K_1	0.00075			0.75[b]	1.0
Vitamin E	dl-α-tocopherol	0.06			60	18.0
Biotin	Biotin	0.0002			0.2	0.2
Folic acid	Folic acid	0.004	0.02	0.0075	4.0075	1.0
Inositol	Inositol	0.005			5	
Niacin	Nicotinic acid	0.015	11	4.125	19.125	15.0
Pantothenic acid	Pantothenic acid, Ca salt	0.008	33	12.375	20.375	10.0
Riboflavin	Riboflavin	0.005	20	7.5	12.5	3.0
Thiamine	Thiamine·HCl	0.0027	3.5	1.3125	4.0125	4.0
Vitamin B_6	Pyridoxine·HCl	0.0075	4.8	1.8	9.3	6.0
Vitamin B_{12}	Vitamin B_{12}[c]	0.05	0.06	0.0225	0.0725	0.050
			IU/g	IU/kg diet	IU/kg diet	IU/kg diet
Vitamin A			22	8250	8250	2300
Vitamin D			2.2	825	825[d]	1000
			g/kg	g/kg diet	g/kg diet	g/kg diet
Choline	Choline chloride	0.5	1.4	0.525	1.025	0.750

[a] Requirements (growth) from reference.[11]

[b] Surpasses the 1978 National Research Council recommendation (0.050 mg/kg)[12] and equals the AIN-93G recommendation.[13]

[c] Vitamin B_{12} (0.1% triturate in mannitol).

[d] To meet the 1995 National Research Council recommendation,[11] addition of 175 IU vitamin D_3 (cholecalciferol; 400,000 IU/g) is suggested.

approximate starting volume. The contents are mixed for 15 min followed by 5 min settling, decanting, and addition of Super Q to approximate initial volume; after repeating this step three more times, the supernatant is decanted and the corn slurry is poured into a large Büchner funnel (approximately 32.5 cm ID). Then, approximately 24 L of Super Q are poured through the corn; the flow rate of the Super Q is adjusted to keep the corn submerged during the rinsing. Excess Super Q is removed by applying a vacuum to the Büchner funnel. The moist corn is then transferred to a plastic tray with an elevated top (to allow air convection). The corn is dried in a convection oven for 12 to 48 hr at about 75°C. The dried corn is stored at room temperature in plastic barrels until needed.

The vitamin mix (enough to complete an experiment — typically for 50 to 200 kg of diet) is prepared by using finely ground vitamins weighed into a plastic container containing the glucose. After each addition, the mixture is briefly mixed with a plastic spatula. After all additions, grinding media (burundum cylinders) are added, then the container is capped and shaken by hand for about 0.5 hour. The vitamin mix is kept in a plastic container at –20°C. Unused vitamin mix is discarded at the completion of the experiment (usually 8 to 10 weeks).

The mineral mix (enough to complete an experiment) is prepared in a porcelain ball mill to which the ground corn is added. After the addition of each mineral, it is mixed into the corn with a plastic spatula. After the addition of all minerals, burundum cylinders are added, the mill is capped and then rolled for 3 to 4 hours. Unused mineral mix is kept in a plastic container at room temperature.

Ferric sulfate is prepared by using iron powder (99.999%, Alfa/ JMC, Ward Hill, MA) and reagent grade sulfuric acid. The resultant ferric sulfate is analyzed for iron content and an iron-glucose mix is made; 1.0 g of the ferric sulfate mix/kg diet provides 35 μg Fe/g diet. The iron-glucose mix is made similarly as the vitamin mix. The unused ferric sulfate and ferric sulfate-glucose mix are stored at room temperature under vacuum.

The basal diet is prepared in a plastic tub. Usually 8 kg of each diet are prepared at a time. Dried skim milk and acid-washed ground corn are weighed directly into the tub; then all other ingredients except the corn oil, dl-α-tocopherol, and choline are added and the contents gently mixed by hand using powder-free gloves. Corn oil is weighed into a plastic beaker and then the dl-α-tocopherol is weighed into the oil and mixed in with a plastic spatula. Choline chloride, dissolved in minimal Super Q is added to the corn oil-tocopherol and the contents are vigorously stirred and quickly poured onto the dry ingredients. The residual contents in the beaker are scraped out with a plastic spatula and added to the diet preparation. The contents of the tub are then mixed by hand for 30 min. Aliquots of the diet are taken for nickel analysis. Diets are stored at –20°C in plastic containers. No metal is used in the preparation, storage, or feeding of the diets. A typical basal diet, prepared carefully, will contain less than 25 ng Ni/g diet.

Nickel supplements are usually in the form of nickel chloride. A nickel-glucose mix is prepared similarly to the vitamin mix. The typical nickel supplement of 1.0 g nickel-glucose mix/kg diet will add 1 μg Ni/g diet. To avoid cross contamination, diets containing the nickel supplement are prepared after the completion of the basal diet. Sprague-Dawley rats (Sasco, Omaha, NE) fed these diets grow well; weanling male rats generally reach 300 g in body weight by about 70 days of age. The fat content of the diet could be decreased; for example, the National Research Council suggests 50 g fat/kg diet for both maintenance and growth[11] and the AIN-93G diet contains 70 g soybean oil/kg diet.[13] Variations of this diet, including alteration of fat content, have been published.[14–16]

Anke,[7] who has studied nickel deprivation in goats and minipigs, used a diet based on potato starch (48%), beet sugar (32%), casein (10%), urea (3%), and sunflower oil (3%), and added minerals and vitamins. This diet contains about 100 ng Ni/g diet and could be adapted for use in rodents. Kirchgessner (see Nielsen[3]) used a casein-based diet for his studies of nickel deprivation in rats. However, to reduce the nickel content, the casein was treated with EDTA, possibly confounding the experimental results (see Nielsen[3]).

All-plastic caging for the animals has been described by Nielsen and Bailey.[17] Any type of all-plastic caging can be used. Food and water cups are also plastic. Environmental conditions have also been described.[14] The use of HEPA filtered-laminar flow racks for animal housing in nickel deprivation studies is recommended but not a requirement.

In studying trace element deficiency, some researchers feed all the animals in a study the basal (deficient) diet but add the trace element under study to the drinking water of the control animals. This perhaps could be done when studying nickel deprivation, but the researcher must remember that studies with humans have shown that nickel is absorbed to a greater extent when given in aqueous solution than when given with food.[18,19]

III. NICKEL STATUS INDICATORS

Although strong evidence suggests that nickel plays an important role in vitamin B_{12} and/or folate metabolism,[1,20] the exact physiological site of action is not known. Because of this, the most reliable indicator for nickel status is tissue nickel concentration. Routinely, nickel concentration in kidney, leg muscle, brain, and bone are ascertained (Table 7). Nickel concentration in these tissues is determined by conventional wet ashing and electrothermal atomic absorption spectroscopy.[21] At a dietary concentration of 1 μg/g, nickel does not accumulate in the kidney. Kidney nickel concentration, however, is not always a reliable and sensitive indicator of nickel status. Nickel content in bone has been the most consistent status indicator. Recently, however, the concentration of nickel in erythrocytes has been shown to be an excellent marker for nickel status (unpublished results). The suitability of bone and erythrocyte nickel concentrations as status indicators could be indicative of nickel's proposed function related to vitamin B_{12} and/or folate metabolism. Typical values for nickel concentrations in various rat tissues are shown in Table 7.

TABLE 7.

Nickel Concentration in Sprague-Dawley Rat Tissues

Tissue	0 Ni[a]	1 Ni[a]
Erythrocyte	0.0027–0.0120 (0.0073)	0.0029–0.0194 (0.0135)
Leg muscle	0.013–0.026 (0.019)	0.016–0.031 (0.021)
Bone, femur	0.059–0.127 (0.105)	0.109–0.252 (0.162)
Kidney	0.368–0.437 (0.402)[b]	0.355–0.512 (0.405)
Brain	0.395–0.456 (0.421)	0.417–0.442 (0.433)

[a] Amount (μg/g) of nickel supplemented to basal diet; basal diets contained less than 25 ng Ni/g.

[b] Range followed by average; all concentrations μg Ni/g tissue, dry weight basis. Values taken from numerous published and unpublished experiments. All values are from male rats about 80 days old; the rats were on respective nickel-deficient or nickel-supplemented diets for about 60 days.

Determination of erythrocyte nickel follows (R.A. Poellot and T.R. Shuler, personal communication). Rats are anesthetized with ether and blood is collected from the vena cava by using a heparin-coated syringe and needle. Nickel contamination has not been seen when stainless steel needles are used under these conditions. Also, in a human study in which the first 5 ml from a 26 ml total blood collection was discarded, Bro et al.[22] saw no nickel contamination when collecting blood with a steel needle. To prepare erythrocytes for nickel determination, 7 to 8 ml of heparinized whole blood, in a 15 ml plastic conical centrifuge tube, is centrifuged at approximately 2000 × *g* for 20 min at 4°C. Plasma and buffy coat are removed with a plastic transfer pipet. The packed cells are gently suspended in 7 to 8 ml of cold saline (0.89%) and transferred to a clean tube. One ml saline is used to rinse the initial tube. The tube is centrifuged for 10 min at 1400 × *g* at 4°C. The saline is removed, the cells are suspended in another 5 ml of saline and centrifuged again at 1400 × *g* for 10 min at 4°C. After the saline is carefully removed, the tube is capped and frozen at –20°C. The frozen cells are lyophilized before being analyzed for nickel as above. Nickel concentration is expressed on a dry weight basis or, if hemoglobin is determined, on a hemoglobin basis.

B

REFERENCES

1. Nielsen, F.H., Other trace elements, in *Present Knowledge in Nutrition. Ed. 7*, in press.
2. Nielsen, F.H., Nickel, in *Biochemistry of the Essential Ultratrace Elements*, Frieden, E., Ed. Plenum, New York, 1984, 293.
3. Nielsen, F.H., The importance of diet composition in ultratrace element research, *J. Nutr.*, 115, 1239, 1985.
4. Nielsen, F.H. and Sauberlich, H.E., Evidence of a possible requirement for nickel by the chick, *Proc. Soc. Exper. Biol. Med.*, 134, 845, 1970.
5. Nielsen, F.H. and Higgs, D.J., Further studies involving a nickel deficiency in chicks, in *Trace Substances in Environmental Health-IV*, Hemphill, D.D., Ed., University of Missouri Press, Columbia, MO, 1973, 241.
6. Nielsen, F.H., Myron, D.R., Givand, S.H., Zimmerman, T.J., and Ollerich, D.A., Nickel deficiency in rats, *J. Nutr.*, 105, 1620, 1975.
7. Anke, M., Hennig, A., Grun, M., Partschefeld, M., Groppel, B., and Ludke, H., Nickel - ein essentielles spurenelement. 1. Mitteilung, *Arch. Tierernaehr.*, 27, 25, 1977.
8. Spears, J.W. and Hatfield, E.E., Nickel for ruminants. 1. Influence of dietary nickel on ruminal urease activity, *J. Anim. Sci.*, 47, 1345, 1978.
9. Spears, J.W., Hatfield, E.E., and Forbes, R.M., Nickel for ruminants. II. Influence of dietary nickel on performance and metabolic parameters, *J. Anim. Sci.*, 48, 649, 1979.
10. Spears, J.W., Smith, C.J. and Hatfield, E.E., Rumen bacterial urease requirement for nickel, *J. Dairy Sci.*, 60, 1073, 1977.
11. National Research Council, *Nutrient Requirements of Laboratory Animals, 4th Revised Edition*, Washington, D.C.: National Academic Press, 1995.
12. National Research Council, *Nutrient Requirements of Laboratory Animals, 3th Revised Edition*, Washington, D.C.: National Academy of Sciences, 1978.
13. Reeves, P.G., Nielsen, F.H., and Fahey, G.C., Jr., AIN-93 purified diets for laboratory rodents: Final report of the American Institute of Nutrition ad hoc committee on the reformulation of the AIN-76A rodent diet, *J. Nutr.*, 123, 1939, 1993.
14. Nielsen, F.H., Zimmerman, T.J., Shuler, T.R., Brossart, B., and Uthus, E.O., Evidence for a cooperative metabolic relationship between nickel and vitamin B_{12} in rats, *J. Trace Elem. Exper. Med.*, 2, 21, 1989.
15. Nielsen, F.H., Uthus, E.O., Poellot, R.A., and Shuler, T.R., Dietary vitamin B_{12}, sulfur amino acids, and odd-chain fatty acids affect the response of rats to nickel deprivation, *Biol. Trace Elem. Res.*, 37, 1, 1993.
16. Uthus, E.O. and Poellot, R.A., Effect of nitrous oxide on nickel deprivation in rats, *Biol. Trace Elem. Res.*, 38, 35, 1993.
17. Nielsen, F.H. and Bailey, B., The fabrication of plastic cages for suspension in mass air flow racks, *Lab. Anim. Sci.*, 29, 502, 1979.
18. Solomons, N.W., Viteri, F., Shuler, T.R., and Nielsen, F.H., Bioavailability of nickel in man. Effects of foods and chemically-defined dietary constituents on the absorption of inorganic nickel, *J. Nutr.*, 112, 39, 1982.
19. Sunderman, F.W., Jr., Hopfer, S.M., Sweeney, K.R., Marcus, A.H., Most, B.M., and Creason, J., Nickel absorption and kinetics in human volunteers, *Proc. Soc. Exp. Biol. Med.*, 191, 5, 1989.
20. Uthus, E.O. and Poellot, R.A., Dietary folate affects the response of rats to nickel deprivation, *Biol. Trace Elem. Res.*, 52, 23, 1996.
21. Nielsen, F.H., Zimmerman, T.J., and Shuler, T.R., Interactions among nickel, copper, and iron in rats: Liver and plasma content of lipids and trace elements, *Biol. Trace Elem. Res.*, 4, 125, 1982.
22. Bro, S., Jorgensen, P.J., Christensen, J.M., and Horder, M., Concentration of nickel and chromium is serum: influence of blood sampling technique, *J. Trace Elem. Electrolytes Health Dis.*, 2, 31, 1988.

APPENDIX

Appendix

SOME ELECTRON MICROSCOPY MATERIAL SUPPLIERS

Bal-Tec Products, Inc.
984 Southford Road
P.O. Box 1221
Middlebury, CT 06762
(203) 598-3660

Electron Microscopy Sciences
321 Morris Road, Box 251
Fort Washington, PA 19034
(800) 523-5874

Ladd Research Industries, Inc.
P.O. Box 1005
Burlington, VT 05402
(800) 451-3406

Polysciences, Inc.
400 Valley Road
Warrington, PA 18976-2590
(215) 343-6484

Energy Beam Sciences
P.O. Box 468
Agawam, MA 01001
(800) 992-9037

EMITECH, USA
5206 FM 1960W.
Suite 100
Houston, TX 77069
(713) 893-8443

Ted Pella, Inc.
P.O. Box 492477
Redding, CA 96049-2477
(800) 237-3526

INDEX

INDEX

C

D

E

F

G

I

K

L

M

N

O

R

T